A Tom Luong Novel
The Tomorrow Collection

THE JUPITER PLAN
ANARCHY AT AMALTHEA

JULIAN PHILLIPS
TOM LUONG

Cover art "Starship" by David Schleinkofer

This book is the property of Luong Films and Tom Luong
www.luongfilms.com | 951-660-6010 | tvluong1@hotmail.com

DEDICATION

'The Jupiter Plan: Anarchy at Amalthea'; is dedicated to my son, Preston Laverne Phillips: "When the wings of Icarus wax wildly, and melting you into the light, my son, love befalls us only mildly, all with your world is right." -Julian Phillips, May 2014.

From Tom Luong, I like to dedicate this novel to my parents, Mike Manh Van Luong and Nancy Ngat Thi Le, both from Vietnam.

CONTENTS

CONTENTS

CHAPTER 1: Dead All Over Again

"What is wisdom? Whatever works, when it's needed and right, or possibly kind, or compassionate, sometimes not, though. Basically, whatever works, is wisdom, for me anyway. Unless it's not."

---Daniel Deveroux, Planetary Program Proficiency Assignment-Researcher, 2,412, A.D./C.E.

"It is my sad duty to inform you, Mr. Daniel Deveroux, of 3344 Oakland Drive, Seattle, Washington, that you are dead. Time-of-death was 3:15 p.m., Wednesday, November 14, 2,412, Common Era, recorded by the Saint Mary's Community Hospital Emergency Room. Cause-of-death was by fatal automobile accident. Please contact the agency for more details concerning your demise, funeral arrangements, disposition of your estate and personal wealth-accounts, residence and loved ones, and your next assignment. Thank you. PPP Human Resources: Announcement 532/Verified."

Such was the Voice-Message recording that Danny Deveroux found on his telephone-system, by Friday of that week, there at his home in Seattle, Washington. It was a dreary day, not a 'good day to die' at all, as the Native-Americans say. As if any day one might be found among the dead, was truly a 'good' day, personally, or even might be said to be a 'day' at all. True enough, having died before, once or twice in his lifetime, Daniel knew that the lines between black-and-white, night-and-day, or alive and dead, could sometimes be blurry. Grey area. Melting into One.

And it was grey, like water-color clouds of moisture and mist, then rain and drizzle, above Seattle, often a rainy town, on the Northern-Western US land-mass, mostly green and verdant, and by that year of 2,412-CE, one of America's largest and most successful cities. 'Home', at that time, but not for long, as usual. Daniel enjoyed Seattle, and often stayed there if he could, instead of New York, or Washington, D.C., like many of his co-workers with Planet-Proficiency. Work is work, and pride or ego-trips about surviving his own death or deaths were not really very funny at all. So, his thoughts were grey, then, like the skies overhead. About 'real' death, his own and that of those he cared for, here on Earth. How and when and why he might die, in 'real-life', biologically, or, perhaps, other kinds of death, such as the death of reason, or the death of soul. The death of reason might indeed be the worst of any of them, as far as deaths one might experience. "You shall not surely die," he thought, from the Bible. But, oh yes, forgot to tell you about the other deaths, "Ooops". And what is death but forgetting? And the death of wisdom, that too. Gone, gone, gone, all gone at last. Buddha's cup was empty, turned over. Oh well.

He had arrived back home at Oakland Street via mass-transit, and then a three-block walk into the tree-lined neighborhood, up a small hill where the road turned, somewhat private, but not isolated or secretive. There had been no automobile accident, and he was in fair health, overall, as much alive as he had ever been, no more, no less. Can life be added to, while living, with more of itself? At that time, Deveroux was a thin and muscular man, fit-and-trim for his age at 42 years, dark-haired and with skin like shades of pale wood, clean-shaven of his face and chin, with a hawkish look. At that time he was also working at the local University, part-time as an instructor or professor, teaching data and research skills to young students half his age. The small private home was more like a wooded-cabin, with pine trees around and birds singing. He moved up the steps to unlock the door, and went inside, setting aside his briefcase and school-papers, and then his coat, and finally tapping the play-button on the answering-system of the house phone. There were two other messages aside from the recording from PPP (his employer). He made coffee and listened from the kitchen.

"Daniel, are you there? It's Veronica, at the hotel. Are you coming over like you said on Saturday? Because if you're not, I'll leave the circulator thing back in San Francisco, because I'm not even supposed to have it with me, and it's heavy and it takes an hour to set up right. So let me know, sweety. Okay? Seriously, I know you're busy, but this thing is only any good after you try it out for a while first, so don't let me down. You have my number. Thanks, hun'."

There was a tone or beep. Then another message played. Veronica was a woman with PPP from San Francisco. The so-called 'circulator' was a spectacular new sex toy for two, installed in a bath-tub, or hot-shower. Supposedly fantastic, but now he'd never know, being dead. It was her idea anyway, they had been romantic for a year or so, off-and-on. Too bad. The coffee was by then hot and steamy as well as his thoughts, and he sipped as the second message played.

"Mr. Deveroux, this is Officer Morton with Seattle Codes-and-Compliance. It's my understanding that a private aircraft, a small Pensacola-jet, twin-engine model with under-wing anti-gravity assist, Aircraft ID Number 5634K, silver and blue, registered to you last year, out of San Francisco, California, has been unattended for three weeks at the Seattle Bungalows Community Airport. Sir, this piece of equipment is now in violation, with serious fees and costs to you. Please contact my office in Seattle immediately if possible, phone is Regional 687, direct line is 234-009-764. Again, I am Officer Morton with Codes-and-Compliance. Thank you, Mr. Deveroux. Cid Bixi Mimim to you, sir."

Damn, he thought. That stupid airplane. Even worse than the news that he had died, or the loss of a hot-night with Veronica. It wasn't for another moment that the PPP recording replayed again automatcally on the Voice-Messaging system then, as he continued with his hot coffee, relaxing after a day's work. But, there it was. If PPP was declaring him 'homo non-viro', for their own reasons, any contact with either Veronica or Officer Morton, was a very poor choice, if not forbidden. Veronica would probably understand, being with PPP herself. But not the Seattle Bungalows Community Airport. Walking around at the hotel, showing ID-papers or running his money, all would expose him as 'not dead'. Death was more trouble among the living, as usual, but in his case, a convenience that PPP reserved for screwed-up cases he and other researchers were involved in. As did many other global agencies. Standard, really. The Pensacola jet was a nice bit of gear, and he hated to lose it. Oh well.

Deveroux worked for Planetary Program Proficiency, and had worked for them since 2,398-AD/CE. The PPP was a low-level global agency associated with the world-government of that time, such as it was. Their task was simple enough: oversight and review, research and remedy, for an array of complex high-end regional or international, or inter-planetary programs and systems, or agencies, military, industrial, labor, political, you-name-it. When something screwed up at a weapons-lab for non-violent crowd-control systems in Edinburgh, Scotland, if the strings-attached were big enough, PPP would end up looking into it. If the Chinese moon-base created in 2,385 AD/CE, was found to have trouble with food-preparation, linked to crop-disease in Malaysia, and making visitors sick, PPP would find out why, and what the 'broken-pipe' connection might be. Large-scale deep-sea fish-farming operations feeding millions of hungry poor-folks in Africa, with on-going small-submarine failure due to legal drug-addictions supplied to ignorant under-water workers (for stamina and oxygen decompression comfort by South-American illegal chem-labs)? PPP would eventually figure it out. The endless variety of PPP assignments, in this future-world period, was indeed stunning. But not the work involved for man like Deveroux, basically a highly skilled and flexibly unattached researcher. It was a great job, with high pay, and status, also mostly secret to the world-at-large, and also patriotic. But the supposed glamour was rather a joke to anyone working there. PPP was an administrative facility at the world-government level. All they really did was 'make things work', or find out why they didn't. But they had a reputation, it was true.

So, why had they suddenly declared him among the deceased? Daniel and his long-term assignment partner, Angelo 'Angel-Face' Martinez, had been called in to review California's new deep-Earth heat-based energy system. Over some 40-years, regional authorities had installed hundreds of very deep heat-transfer 'rods', that went down several miles to where the Earth's natural heat would conduct upwards to common electricity-generator stations. It was a clean-efficient way to provide power to the nation's most populated and busy region, by 2,380AD/CE or so.

California by then was home to almost half a billion people, all needing common power-sources for modern living. Trouble was, the system was being systematically dismantled, from within the responsible energy company itself, over several years, basically to provide easy big-money contracts to the second-level materials and tech-companies they worked with, and a huge slate of investors and money-stake holders. The contracts were worth many, many vast personal fortunes, enormous wealth for a select group of individuals. So it meant nothing to the energy company to 'screw up' their own systems, they would be repaired anyway, at high cost. But it was also a very corrupt practice from a public-ethical view, and the way it worked out, many of the electric-systems started to fail for long-term reliability, and there was also associated property crime, and even attacks, beatings and a murder or two. And now Daniel would need to disappear, declared dead himself, as a result of the intrigue.

So, by the time Deveroux and his co-worker Angel-Face un-tangled what was happening, his own life, as an investigator, was threatened. The tech-materials unions and big-money boys knew they could 'rub out' a low-level global-agency whistle-blower, and hide away their deeds even longer, while raking in billions of units in peronsal wealth. So, PPP would declare Deveroux had been killed in an automobile accident, on Wednesday at 3:15 p.m., to save his life. And not incidentally, to rescue the assignment. He wondered if Martinez had been declared dead as well? Maybe give him a call.

"Hey, Al, you dead too? Probably for the best."

The records, police-work, any actual car wrecks, his funeral, the medical-emergency doctors, the minor news-media, and a spare corpse that resembled a man his age, and so on, would all be gracefully falsified such that Deveroux's death would become an official matter. The University and his students would start over with a new instructor. Within a short time, he'd be relocated to a new town or city, under another name. And of course he was off the case in California. And so on. Part of the job, but no one's favorite. It made his life creepy and sad, he was always a stranger, it seemed. Public service. maybe even essential, but why? Sleaze-bag union-deals were in no way worth his life or anyone else's. But, the whole thing sucked and he knew it. Just the way it was.

So, along with the sad news, there was a lot for a dead-man to tend to. The private jet, Veronica, his comfortable home here and his personal belongings, all the usual life-stuff, had to be dealt with 'under-wraps', making things even more difficult.

"What's a corpse supposed to do?" he said to himself, joking. He decided to check in with local PPP management, to see what was next. And if at all possible, he was inviting Veronica over for the night, with or without the circulator-thing. She was a peppy gal, really cute and affectionate. She worked at a higher-level than he did, as far as PPP was concerned. Her job involved deciding what work he would do, or the other Researchers, of which there were many. Assignments were chosen by committee, and she had a permanent seat on the one he worked under, Western-Regional. Kinky, Vera, he joked under his breath. *Intimate relations with a dead employee? Hmmmm? How's that going to look?*

He spent an hour so by private-communications to his immediate sources with PPP. More coffee, then a hot bath-shower, then down-time viewing local and national news-media, none of which he believed to be telling him or anyone else anything remotely truthful about the way the world was then, at year 2,412AD-CE. No one else did either, but it was entertaining in a way, feeling above it all. Nothing bad will ever happen to me. But that other poor bastard, my god!!

By dinner-time, he met with Veronica. Things went well from there.

"Daniel," she said. "What do you know about Jupiter?"

"The planet? Not much more than anyone else with a college education," he responded. "Why?"

"Well, head's up," she said, enjoying a glass of wine, with a fish-dinner they shared, as the evening went by, like a dream. "The agency is looking at something going on with the Jupiter space-program, the whole thing with the ships and transport and giant hydrogen cargos. Your dance-card is now officially empty, for a while. I hope you don't get space-sick. Just FYI, nothing's certain at this point."

Daniel lifted his wine glass, touching it to hers. "Well, here's to wisdom about it, then," he said.

CHAPTER 2: Mendoza's Mouse

"Our programs on Jupiter have never been in any jeopardy, and are likely to continue to create wealth and materials resources at home, for a second hundred year period. There is no conspiracy or serious failure with the Jupiter program."

-Dr. Wu Menuda, to the General Assembly Earth-Space Program Authority Council, Geneva, Switzerland, January, 2,398 AD/CE, three months prior to the disastrous and unprecedented crash of the space-transport ship, 'Ferrous-2', and related loss of life and equipment.

"They had to kill me, I guess. I got in the way. Part of the job," Deveroux was saying. "But, I'm still here."

"How could it be otherwise?" said his partner, a husky Hispanic man named Angelo 'Angel-Face' Mendoza, about age 38-years, and given to fits of opinion about almost anything. Angel-Face was quite intelligent, moreso than many, but it was of a sort that was enjoyable or versatile, and Mendoza had proven himself time-and-again, to be dependable for the work involved as a Field Researcher for Program Proficiency, working with Daniel for as long as three years, on about eight major investigations. He was tall and brown, with dark hair and romantic brown eyes, fluent in several languages, and an expert in martial-arts, as well as a wide variety of high-technology machinery and devices.

Following his 'death', Daniel Deveroux tended to his personal affairs, as neatly as he could. California-based Thermal-Energy Systems providers were vendors to a regional system that supplied electricity to almost 500-million people, and thus enormously wealthy, and likewise 'probably corrupt'. When things got too hot with the PPP work on the case, lower-level sinister thug-type labor class stake-holders, were activated to instigate a 'hit', and Deveroux was the target. His cover had been blown, and for whatever reason, they wanted him 'rubbed out', to conceal their secrets. The hard-core unions were not even associated with the actual energy-company top-leadership at the Corporate-Level, or the World/Western money-traders. It was a sub-contracted parts-supplier network, with lesser resources, and hungry for big, fat long-term deals to supply all kinds of Thermal-Energy Underground repairs of equipment. But the systems were breaking down, or needing costly repairs, mostly as part of a planned sabotage by which insiders could personally profit almost endlessly, unless they were found out in their scheme.

The two men were aloft and riding on the cold winds and shifting cloudy air-mass, about 300 miles East of Seattle, at an altitude of 15,000-feet, in Deveroux's Pensacola private jet aircraft, successfully liberated from the authority of Officer Morton and Seattle-area 'codes-and-compliance'. It was an elegant and superior aircraft for its time. One reason Deveroux enjoyed this jet model, was the anti-gravity under-wing assist 'pods', creating a lightness and very fast air-speed that a personal aircraft of prior eras could never have achieved. Anti-gravity technology was very new, even then, but held great promise, and Dan's Pensacola twin-engine jet could easily travel at more than the speed-of-sound, coast-to-coast above the US land-mass continent in about 45-minutes. From within, only moments after take-off from Seattle, Deveroux was at the controls with Angel-Face as co-pilot. The interior was comfortable and efficient, and large enough for whatever they needed. Weather conditions were mostly favorable, with light rain below and mild winds.

"It wasn't the money, you know, Angelo," Deveroux was commenting. The jet-engine sounds were dulled to a low-streaming rush and hum, the control-boards glowed with digital systems-detail and controls. In another ten minutes, they would put the aircraft on auto-pilot. "It was the sabotage, and the fact was that California was going without power for months on end. They cared nothing if hospitals, homes, road-work, or mass-transit, were without electric. The money really was only important to my killers, I guess."

"Well, I'm glad you're dead, then, I guess," said his partner. "But please fly the plane with the greatest caution. Thank you. "

Then Angelo just hummed softly to himself, an old Mexican traditional song, called 'Posada'. He closed his eyes, drifting into a space within he liked to keep clear of confusion or messed-up thinking.

In his own life, he cared very little, personally, about the California energy system needs. It was just a job. He didn't live there, or not very often. He had no authority or office as far as decisions to make, and he represented no constituency. Only a few scattered family-members (Mendoza), were California residents. While Daniel was working through the recorded failures and logs at various Thermal-Plants, Angelo had been leading a mobile maintenance-crew, in the Central-Valley, serving several of the deep-earth heat-transfer plants, as a 'new-hire boss', totally under-cover, learning all he could regarding the allegations. From the Pensacola's co-pilot's seat, Angelo hummed softly to himself, eyes-closed, not answering or commenting. After a bit Deveroux settled in for the flight, and they were both silent.

The lands and mountains, West to East, were as-ever dark, vast, multi-colored and layered of endless variety of flora and forested green, or hundred-mile swaths of crops, and of course the cities and towns. The 'future' came and went, then passed behind to whatever it had been, to whatever it was to become, and although the people and folks of 400 years-past were now certainly gone, the Earth itself, somehow, had not changed much at all. The powers and leadership finally recognized that the future of the Ecology in general was likely more valued, even during the temporary passage of one generation or another, than the greedy and reckless ruin thereof.

Large forests were coming back, urban qualities had been re-invented for long-term sustainability, pollution from carbon-based fuels was now very, very low. New industries and policies made it possible for a general restoration of much that had been spoiled by the wasteful practices of the past. Global warming and climate-change had been mastered, by use of vast deep-ocean pollution-sucking bio-engineered bacteria 'ponds', in-place now for 100 years. The deep oceans of the Earth had been pioneered for super-abundant food sources, around year 2200, utilized of bio-technology as hyrogen energy to restore global Terra in thousands of new and highly effective ways.

War, among nations, had been outlawed, about 100 years after that. Of course few if any of the local nations and peoples kept those agreements perfectly, but it had permitted of a shut-down facility in international forces, that easily and painlessly disarmed any serious conflicts. If any nuclear weapons or bombs remained active or functional, anywhere on the planet, Deveroux did not know about it (and he in particular better informed than many). The approach by then was non-violent regional control, and cooperative international wealth-and-resource management, backed up by global protection forces, similar to military, but almost exclusively adopted or employed for disaster-assistance and conflict management. Thus, 'no war', and a satisfying prosperity. Both men, there in the aircraft high above the Rocky Mountains by then, knew, or suspected, that things could always change very fast, as far as Earth-powers were concerned.

One never really knew anything. And at the same time, by that year or era, only 40-percent of the world's population of 12-billion people, could procreate, or have any children at all. Less than 40-percent. Science finally could blame the AIDS virus with its very own Extinction Level Status, but human fertility had dropped dramatically, from about year 2,150. The remedy applied had brought this unintended result. The African and then global AIDS crisis was finally stopped cold, by use of a Universal Prophylactic, widely in use for centuries (by then), but with the unintended consequence of creating billions of completely infertile sexually active people, both male and female, who could not produce off-spring. But, they didn't get AIDS, either, and that plague, and the billions of lives that the virus had taken, over about 175-years, was overcome. A peculiar justice to it all, many thought.

The anti-gravity assisted jet-aircraft glided effortlessly through the windy high cumulus clouds, a silver bird of tomorrow's dream. Deveroux had the help of the PPP to settle out his life in Seattle. They didn't want to lose him as a pawn in some bloody hit-job for a big power-company union dispute. So, they used a synthetic-corpse (with real lab-fresh human tissue!), faked a highway accident complete with wrecked cars and medical-emergency response, floated a lie to the media with total authenticity about the tragic death of a Seattle-area University professor no one really knew much about, and cleaned out his cozy home there on Oakland Street in a matter of hours. Now a non-person, Deveroux hooked up with his assignment partner (Angel-Face) within a day or so, and at the advise of his San Francisco regional PPP task-council officer and sometimes lover, Veronica Signo (also known as 'V'), they were soon scheduled for the flight out to New York.

Medical records, employment records, tax records, transportation documents, passport, and other evidence of Deveroux's existence, were neatly cleaned up in complete secrecy. Glamorous? A future-world CIA or FBI? Daniel and other PPP researchers loathed such a comparison. They rarely used weapons, and the military or police powers avoided the PPP with great prejudice. Individually, men like Angelo and Deveroux had almost no real power, no important votes or laws to pass, no big-money decisions or inventive new ideas. They were truly 'researchers'. Their only real power was the one thing many of the Earth's corrupt and selfish king-makers feared: truth. So, they were very careful about the details and types of factual information they were moving. Lives sometimes depended on their work, including their own. But James Bond and company were thought to be worthless myths born as cartoons for children.

"It was a mouse, a field-mouse, just a small rodent," Angelo was telling Deveroux, as the jet began its programmed descent to an up-state New York airport, for their landing and disembark. "A mouse, Daniel. Or, I guess, a family of mice, maybe a few dozen, in a hay-field, south of Fresno."

"A mouse?"

"I don't expect you to believe me, but I think it's true. I was under-cover as Project Manager for Scheduled Repairs, at Geo-Thermal Fifteen, back a year ago. It's on 1,000-acres, nothing but hay-fields and a few cows. The men were doing their normal work, mostly checking the plant heat-sensors and heat-efficiency, and then re-installing whatever was not up-to-standard. One of the workers reported that he was going to need an exterminator, to destroy a family of field-mice, living under the wall of a storage-area. They were getting inside nearby sensitive equipment, eating through wires, and opening the dirt-wall floor beneath the storage, to cold and wet and dirt, or dust. So I went to look, before he killed them. Inside was all the evidence I needed, I took pictures, I made recordings and measurements. The storage was full of parts and pieces from the thermal plant, that had been deliberately sabotaged, removed, broken up or pulled apart, or basically just pulled out from use at the plant. That was how they pulled off the trick, I guess. They just yanked out expensive gear, ruined it on purpose, stored it temporarily, and then made millions of units fixing what they had broken themselves, year-after-year, as sub-contractors. I logged it all. But far more of interest, was the fact that a mouse could in some way have been instrumental in bringing down the union-scheme, you know??"

"Mother nature wins every battle, Angelo," Deveroux said.

He landed the jet successfully, and they were later on their way into the city. New York by 2,412 AD/CE, was a regional authority or self-governed nation-state, on its own, and no longer truly responsible to the laws and 'federal' government of the US lands and people of that period. Population: 60-million for the City of New York alone. Somewhere down in that human zoo, the Planetary Program Proficiency facilities were tucked away, almost unknown except to those involved. Deveroux and Angel-Face now needed to discover what it was that Veronica was really talking about, before they left Seattle.

Jupiter program? Space-travel? All the PPP researchers had a choice to accept or deny any of the cases the PPP came up with for them. So Daniel and Angelo would spend some time going over what was being proposed. A single case could take even more than two or three years, and the Jupiter Program was immense in scope, more than 100-years of successful raw-materials harvesting and transport, from the planet Jupiter and its moons, back home to Earth, and to other near-planetary resources.

" I don't think I can do this, Daniel," Angel-Face said, on their way into New York by railway, a day later, after a short stay at a hotel near the airport. "I don't want to travel into deep space. I haven't done much of that. I never liked it."

Daniel shrugged restlessly, and continued reading the local video-feed for area news, as they rode together on the pneumatic-force people-mover system, a glass-and-plastic wonder that floated on magnets smoothly into the endless urban realm. He didn't really care for the idea, either. He was older now, and had done too many PPP cases. He wasn't their best, but he and Angelo were known for their loyalty and skilled work, and also had many friends. But they were near burn-out. The pay and perks were great, and in fact, by the time they retired, PPP would provide them with quality lifestyle needs for the rest of their lives, including money, and homes. But space-travel was strenuous. The Jupiter Program was the sort that Earth was proud of, and shared as a great boast, even on other worlds, according to what he had heard. Deveroux preferred to work in areas he at least had some knowledge of, it helped in many ways. He was not a space-man, and neither was Angelo. But like many PPP cases, it was their status as 'outsiders', that supposedly gave them a fresh view.

"Maybe it's not important," he said, with Angelo now gazing out the window. "Maybe the Jupiter Program doesn't need us, or maybe we can just walk away, let someone else do it. Fine by me. I have no interest in space-travel, either."

They moved into the city like ants inside a dusty Christmas HO-scale train-set from 1952.

CHAPTER 3: With Me, With Wu

"Are you there? Can you hear me?"
--Captain Martin Brendeis, pilot of the deep-space Jupiter-program transport 'Down', to Jupiter Base-Alpha, via ship-to-ship radio, mid-year 2,410AD/CE.

Jupiter's Alpha-Base, in that late period, the far future, or the endless tomorrow, beyond year-2,400, was a dream of Mankind's improved awareness. This home-a-way-from-home, at such distance and of such a nature, that one might only suggest it as an impossible place that we needed, and built and created for ourselves, to explore and exploit the near-Earth planetary system. This was by that time, more than 80-years in-operations, the base itself, and headquarters for Jupiter-program management, as far as deep-space support and command. Vision, and the spirit of man, had learned a thing that could never be reversed, by this time of the human passage of us all, that our salvation depends on ourselves. All we needed was at our very door-step, from the circle of planets nearest the Earth, now within reach.

The Alpha-Base was one way this idea had become real, the physical form and How-To Reality, where people lived and worked and played, and were born and died, and all in-between, a bubble of humanity, more than a space-station, like a world unto itself, visited by few, a place with its own rules and ways.

"Welcome home, Commander Wu," a robot-voice greeted a certain space-voyager, known as Menuda Wu. As he finally reached his personal quarters at Alpha-Base, following a visit to Earth, it had been a journey there and back of nine months. Wu was a Jupiter-Program big-shot, essentially the Program Director. Not a king or dictator, not demi-god or Ceasar, no prophet-magus, no sky-bound CEO, no grandiose extreme Uber-Mind, and not granted such as life-or-death power, or to keep himself slaves or concubines.

Wu was instead only a man with a job to do, happening to be management for Earth's Jupiter Program Off-World Division. Once the men and ships and robots and equipment left Earth, he was the responsible final decision-maker. And of course he in-turn answered to other authorities, councils, and Earth-based management. Top-dog, boss, Caller-of-Shots, chief, head-honcho, show-runner, or ring-master, Menuda Wu was a man of his times, no Siddartha at age 53-years. Wu was a person bored and restless with a burden of responsible service, somewhat trapped in the high-technology loneliness and isolation of it all, and seeking comfort where he could find it. Thus a reputation as a considerable sleaze-bag followed his entire career.

Wu moved through the entry-way of his private rooms, to be found in one of the more elite housing areas at the base. Jupiter's Alpha-Base was built on a very large planetoid, or asteroid, a huge floating iron-granite rock, called Almathia, a planetoid, irregular in shape, not a globe, but in orbit around Jupiter Main the same as her moons.

Orbits including Doctor Wu. Like most men, Wu was better appreciated as-advertised, than in-person, given his position and role, and in his own self-esteem or idea about himself. Maybe it was all the machines and sleek, beauteous ships and space-stuff, the fancier-than-thou life-or-death air-breath dependency, the Star-Trekkish appeal of it all, un-naturally man-made in every detail, separated by necessity from the dirts and dusty soils of Earth, the gritty life-giving filths and hills and valleys, back home. The complete contrast crept into their souls, and persona, with long years at work in space.

Wu was was somewhat unpleasant, to some people. He had an effete-snobbish luxury about himself, not strong or athletic, but not very weak or very obese, not an outstanding handsomeness or appeal, but not really rugged or craggy, and in his dress and outfits, he was not a fashion-plate, not really creative or outlandish, but not either very conservative or totally in-uniform or conforming, but not sexual, not really very aggressively male, but suggesting a willingness to join in, give-it-up, or find a secret way to indulge that no one would find out about, because he had a high-profile, even with himself.

Conceited, in a word. Some space-workers found him even detestable, but he was kind enough, and in every way efficient and in complete control of his assigned tasks. The world's Jupiter moons raw-materials and exploration programs were vast, by era-2,400, just like the man's ego.

Wu's voyage had taken him by people-mover space-ship (rather than the huge raw-materials transports, like the 'Down'), back to Earth, some 700-million miles from the Sun, or about half that distance from Earth. The journey was, as always, arduous and fearful. The ships by that period of Earth-launched local space-travel, had advanced significantly. It was a new era, men of Earth had ventured into the local planet-system, for fun and profit. Journeys to even a single 'other star' beyond Sol, even in the Home Galaxy (the Milky Way), had not even been seriously considered by space-program planners, and seemed to be forever out-of-reach.

But our nearby planetary system, Mars, Jupiter, Saturn, Venus, and the others, had opened up a vast realm of potential wealth and resources that could be of tremendous value to Terra. Often, this was not the case, such as on Mars, or the Earth's moon, where little had ever been discovered that was useful or valuable. In general, it was for raw-materials like the enormous supply of pure Hydrogen on gaseous Jupiter, that the programs were initiated. After 100-years of exploration and development, the Jupiter Program was by that time a routine and systematic harvest of such elements. So, a 400-million mile jaunt back home, for Wu, though it took nine months or so, this too was routine. 300 days of space-travel, compared to the early years, when it took five years to reach Jupiter, it was really nothing.

He lived in space anyway, deep-space was home for Wu, and the 550 residents of the Alpha-Base. The new ships, like the passenger-ship that had ferried him back to Alpha, could attain near light-speed travel, or half that fast. He really didn't care, in the sense of being very impressed. The novelty eventually wore them down, and it was more of the same, astonishing and amazing, dull and dreary, perfectly safe, perfectly terrifying. This is space-travel.

Wu settled into his cavernous personal-quarters at the Alpha-base, or, as they called it, the Rock. Alpha-base was still another 20-million miles from the nearest on-site Jupiter-moon facility, of which there were 12, surrounding the giant gas planet at various points. But it was the center of all their efforts, with large and small ship docking to refuel, life-sustain and energy-charge, communications and telemetry-tracking, information and dispatch or schedules, and many other resources and essential program technology and staff. Hard to call it the center, since it was in orbitude as well, along with all the other ships and loading or harvesting platforms and bases, that is, in-motion. There really was no center.

As he began to relax, home at last, he was able to tend to personal needs for an hour or so, mostly checking out his effects and items he always had with him: his Priority-Communications and Command Module (much like a High-Level one-of-a-kind Palm Pilot, by which he could keep in touch with his power-base and underlings); his toilet-kit and Sundries-Baggage, (very important in space-travel, given that things could always change), and also personal, so that in his case it included durable and softly woven real cotton-fiber clean towels, wash-and-rinse effects for mouth, hair, body, toilet, teeth, eyes, foot-care; also his private (and secret) personal defense 'weapon', a specially-made gun-like technology known as a TASP (Temporary Alpha-Wave Stun Pistol); not deadly, but very effective when he felt threatened for whatever reason.

"All this junk," he spoke aloud, to himself only, there in his private chambers. He briefly checked his sources to learn if anything had changed there at the base that might somehow effect him personally, such as if there had been any odd reason why security may have entered his private quarters, or like that. His priceless collection of rare Earth-born comic books, for instance, was in-tact and un-touched. Wu often felt some of the base-regulars coveted one or two of the titles, sealed in plastic for even hundreds of years, and preserved with a special glass-like substance, so they would remain without decay for a long, long time. But, no matter.

He had traveled to Earth to connect with program leadership groups, concerning the very rare event in which one of the large transports, the 'Ferrous-2', had collided or crashed with an external loading-docking station, one of two in orbit around the Jupiter-moon Ganymede. This had been about a year ago, by the time he found himself there in his room at Alpha-Base again. The fact was, as old as the Jupiter Program was, the very large transports, like the Ferrous-2, were never even close to the docking stations or ports or orbiting stations like the now-antique Monilari, and Molinari-2 (between Earth and Mars).

There had been perhaps three major crashes, in the history of the program, and this one made four, and on his watch, too. Minor mishaps were also very unusual, and obviously in any deep-space work, even small mistakes could be deadly and cost many lives. So, following up on this circumstance was his job. Dr. Wu made the trip home, ending up at Program Control, which at that time was in Europe, far from the actual launch sites or space-ports. Mostly his role was to comfort nervous financial and investment concerns, and to smooth over things with suppliers and the space-industry in general. Again, the Jupiter Program in 2,400-AD/CE was a gigantic enterprise, with hundreds of thousands, if not millions of employees.

Wu sat down with his personal computer-kiosk, and brought up information on one of the transport Captains, Martin Brandeis, of the 'Down'. Brandeis was currently active as a cargo transport Commander with the program. The 'Down' was similar to the 'Ferrous-2'. These were huge ships, about half-a-mile long, with gigantic rear-thruster engines, and a series of holding tanks in-between, up to the Flight-Module Control-Room in front. These big space-vessels, but for the novice, were impressive without a doubt. Doctor Wu enjoyed music as he worked, a popular rock-n-roll tune of the times, sort of a trance-inducing beat, with bizarre-yet-soothing electronic tones. He already knew much of what he needed to know about Brandeis.

"Wu-142, please display current day-date-hour information and details, Captain Martin Brandeis, transport ship is the 'Down', Program vessel ID is Down, Brandeis-Europa, station four. Waiting."

The computer brought up the requested information quickly. As Program Director, Wu could find out anything available, any time, regarding anything or anyone, involved or connected to the Jupiter efforts under his command. Anything true, that is. Dream-states or stomach-contents were a bit more difficult.

There in the display, was an image of Captain Brandeis. He was a husky-looking, athletic, mature astronaut, rugged and ruddy, and like most of the space-workers, he had an intensity and intelligence about him. Good-looking, 'the best of the best'. To pilot one of the big transports, nothing less would do. Martin's face was that of a 45 year-old Caucasian, with a deep brow and brown hair. Nothing unusual-looking.

There was a cue-tone now at the doorway to Wu's apartment. It was a service-robot with his dinner. Wu was assigned to a particular robot for dinner, baggage-carry, room and area gear, house-cleaning, alerts and announcements, social matters such as introductions or escort, minor repairs and reproductions (temporary data transfer), and many other functions. He called the robot, 'Burt'. The bot's personality-program was male, his basic role was such as a butler.

"Come in, Burt," Wu called out, and the room-door opened automatically. Burt's appearance was rather like a rolling cart-thing with a domed camera-eye head-box, and moving arms and serviceable hand-like clasps. He was self-motivated and self-sensing, and there were controls and operators-command buttons. He could also talk.

"Dinner is served, Doctor Menuda Wu," Burt intoned. His voice-tone was as pleasantly non-robotic as possible. All the robots at the base were rather like adopted pets, for the staff and crews. The brainless, soul-less machine moved through Wu's door, and into the area where Menuda normally took meals. Mechanical, jointed limbs and grasping hands moved the dinner plates and food items in perfect manner, onto the table, with flawless etiquitte. The meal consisted of a light pizza-bread with cheese, hot coffee with drinking alcohol, and small candies.

"Thank you, Burt," Wu said. "Dismissed."

"Yes, sir," the robot voiced. After a few moments, the thing turned and tucked its arms away, with small, soft, rubberized wheels motoring by electric-servo, back out the doorway, which then softly shut behind. The Alpha-Base had a false gravity-field throughout most of its working areas for staff, so the robot rolled as if back home, held down by the normal pull of artificial gravity (which included 30-percent real gravity, from the giant asteroid where it had been built). As Wu enjoyed dinner, he spoke by video-phone with one of the Human Resources staff, a young woman he hardly knew. The connection lit within moments with a few deft computer commands.

"Yes, Brenda, I just wanted to chat about one of the transport pilots, you don't mind?" he said, charming as always.

"Of course not, sir," she answered meekly. Brenda worked in staffing and personnel. She had a lot of personal contact with the pilots, mostly as a connection to families and private matters back home, standard for all long-term workers. This included payments, housing, medical, vacations and leave, recreation, sons and daughters back on Earth who needed to reach their dad's in deep-space for some urgent matter, as well as spouses, and many things like this.

"I'm interested in a certain transport pilot, Captain Martin Brandeis of the Jupiter transport 'Down'," he said. "I have all the usual details myself, of course. But, do you know him?"

Brenda seemed to hesitate. "Yes, sir. I handled his account last year, for about six months, then it was turned over to another Staff Concierge. I don't know who, we have dozens. He's a very nice man, sir."

Wu continued with his meal, alone, yet not alone. Brenda, the young Human Resources Staff Concierge, was perhaps many levels away, on-call during regular duty hours. She would normally have filled many other roles in her work, handling the personal affairs of the pilots was just one task.

"Well, nothing like the personal touch," Wu said. "Makes the difficult work for the pilots a bit easier. But this guy, you know, does he seem like any kind of a trouble-maker to you? He looks very intense from his profile, the one I have. The reason I ask, I am told he has a complaint he has been working on, about the big crash we had a year ago. Some kind of private research he has been doing, I guess, or some collection of information or science, something he feels he wants to bring forward. It's not a secret. I just wondered what he's like, or if he has a reputation, you know, whistle-blower, or rebel. What do you think? Have you heard anything?"

Brenda again paused. "Not really, sir. Everyone gossips, the pilots have their own supporters and fans back home, like athletes. Brandeis is not unusual. Real straight-up. I guess he likes horseback riding back home, and there was something about a large horse-breeder operation. He felt it was not managed properly as far as breeding techniques, and it ended up with a big lawsuit, I guess what they call a Class-Action Lawsuit, there were meetings and industry people, defending their medical practices. With the horses, I mean. It ended up in court, but they lost. Things like this. They all have other lives, besides the ships."

Wu listened. "A breed apart, true," he said. They chatted more while he ate. Brenda felt impressed with Doctor Wu, he could be very charming and of course he was a powerful man at the base. She had other information and opinions as well.

"Welcome home, Doctor Menuda," Brenda said after a time chatting with him. He politely ended the call, as the vid-phone shut down. Wu fidgeted with his dinner candies a bit, trying to decide if he would make an enemy out of Captain Brandeis, or not. As far as the fallout from the crash of the 'Ferrous-2', and the on-going recovery and repairs, and how to deal with the effect on his reputation back home as Commander of the overall Jupiter

Program efforts, there in deep-space, there in tomorrow, Wu was elsewhere, not different, not the same, in his mind and thoughts.

CHAPTER 4: The Down

The deep-space raw-materials cargo transport ship 'Down',
also called a Whale-Hauler, by Jupiter-Program regulars, was
3,532-feet long, just about two-thirds of a mile. Without the pull
of planetary-gravity, a gigantic ship like this, could cruise at very
high-speeds. No friction like an air-ship, no escape-velocity to
enter orbit, no lift-off or launch, and nothing but emptiness itself,
to inhibit forward motion. The new generation of thruster engines
could approach light-speed, even between Solar-planet system
objects, such as Jupiter and its moons, the asteroid-belt and the
Alpha-Base program headquarters at Amalthea. The greasy-brown
epidermal skin and somewhat atrophied muscles of Program
Director Doctor Menuda Wu, currently sweated salt and toxins, in
the gymnasium he used, mandatory for many space-travelers,
given the overall stress and demands of space-work. He was
thinking, and it showed.

The 'Down' was much like a giant whale of a space-ship, huge
to gaze upon, as tourist-voyagers and students would do at times.
Two-thirds of a mile in length, and the width of the thing varied,
much wider at the back-end and main thruster-engines, and then
very long in the middle, with a series of holding tanks and large
cargo bays, and other needed structures, and then in front was
the Flight-Command and Navigation Deck, and also staff and
quarters, life-sustain. It wasn't a very attractive ship, not sleek or
appealing for beauty, or sharp-looking as if a fighter, or a
pleasure-ship for passengers.

Brandeis had been the Captain of the 'Down', at that time, for
about three or four years. Martin was among an elite type of
space-worker, some 300 years into the new era of space travel.
Given that Venus, Mars, Jupiter, Saturn, and the outer planets,
and also the inner planets, and the Earth's moon, had opened up
into a period of exploration, it became possible for those planets
to be reached by travel for passengers and workers, in an almost

routine way. For the very large transports, like both the 'Down' and the 'Ferrous-2', the program had about 40 of these types of ships, and twice as many capable pilots or captains. Of course a qualified team of five or six capable commanders were required to operate each ship.

The program was thus a very successful Earth-industry, only to build, maintain, operate and staff, these voyagers to the local planets. The same industry, better described as an empire, manifest itself 700-million miles away, on-site around Jupiter. Basically the whole of the Jupiter Program, along with governance, science and technology sources and learning, and much more, was here. There was no other way it could be done, but it had taken several hundred years. The Jupiter Program was more valuable as a new, highly productive industry back home. Planners had found many off-world resources, mostly vast quantities of useful fuel types, such as Helium and Hydrogen, and H20. Thus, a man like Martin Brandeis, and other people with his skills and approvals, were an important key in the chain of ships and cargo, back and forth to the spinning giant, and home again. He was an important man. He had clout.

The 'Down' and other ships like it were fairly simple in design, although also of course very complex. Their function was to make the journey back-and-forth to various loading-points near Jupiter, and receive the raw materials in the big tanks, and return to Earth safely and unload them. An extraordinary task, so the planners and designers wanted the ships as large as possible. The weightless or gravity-free environment in space made heavy loads light, and reduced travel times. So, giant ships meant the trips weren't wasted, and they could haul as much of the 'stuff' as they could handle, millions and millions of tons in one trip.

In a way it didn't matter how large the ships were. But there were limitations, especially command-management and navigation. Design and technology had been developed over 300 or 400 years of modern science. Like almost everything else modern societies enjoy, it was from the past hard work of inventors, students, philosophers, and scientists. On the shoulders of giants, they were able to create and design marvelous wonders like the 'Down' transport, and the other Jupiter Program features. All of the program features were a group effort by humanity in general, there was no super-genius designer, or Space-Messiah, telling everyone what it was all about, what was going on, which way home. The program was strictly a commercial-financial-exploration effort, intended that way to benefit the people of the Earth. After 100 years or longer, the routine of operating these large ships was successful, year-after-year. No room for errors, so there were none.

In viewing the 'Down', Martin Brandeis' ship, one might start with an external view, such as a University student tour-ship group might enjoy to examine and review the workings of the ship from the outside. Of course, outside was the abyss itself, and instant death to any living creature. When ships like the 'Down' were in-orbit around the Earth's moon, on return from the transit, or in-between voyages, the World Student Assembly, in that year of 2,410 to 2,412AD/CE, would use study ships to do just that. What they would see was a very large, 2/3-mile long deep-space transport ship. Luna was a natural orbital point for the big ships to park, as large as they were. The Earth is about 28,000-miles in circumference, 250,000-miles from the moon.

The Jupiter program had many of these large ships by this time, and they had to maneuver, unload, navigate, dock, etc. Even at 2/3-miles long, the 'Down' was less than a fly speck in the Solar System at large. It was the same as they moved around the moons of Jupiter, and the upload stations and docking platforms, and the Alpha-Base Amalthea Asteroid-constructed headquarters, where Program Director Doctor Wu Menuda was still sweating-to-the-Oldies. He was thinking about the 'Ferrous-2'.

From the back or rear of the space-ship 'Down', students aboard a small study-ship (that might hold 20 or 30 high-end science-tech learners), the large thruster-engine were quite a sight. It wasn't so much their size and position on the frame and structure of the ship. The engines were a new type that were suitable for very high speeds and very great distances, in the local-planetary system. It's hard to conceive just how far the planet Jupiter really is, and yes, if you had to walk it, you'd still be walking.

The engines were known as Huschcroft Ion-Accelerator-types, from a concept developed by a Fifth-Grade science teacher 150-years previously. True enough, the concept-creator, William A. Huschcroft, had no real credentials in space-engine design. But it was a popular hobby for many science fans and students, at that time, to toy with, or 'paper-out', some of the physics and ideas of what might serve well in those environments. The Ion-Accelerator thruster engines worked on the principle that a big ship in deep space could accelerate almost endlessly, by amplifying action-reaction thrust, as a repeating-cycle of small-energy output bursts. There was very little ionization involved, that only referred to the type of fuels used.

A common mistake in understanding high-speeds for big ships in null-gravity, was that one simply used huge, high-energy, Almighty-level power-thrust, like gigantic simple rockets, but this was not the case. Big push, fast ship. The ships needed to speed up to tremendous rates of speed, (almost one-tenth the speed of light, or even faster) in order to be useful in reaching Jupiter, and returning, without the crews suffering from old age. How would it be done? The Huschcroft-Accelerator concept negated the image of increasing levels of rocket power as the ships would blast through space with smoke-and-fire flaring.

Instead, repeated small-level thruster action, over-and-over, were applied, increasing the weightless ship's speed as a constant rate, or velocity amplification. This was only a theory about the year 2200AD-CE, but no one had really figured it out for actual use or application. Great gas-mileage, and they could still reach high speeds needed. There were more traditional forms of propulsion the 'Down' used as well. The pilot would move into his navigation-track, and ramp up to a basic speed, for a 300-million mile run, then he would move the large ship through a common-propulsion series, and then drop into the Ion-Accelerator Drive. In about ten days, they could achieve about 40-million miles per hour, with the speed-of-light as a constant 186,000-miles per second, or more than 600-million miles an hour. These are only very general layman's approximations.

As the planets moved in their paths, the longer distances changed depending on the various positions, much like astrology. As the smaller student-ship carrying World Student Assembly students, moved keenly and with high-maneuverability near and around the massive 'Down', in a parking-orbit around Luna at that time, what they would see were the engines of this type and their output vents or thruster-stream cowlings and the external housing, for all the plumbing-like tech. It was all quite impressive, the 'Down' was perhaps 1000-feet across, or by width.

The moon below them was so small compared to the gas-giant Jupiter, where the crew and staff of the 'Down' spent months at a time, Brandeis and others from the Command Deck may have mocked at little Luna's stature among the planets, as hardly more than a tennis-ball hung by a thread, by comparison. Jupiter has a mass-density of more than 300 Earths, and an inner radius of at least 12 Earths, side-to-side equal to 24 Earth-diameters. For workers and laborers on-site near Jupiter, the thing simply filled the entire sky, with dark, multi-layered cloud-forms of colors and shapes, and the hellish, legendary 'red eye', moving past them in a daily orbit, spinning like a top at only ten-hours per complete planet rotation, despite it's massive size.

The students in the small tour-ship could now view the entire motor-engine display or array. Their little ship was scooting around the whale hauler like a sardine interested in a aircraft-carrier. The spine of the 'Down' was for the cargo, and consisted of row after row of very large tanks, and also other types of containers and holding cells. There were three sections, one with about 20 fluid-type containers, each with a capacity of many millions of cubic yards. These were also of course sealed and protected for deep-space transit, and also filling, and un-filling, ship-to-ship; they resembled big dark tubs, with huge rings of strong materials, and long ribbing, and many openings and access points. Somehow reassuring.

The second section, moving towards the front, was not specifically a cargo-hold, but was occupied with many connections and system-technology, also long-form communications-tech, energy-solar collectors, smaller ship docking points and loading points, ship-based resources and fuels, generators, a maze of high-tech wonders, all necessary, all man-made, all fallible, all backed-up by redundant systems. The 'Down' also had small shuttle-ships of her own in this portion, tucked inside parts of her hug belly. This section also had holding cells for other types of materials, tanks and inert-materials and substances containers, sealed and prepped for long-distance vacuum.

Forward towards the command deck, one-third again of the mid-part of the 'Down', was reserved for H20, mostly from Jupiter's ice-moons, like Io and Europa. The pure water was harvested as frozen ice, and loaded, and remained frozen in the cold of space. There was some controversy about collecting H20 in this way, and whether or not it was a resource Earth needed in such volumetrics, and with such on-going regularity. But it was a dedicated cargo for these ships, and over many years was very useful back home, where natural environmental changes made pure, fresh, elemental H20 in large supplies, a valued item, shared by all for life itself to flourish. It was traditional in that sense. So, this part of the ship looked different, the H20 blocks were not liquid, or gas, and were handled in another method. So this section resembled a series of very elongated front-to-back tube shapes, with portals and loading entry. The spine itself traversed the entire ship, and connected to the Command Deck or Flight Command part, in the very front. Workers could move anywhere on the ship via the spine, and there were other passages as well.

The Flight-Command Module was in appearance much like one would expect, with windows, passenger-ports, a sort of upward-tilted 'nose' shaped aerodynamically, if only for the look of it, perhaps psychologically motivated by the designers. It was also very large, big enough for all crew, crew-housing, control and command functions, telemetry-tracking, docking, life-support systems. It was like a very large, exotic, deep-space high-tech mobile airport, where Captain Brandeis was also able to operate the ship, along with his teams. And true enough, they had the best of everything. Gone were the days when space-travel was a painful and arduous struggle to survive against-all-odds, with null-gravity environments and Spartan lifestyles, moist or dull CO_2 contaminated passageways where life-support was not quite cutting it, and fearful shortfalls or temporary systems failures.

Inside, Brandeis was busy with some routine work, in this case prep-work for their departure from Earth-Moon orbit, and orders for their next run, destinations, loading points, cargo types and amounts, planning their navigations and changes. This was done as a team in a work-room with all needed computer-based systems info kiosks and command-control portals, under the ever-dim central lighting and wall-screen info. The surrounded themselves with emotionally supportive music, beauty-images, sexy entertainment, and a run of all kinds of details and closed-circuit ship-board only media. The boring work had to be very properly done and double-checked. Much of their work was automated, but never the real choices and decisions.

"Martin, it never really dawned on me that your research about the crash of the Ferrous was going to cause a big stink on Alpha-Base and back home," said his Second Command, a woman, whose name was Menima Pearl, a demanding and authentic space-commander in her own right. She had dark hair and big eyes, athletic but abundant, not much for the jokes. "You haven't released all that yet, have you? I mean, your findings are not public to the Program, at this point? How did Menuda find out so much of what you were trying to do, so quickly?"

A long series of computer calculations were flashing before them on a vid-display, moving through planet pathway projections and waiting to create a secondary navigation they would hold for possible back-up usage many months ahead. Menima had initiated the basic plotting, as she usually did this chore, but it could take hours for the results they wanted, and was based on actual current-hour data about goings-on in the solar-system, gathered by their tracking and telemetry machines (telescopes, radio-antenna, radar, this sort).

Brandeis gazed dolefully at her a moment, looking up from his work on a hand-pad. He didn't want to broach the topic about the 'Ferrous-2', not with her, not at this time. Like any command-deck relationship, they were jovial and friendly, and even close. But they kept a very strong distance emotionally, more than at arm's length. It could be no other way. And yes, it was quite true, as a leader among the Jupiter-program commanders for the transports, Brandeis had taken it on himself to look into the entire case, and gather whatever he could find out, as a documentation, for whatever purpose might be found. He had been working on it privately since the accident.

"For a big stink, Pearl," he said. "You ask Life-Support. Otherwise, I do not wish to discuss it now. I imagine the Program Director has all sorts of mysterious ways to spy on me. Big deal, nothing new there. They are very sensitive about the crash of the Ferrous-2, that's all. Not surprising. I will probably have to retire."

He smiled and Second Commander Pearl laughed just a bit. After all, the 'Down' could always smash the corner collector-panel array of a Ganymede loading station, in a similar way, and they needed to understand the details of such a rare incident, to learn from. But there was much more going on, and they knew it, a sly sort of un-spoken agreement between them, and something else.

CHAPTER 5: The Predictable Path

"The planet Jupiter represents wisdom, like the old Roman deity Jove, or Zuess, king of the gods. Not that there's anything wrong with that,"—Captain Martin Brandeis, Commander of the deep-space transport, 'Down', 2,412AD/CE

Astrology, or the practice of figuring human fate and personality, the love-relationships and annual ups-and-downs in our lives, and so much more, was always a faddish thing. Among the Jupiter-Program workers and staff, and families connected to them, it was sort of a popular game. Capricorn, Aries, Taurus, Gemini, Aquarius, Pisces and so on. Linked in ancient times to the paths of local planetary objects, that is, the planets, it made sense culturally, as little more than an amusement. Space-workers in year 2412 might contemplate this planet or that, favorites, personal birth-date relatedness to planet-paths, and historical legends. Serious atsrologers might take note that with physical birth and death now at least somewhat common elsewhere than on Earth alone, the maps needed to be changed, for any gravitational influence on any particular human person.

"Cid Bixi Mimim," was a byword-phrase, or greeting, used in social interaction. It had no meaning at all. When at meals or recreation, in those times-to-come, for people to exchange the pleasantries of belief and superstition, they may say, *'cid bixi mimim, sir',* as teams or like-minded groups (such as the star-signs), in particular in regards to an individual's name, and ESP (Extra-Sensory Perception). The system's largest gas-giant planet, Jupiter, was thus a prominent player in this game, with a certain esteem and clout, in which the program employees took pride. More like a dance than anything else, so when moving about in the Program, 'Cid Bixi Mimim' was a courtesy, and polite introduction to the strangeness of the circumstances. Language-science researchers had studied origins of the term, but it was inconclusive other than sourced to some ancient Greek travelers

and their languages. But it was very popular, like a mind-reading guessing game.

They had called the facility at Amalthea, 'Alpha-Base', early on when the program was young, as a convenience or simplicity. There was also a more formal name for the Jupiter-program HQ, that being the home of folks like Menuda Wu, and so many others, like Brenda in Human Resources. The people there called it other names, like 'the Rock', or sometimes 'Minerva', the Roman mythological wife of Jove, or Zeuss (also called 'Hera'). Or it was just home, even when back home was Earth, but not for a long while, anyway, for them working and living there. By this period in human development, the Jupiter-program head-quarters at Amalthea, was a marvel of space-survival technology and achievement, there at the rim of Jove's many moons, structured safely and with a great sense of permanence, on and into the iron-granite form of a very large planetoid.

Why was it built, what was it for? Planners wanted to expand Earth's reach into space, and local planets were the next step, given the stars were truly out of reach, probably forever, even within the Milky Way galaxy. But not the planets, by that era. Jupiter's Alpha-base was needed for work at Jupiter. It wasn't even really near Jupiter, by spatial proximity. The planets still turned in their annual orbits, and the asteroids between Mars and Jupiter glided along like a bunch of stone bees on their way to some giant, sweet flower. And yet never arriving.

Many thousands of stones, probably from yet another Earth-Solar planet from times-before-time, somehow shattered into a million bits, and left to float in gravity-bound circular paths. Big asteroids, small asteroids, big planets, small planets, planets with two dozen large moons the size of Earth itself, ringed planets, dead planet, and the Silent Planet, Terra, Earth, Urantia (cultic), the place of cool green hills, food, Eden, etc. Amalthea-base was a practical and utilitarian facility, but in the legends of Man, it took on the stature of the pyramids of ancient Egypt, grandiose and ever-lasting, mysterious, even dark, as one of our first extra-terrestrial colonies of any long-term function and intentional service. *"Made to last"*.

Doctor Wu was Program Director for the Amalthea home-base. He hated his job, he hated himself. He was good at his work and seemed to please everyone, somehow. The hierarchy was much like a corporate-business would be. Decisions and choices were very few, actually, because the science never changed. The moons and the gas-giant were big raw-materials sky-bound supermarkets for Earth's many needs: helium, hydrogen, H20, and other substances. The basic ingredients they could harvest regularly, in vast quantities and significant purity. Drilling and mining, or collection-gathering sub-stations orbited Jupiter, along with the moons, like Io, Europa, Ganymede, and Callisto.

The four Galilean moons each fill over 1,900 miles in diameter. The largest is Ganymede, the ninth largest object in the Solar System, after the Sun and seven of the planets (Ganymede being larger than Mercury). All the other Jovian moons are less than 160 miles in diameter, with most barely exceeding three miles. Orbital shapes range from nearly perfectly circular globes, to highly eccentric shapes and inclines. Many revolve in a direction opposite to Jupiter's spin (a retrograde motion). Orbital periods range from seven hours (taking less time than Jupiter does to spin around its axis), to some three thousand times more than that (almost three Earth years).

And at each station, space-workers (among the toughest and most highly sought-after Earth-men around), did the grunt-work of prepping the high-tech transport of such materials. The raw materials were drilled or mined, (for lack of a better term), and shaped or converted into forms suitable for loading and travel. Just how hard is it to suck up 1-million tons of gaseous pure helium from the clouds of Jupiter? How much trouble is involved in chipping off half a million tons of H20 'ice' from the Jupey-moon Io, and loading it safely onto a transport like the 'Ferrous-2'? Impossible, improbable, a challenge, and then after many years, a reality.

There were many other highly-valued raw-materials at the Jupiter-Program sites. Fuel-sources, Earth-human usable life-sustain matters like water, and O2 (oxygen), were chief among their usefulness. No food, no DNA, no life-forms, no minds or consciousness, no artifacts or archeology, were ever found, by that date. Still they dreamed. The program made sense and 'worked'. Back home, it represented a economy on it own, huge, employing many millions who would never travel into space at all. It also meant new knowledge, new powers, and new sources of these rare materials, that could be applied to troubled Mankind, and the aging, corrupted Earth. If Doctor Wu was somewhat loathed among his sub-ordinates, it was only *'cid bixi mimim'*, as ever, among men.

Transport ships, (the whale haulers), were managed by a committee or team of ten or twelve specialists, led by a man named Confey Gomrah. The schedules, crew-staffs, the navigations, the loading and back to Earth, problems with technology and maintenance, safety and life-support, communications, were all directed by this task force team, and they answered to Wu. There was a similar management committee, back on Earth, that did not answer to Wu. At Alpha-base, there was another group in charge of affairs dealing with the drilling and miner-collecting sub-station or sub-bases in orbit around Jupiter. The task force teams were basically only management.

Work at the sub-stations was a lot more complicated and dangerous. It's one thing to run a giant automatic space-transport ship like the 'Down', back and forth to load, travel, into orbit, unload, and repeating. Captain Brandeis had a kush-job, compared to the drilling team 'miners'.

It was truly a thrill to find yourself responsible for ships and men, much smaller machines and lesser ships, that would dip down like bizarre silver pelicans from 'above', to skim or scoop across the fractional surface-level gases of Jupiter, to set up collectors for helium from giant pools. Some were more vast in areas known to contain helium concentrations, than the Atlantic ocean on Earth, and that a small portion.

Or the ice-workers. Planet-moons like Europa had surface ice-structures (H20) covering almost the entire planet to many hundreds of miles deep. Pure, clean, frozen water. These teams and sub-platforms or sub-colonies used high-tech means to simply melt away blocks of the ice in lines, and then lift them out of the planet moon's gravity-well, into orbit, and then for loading or transport. Management for these programs and people were also found at the Amalthea Alpha-Base.

So by year 2412, there were 12 main sub-bases or sub-colonies, at various points in the Jupiter-system. And more structures being built, or being disassembled after temporary use. Jupiter, at more than 300 times the size of home-planet Earth, has 63 significant orbiting bodies. Only three or four of these stellar-objects are as large as Io or Europa, or, similar in size to Earth (smaller). Many of the program applications or labor was performed by robotics. These were large-scale automatic machines that did the impossible work, of gathering and collecting, identifying and measuring, drilling and mining, and lifting into space, and loading. On-going science-technology research and exploration was included. After 100-years in the effort, the thrill had depleted and new discoveries were slowing down. Commonplace astonishment, boring-dull amazement and wonder, gigantic smallness and mind-blowing simplicity. It was truly a new era, full of adventure and hope.
Just another day in the abyss.

Down-the-line, an ice-block the size of a modern sports stadium arena, had broken free of its rigs and was drifting away off Europa, one of the most successful water-ice operations. Smaller than Earth's moon, the planet is covered with ice, frozen H20. The surface-teams operated the robot-machines, as each block was cut by precisely measured melting lines, using heat-lasers. As small as Europa was, launching the ice-blocks into orbit was easy, as far as the moon's gravity-well strength. They did this by attaching thruster-rocket pods at strategic point to the ice-block itself, calculated carefully for a successful lift to orbit. Once the block of ice was in orbit around it's source-world, it could be manipulated into loading position, for either the large transport whale-haulers, or parked in orbit to await pick-up.

It was common for many of these frozen ice-bergs to be parked in orbit, even for a year or more, until the transport pick-up was ready, arranged and then successfully loaded. In this case, one of the new ice-blocks was not behaving properly. And it could be a problem, from tracking-telemetry, because the thing might accidentally collide with something more fragile than itself, such as a ship, or a robot-machine they used.

"Europa, this is Alpha-Base, please respond?" the human voice of an Alpha-Base communications-monitor traveled across the Abyss, by common radio-waves. Communications between the various Jupiter-program stations and sub-stations was essential and vital to their success. Each ship, base, or sub-base, robotic device, and human worker, was in constant-connection to some source of information and authority, mostly from Alpha-base, or Earth. The Amalthea-built HQ had better tracking, better overall data and science-uses, superior navigation-planners and teams, and a better view of the program's day-to-day schedules and movements, and staffs and crews, technical resources, availability of rescue ships, fueling, and more. Even at a great distance, many millions of miles from Europa, Alpha-base could still link with on-site teams, to help with such as this off-course ice-block and it's resolution.

The Uhuru here was a young man, his only job was to handle the radio-calls, and relay messages or commands and instructions. The communication was from Alpha, in the communications-center, only one part of the base's many facilities and chambers. He was a slightly-built, athletic Indian-Hindu man, about age 30-years, in top health, with complete understanding of his equipment and task.

His boss, Confey Gomrah, the transport-ship manager, was there with him at that time, with his own information and sources. The kiosk-room they worked in looked like a hallway ending in a mirror, which was actually a mapping screen they could use in many different ways. The radio-kiosk was there, and they were surrounded by data-base computers, other tech-workers, decorations, seats and chairs. A service robot similar to Menuda's servant 'Burt' was also nearby.

"Wake them up," said Confey. "Do they answer?"

"Ten-minutes between intervals, sir," the young radio-man answered him. They could hear the buzz-and-hum of the communications radio-system by metallic speaker cones in the walls.

"Alpha-base, this is Europa, Charles Benway in command here, I'm at the sub-station, please confirm," a voice then was heard.

"This is Alpha-base Monitoring, Europa, how's your planet looking today?" the young operator joked. By this time the ten-minute delay was compressed to the rate of a near-normal conversation as the signals were matched.

Benway was in charge of the Europa water operation. He was a glad-handed man of wisdom and intensity, common to the work needed, very experienced and trusted.

"Nice, nice," he said. "Nice day out here I guess."

"Handing the call to Confey Gomrah, transport-ship management, Commander Benway. Over to you now, please."

A pause, as Gomrah handled the call, which was not a physical object at all, and was automatic, not like a microphone or telephone hand-set, operating directly from the communications-console.

"Benway, this is Gomrah with transport, Cid Bixi Mimim to you, sir."

"Thank you, Gomrah. Go ahead. Cid Bixi Butthead. Ha, ha!" Gomrah had a portable data-tablet, that he could use to confirm his information, hand-held and very powerful, linked to much-larger data-base memory-computers and analysis, elsewhere at the base.

"Thank you, Benway. Okay, then. We have you with a loose block of ice. I have a transport loading in three weeks. What do we do? Give me the down-low on what you know. My whalers are not cruising into to load a single ice-cube for you, with a loose block of ice out there, understand? We have about 900 blocks about now."

A pause, the radio-link slowed down again, as the signals were matched, about 3-million miles between them. Benway sounded somewhat garbled or mechanical, his natural voice-tone reduced to a wave of energy.

"Hell no, Officer Confey," Benway said. "My ships neither. It's worse for us here. Here's your basics. This is Block ID-32A-Europa, Lot 63. I have 28 blocks like this in stationary orbit right now. My schedule says your next transport will pick up 14 of them. This is an old one, been floating around more than six months. One of my loader-ships passed too close, the thruster-engines had an unexpected effect on the object, and it started to move out of predictable path. That was two weeks ago. It hasn't hit anything, but it might. All the other blocks orbit just fine. There's too many, basically, your transport-schedule is behind, and I'm backing up. So when your ship gets here, 32A might not be a known pathway, by then. That's about it. Personally I recommend we destroy it. Benway out. Please respond."

"Good idea, Europa. How?" Gomrah answered by radio. The young dispatch-monitor felt privy to be a part of the conversation, but it was really a small thing for the program overall.

"We have ways, Alpha-base," Benway answered. "Probably approach with a couple of operations-utility ships, and just melt it. Probably particle-beams, the same we use on the surface to cut the melt-lines before we lift. We could also just break it up into small pieces. Same thing."

"All right, you have my go-ahead, Europa. How long will it take? The transport Osiris is 20 days from you, under Captain Rice. How long? I want 32A gone before they arrive. 32A is now a hazard status object, please log this designation for Europa and all stations right away."

A long pause. Benway then answered. "Give me 24-hours to calculate an accurate answer, Alpha-base. I need to look at my available utility ships, and crews and equipment. My staff will figure what I can to plot the work for safe movements, and hours involved. The particle beams are also not usually used in this way, so they have to be fitted. Supplies are limited. Let me get back to you. 24-hours, I can tell you accurately."

Gomrah nodded, pleased that his man at Europa was covering his ass, instead of guessing, when others depended on him. "That would be fine, Europa. We can do it. Hate to lose tons of pure H20, but I guess you have plenty. 24-hours to you, 20 days to Osiris. Confirmed, go ahead, green light to you, sir. Alpha-base out. Thank you Europa."

"You're good, Alpha-base. Thanks. Benway out, Europa docking. Peace. Cid Bixi Bozo, or whatever."

And so it went.

CHAPTER 6: No One Needs The Water

*'Mirror, mirror, on the wall, up-side down, weightless is all.",
Angel-Face Mendoza, Planetary Program Proficiency researcher,
regarding space-travel.*

Mendoza's observation was among the most commonly-held trouble-points with space-travel in that era. This was especially true with beginners, and both Angel-Face and his partner Deveroux, were very inexperienced in local space travel, prior to deciding to take the assignment to research affairs with the Jupiter-program. Once a person left Earth, rightside-up and upside-down were relative, at best. It wasn't that types of artificial-gravity were not working. Early space-travel efforts had no way to create artificial gravity on deep-space trips, beyond the use of centrifugal-force 'spinning rooms', where folks could exercise, or relieve from the strangely stressful sense of aphasia and atrophy that came with null-gravity.

But Angelo might be pleased that there had been advances, notably the work of a Russian physicist named Podliakov. It was much-later possible to simulate 'real' gravity, on board the deep-space ships and stations, using an incomprehensible system of interlaced unique alloys and highly charged ceramics, that would spins in close-proximity like discs. High-energy types were manipulated passing through these gravity engines, beneath the floors and levels and at strategic locations within the design of the ships and space-stations. Podliakov had found ways to reduce gravity-pull on Earth, by the smallest fraction. In this way, and later, for deep-space, increasing gravity was easier, as energy-systems always run downhill. Somewhat like a modern refrigerator does not increase coldness to produce an ice-cube, but decreases hotness, through imbalanced heat-conductive chemicals.

But for space-voyagers, on their way to work at Mars, or the Jupiter space-workers, false-gravity didn't change the fact that the ships or space-stations held no conceptualized 'up' or 'down'. They may as well be always facing what the traveler thought of as 'up', or even 'forward or backwards'. It meant nothing to the way they worked, navigated, and traveled. But, this took some getting used to.

And if anyone was ever used to it, it was Charles M. Benway, the man-in-charge at the Jupiter-program Europa orbiting raw-materials/resources ship-to-transport operation. The problem with an out-of-control H20 ice-block that had floated away from its moorings in storage-orbit, was now his problem. Benway was man who had a job to do. He knew what was involved, he enjoyed command, and he was methodical and deliberate. Charlie was just a good-old boy space-guy. During this period, 300 years into tomorrow, much as it had always been, especially with men whose jobs and work involved both high-tech science skills, and very demanding physical labor. He was a lanky, heavy-set man, about six-foot tall, at 220-pounds, very fit, but large. Charlie was age 45-years, Caucasian, and sort of pasty-skinned with freckles, much due to so much time working in space. He was well-liked, feared by some (mostly the Unions). He kept two wives for himself, also Jupiter-program workers, both African-ethnic ladies, both somewhat younger than himself, and both deeply affectionate with him.

This sort of thing was common. The space-workers needed and wanted to maintain their vigor and enthusiasm, and psychological morale, so two wives was nothing at all unusual for Charlie, and other men in his position in life. True-to-form, even in 300-years, technology and capacity or travel-abilities had changed dramatically for people of the Earth. Yet the people themselves, never really changed that much.

Daniel Deveroux, the lead on the PPP assignment to review Jupiter-program standards, was probably the great grand-son of a man named Alex, known to the US government and world governments at that time, to be a perfectly humanoid person from another world, an extra-terrestrial humanoid immigrant, brought to Earth around 2090, to work on anti-terrorism efforts, as global warfare and chaos finally began to slow down and simpler controls were set in-place. So, the new normal was normal enough.

The strange and unusual were generally accepted, given they were peaceful and did not interfere with normal affairs. Charlie Benway was a very loving man, but the work at Ganymede and Europa was hard, far more demanding than to command the transport ships, or run things from the comfort of the Amalthea Alpha-base. And yes, Charlie had been there himself, and only attained command status after years down there, working the raw-materials transports, lifts, orbits, melt-lines, drills, etc.

"Charlie likes whiskey, Menuda Wu likes wine," the men would say.

The job collecting very big blocks of pure H20, frozen as ice, from the surface of Ganymede or Europa, was a swan song. These were Jupiter's largest moons, and one of them was the ninth-largest object in the Solar System. Raw hydrogen and helium, and other easily-available large-quantity, very pure chemicals and elements, were also hazardous. Useful? Practical? Easy? Profitable?

In any sensible scheme, profit-and-loss would not even be a factor. The main idea was that money is no object, certainly when the program was started, 100 years past (by then). Actual money-exchange on such a project was an impediment or block to real progress. Benefits or success, or overall wealth and blessings from the program back home, depended on much more than mere chemical elements, even in large quantities.

For agents Deveroux and 'Angel-Face', this question was totally meaningless. PPP didn't want them to analyze profit-and-loss, it would have been senseless to try. Far more difficult perhaps, the two researchers were intending to find out if recent failures were some kind of sabotage, including mysteries like potential off-world alien activities. The crash of the 'Ferrous-2' was such an incident. Benway and his teams functioned like miners or drillers, much like the old-era oil-workers, or coal-miners, that sort. And by year 2,412, this was being done in deep space, in space-suits, using ships both large and small. The harvest operations were a critical part of the program. There were others, about 12 sub-stations in all, each with a man like Benway in charge. Any of the platform-based orbiting local-moon Jupiter collection stations had a lot of influence. They were important for everything else that was happening with the ships and schedules and workers. The end-point value of it all. So, if their needs were not met, work would slow or fail, and the Unions could make all kinds of demands, and get what they wanted. It was hard, dangerous work, so it all made sense, as far as the perks and status.

The former-era NASA moon program, or the space-shuttle programs, the earth-satellite network and launches, and the international space-station work, Hubble-telescope, and many advances like these, may help to imagine a planet, much like the Earth's moon, about the same size, for comparison, that is covered with frozen ice, and other substances. A permanent orbiting platform-type station is there. This is the hub for the workers and so on, and also Benway's home for even years at a time.

Perhaps such as *'The Little Prince'*, (1943), by Antoine de Saint-Exupery, as far as Charlie was concerned. Just a small, barely suitable, hardy even comfortable rock in the sky, that he ruled. The platform-type HQ was serviced by its own small fleet of utility-ships and service-vessels, and some specific high-tech gear they used. Such as the particle-beam laser rigs they used for melt lines, the attachable launch-rockets and thrusters that would move the blocks from planet-gravity to orbit, and then specific rigs for specific elements, such as giant lakes of helium, or hydrogen, and so on.

They knew what they were doing, it was all very detailed, calling for close-quarters navigation and space-flight, and open-to-space life-suit labor. Ganymede has no suitable atmosphere, and all work near Jupiter was far enough from the Sun, that is was intensely cold, almost without any relief at all, from Mother

Nature. And of course the Abyss itself was instant death for any space-worker, without protection. No air whatsoever, intense cold, exposure to a total vacuum, anyone would be dead in a few moments, without the life-suits, or the ships, or the bases.

Could it be done, this kind of work? A typical assignment might involve harvesting 300 tons of liquid gases. The pooling sources of the substances on Ganymede and elsewhere in the Jupiter-moon system, were well-mapped, and could be pin-pointed for navigation. On Jupiter itself, a whirling hell of fast-moving streams and cloud-forms, an eternal storm or elements. But the pilots could navigate down to the tiny planet's surface, like Europa, or Io, working with very strong, new-era ships of great specific capacity use, and versatility, and also very safe to work with. Far-advanced of the old NASA ships, and moon-shots, shuttles, etc.

They didn't actually land ships on the moons, like Ganymede. The different moons at Jupiter had different qualities. On some they could actually reach the surface. It might take two or three ships a week to do the melt lines'on a large H_2O ice-berg, in this way. And then another three or four days to launch it into orbit. The very small Jupiter-moons, of 63 orbiting objects, had only tiny gravity pulls. So, for any of these, the labor and process, resembled these ships moving in a simple orbit path, like a sea-born Pelican over the poison lakes, sending down sky-hook riggers, or diving-bell type rigs, to set the rocket-thruster pods, attached strategically, and functional to lift the things upwards.

The silver-bird mechanical pelican-scoop space-ships, maybe the size of a common football arena on Earth, would drop away on schedule from space-dock at the smaller orbiting bases, at command of Benway and others. Following a calculated curve or path, dropping easily into a pre-programmed orbit, then moving into a stabilized holding pattern, or semi-stationary constant velocity. Liquid gases, unlike the ice-forms, were handled differently, some substances froze at one temperature, some at another, some could not be frozen into ice-forms at all. The main-platform base monitored and tracked each task.

The beautiful mechanical space-birds, with long umbilical-line assemblies, working much like a scoop, funneling upwards, hovering and moving, like giant mosquitos. With proper handling and regulation task-protocol, they could move 300 tons of pure hydrogen into orbital storage within a relatively short time. Each 'run' was a thrill and a challenge, and the men and pilots celebrated back at the dock-station, when done.

The other role the collector-stations played was to hook up with the very large inter-planetary ships, (like Brandeis' ship, the "Down"), for the cargo-run back home. The large transports, again, at a mile long, arrived to haul very large amounts of materials. They moved into position like gigantic space-whales, or versions of Ezekiel's wheels, fed from small utility-tugs, equipped to connect the docking gear, for whatever they were loading. Docking, loading, moving the materials and the ships, in the harshest environment ever, in complete safety.

Benway knew needed an estimate on rounding-up the lost ice-berg he was discussing with Gomfrey, the transport-cargo manager. Here was one of the ice-bergs, parked in orbit until pick-up and loading, then back home to get someone an iced-tea. One of the utility ships had passed too close, with a sort of 'wash-effect', and the ice-berg moved out of its path. The parked orbit paths needed to be a reliable navigational constant, for the safety of other transports. Gomfrey wanted to know how long it would take. Another radio-link connected the two men, many hours later, after Charlie had a good take on the matter.

"Go ahead, Ganymede," Gomfrey was talking, from the Alpha-base, ever-conscious of the looming hierarchy of the program pilots and managers and greedy, selfish, opportunistic and enormously wealthy and powerful owners.

"Back at you, Alpha, this is Benway again at Ganymede, with your info on the one that got away, as we discussed," Benway replied. The radio-link was very powerful, and almost never failed or was garbled or disconnected. Advances by this era in radio were also quite successful. Physical limitations for radio waves had expanded by quantum leaps in stable and powerful communications.

"Go ahead, your details will be logged for review at this time, thank you," Gomfrey replied.

"I'm going to need a week to destroy the damn thing, bottom-line. We have to set up a particle-beam that will work from a utility ship, that's two days just to prep. The lasers are only prepped now for the melts on the planet, this is like ship-to-ship. Then we have to do the navigation, which is a mess, the ice-berg is off-course, and all the other ice-bergs and ships, well, we can't have it hit something. So, we're already starting, but give me 24-hours with my navigators. And then to destroy it, another two days. So, give me seven days, and it's gone. Plenty of time for a clear run for Osiris. Commander Benway here for Ganymede, out."

Gomfrey paused. "Logged and noted, Ganymede. Giving you a green-light. Go ahead for now, follow up with command-approval for the action shortly. Sounds like a plan, as I said, I have a transport due soon, that really what this is all about. No one needs the water."

"Sure, Alpha, sounds fine. What do you hear from the deep? Anything on the aliens? You guys arc full of shit over there, we have drillers praying to the Virgin Mary out here, scared for their fucking souls, cuz' Program won't specify an unusual threat and danger. Seriously, Alpha, you need to be straight with us out here on the realities involved when no one knows why your whale-hauler lost control a year ago. No kidding, right?"

This was all gossip.

Gomfrey paused again, then replied. "Aliens, Ganymede? Yes, we have no aliens. Groucho Marx, mid-century film-comedy. Yes, we have no bananas. Right? I have no idea about all that, it's just a myth, and it doesn't set well with management, because you low-life muscle-men are still shitting your life-suits about the aliens and the conspiracy and all that. It's useless, counter-productive. Gossip."

Then they gossiped more, a favorite past-time throughout the space-program and the new Solar system as a whole.

CHAPTER 7: Planetary Program Proficiency

At the local planet system, where Earth is concerned, you get astrology, as part of the deal, along with Greek-mythology, and ancient wisdom reflecting on the movement of planets and so on. Astrology ranges from a very detailed projected analysis of gravitation of large heavenly bodies, as they might effect the birth-circumstances of citizens of Terra, to parlor games, trading cards, advertising schemes, dating and romance. Most people have some idea of their astrological sign, and enjoy comparing beliefs about the accuracy or truthfulness of the mystery of the planets. And this was still the case by the era of the Jupiter-program, 2,412-13AD/CE. Sagittarius, Capricorn, Aquarius, Pisces, Gemini, Cancer, and the lesser star-signs, generally follow in the public eye as a harmless and enjoyable past-time.

The actual planet Jupiter as revealed by scientific truth, is far more mysterious and impressive, and vastly more enduring and even eternal, than all of the bar-room pick-up lines and pre-sold dating schemes associated with the astrological-review. Did the planet Jupiter make it's physical appearance on the galactic scene, prior to the manifestation of human life on Earth? Will the planet Jupiter continue its physical history, after the human species leaves the planet Earth, or fades away, for some reason?

By about that date, this included the Jupiter space-program workers, pilots, navigators, space-station staff, communications-tech people, and the inhabitants of Earth who might be sympathetic with astrology in general, or mythology, Greek and Roman myths, and more current superstitions, etc. One might have thought that superstition would have faded into the past, by this era, with all the science and space-travel and education, etc. Maybe it was just the overwhelming near-vision view of the Jupiter planet and moons, for the space-workers, that created a need to express a primal spiritual truth, or seek into mysteries of their own hearts and minds. Astrology was the basis of much of those beliefs, for workers in the Jupiter-program, for better or worse, far into tomorrow.

For Daniel Deveroux and his partner Angelo Mendoza, the challenge of deciding to accept or reject the PPP assignment to look into scandalous Jupiter-program affairs, was not very difficult. Bottom-line, any day of the week, they would have turned it down if they could have, or if there had been a better choice. It was a serious and problematic assignment, for them to take on. They took their work seriously, the only way they could succeed when dealing with corrupt institutions and corrupt powers. It would be messy, for the two men, to travel into local space to visit the Alpha-Base, and work there, supposedly in concert with program-director Menuda Wu, as co-laborers along with Wu's teams and staff. It was a trouble and inconvenience, it was demanding and could be dangerous, and might not even be necessary at all for their investigation for the Planetary Program Proficiency research, on the assignment.

They found themselves together again, on their way into a PPP facility in the up-state New York area, where they had flown in earlier in Deveroux's private aircraft. Some time had passed and they had been able to review at least some of what the case was all about. A scheduled transport brought them into the area where the PPP offices were hidden away somewhat, because the program was not public and was not available to the general public. Up-state New York was beautiful that time of year, with many trees and seasonal falling and colorful leaves, and a blue sky above, punctuated with watery-looking, gray-white clouds. Daniel and Angelo walked together across an ivy-covered campus-like environment there, after dropping off from the shuttle bus.

"Veronica is on the assignment committee for this circle from San Francisco," Deveroux was saying to Angel-Face. "She'll meet us in about an hour inside. We're supposed to talk to one of their new machine-things, a data-robot, about the Jupiter program. We have to decide today, if we'll take the assignment. I hope you're ready to choose on that, you know."

Angel-Face Mendoza seemed to take to the whole thing with typical good-humor. "Right, right, a data-bot interview, a decision-making tool. Its sure to be flawless."

They both laughed together a bit at that. No one trusted the machines and robots or computers, despite their usefulness. "It's procedure, Face, please abide," Deveroux said. They were now near the entrance.

"At least Veronica is fully human," Angel said. "And fully female. I'm not too interested in data-base robots asking me a lot of stupid questions. And what about that circulator-thing you were telling me about? I need more details there. Please abide."

"I'll tell your wife," Deveroux joked, now holding open the swinging double-glass doors that showed the way inside.

"Which wife?" Mendoza laughed. The surrounding pine trees, sycamores, birch-trees, and the grassy hills around with large boulders, now behind them, the real world just a doorway beyond, as they entered. The building was old, not new at all, even historic, but obscure, there where the PPP folks worked, happily unobserved. The two men had a choice to make.

Veronica, or 'V', met them shortly in a comfortable foyer where they waited. She was very attractive, busty but athletic, dark-haired, smart. The PPP offices seemed quite inactive. Many of the PPP employees felt that it was ironic that the agency was secretive and aloof, at least as far as their persona. True enough, during that period in the US and the West, many common management functions seemed clandestine and out-of-view, due mostly to the complex and high-end nature of many of the systems. The scandal at the California geo-thermal power grid they had worked on last, before Deveroux was declared dead, was that sort. Many public management systems and large-scale operations-systems, were also staffed robotically. The PPP was a big program, but most of social programs were by this time, in a very complex world-society. These were not actual robots, like R2-D2 or C-3PO, in the movies. They didn't walk or move. They were data-bank systems that could process a lot of information at once, and were operated by staff-level people like 'V', from the assignment committee.

Within a short time, after viewing the sometimes unlikely art-work and decorations in the foyer, they met with Veronica, also Deveroux's hot date, at times (which was no secret). He actually did finally provide more details for Angelo about the circulator sex-toy he and Veronica wanted to try. It was a sort of in-place hot-tub appliance for their pleasures. But, maybe something for later conversations. They said hello, and settled into one of the kiosk-offices down a few levels of hallways, then sat down comfortably. There before them and surrounding much of the room, was a series of computer-displays, voice-activated inter-active inputs, star-maps and navigations, input pads and controls. There was one large window that leaked outside light inwardly from some florid, well-lit balcony.

"What is this machine we're using, please, Veronica?" Angelo asked.

The system was somewhat ominous. It provided information about the assignments in almost limitless detail, with access to all sorts of off-site current data. This including people and staff. 'V' explained that PPP wanted to ensure authenticity on all current or new research assignments, so the new system would log their discussion and choices. In this case it would provide information about the Jupiter-Program, with maps and navigations, details on the ships and captains. The sessions could also be approached as an inter-active voice-controlled question-and-answer source.

"Of course, if the machine knew everything, we wouldn't need you two," Veronica said. "The human factor is far less dependable for this."

"Fine," Deveroux said. "Do we ask it questions, or does it ask us questions?"

"Both," Veronica replied curtly. "Don't worry about it. It's to your benefit. We want to assign the right people to the right jobs. All PPP needs is for responsible and well-trained agents to help look into the Jupiter-Program. If it's not the assignment for you, that's fine. This will help us find out, and you as well."

"We used to sit down and have meetings with flesh-and-blood people to decide. The assignment council? What happened to that?" Angelo asked.

"Cut-backs," Veronica said. "Let's get started."

She tapped a control on the system-board. "Unit on, please, Veronica Ubinucus Signo inquiry session with agents Deveroux and Mendoza for today's date, regarding Jupiter program research assignment. Please activate."

Now the computer and screens seemed to hum and glow, there was a pleasant tone, followed by a female robotic voice from within the automated, mindless computer-brain the thing ran from, somewhere among its endless circuits and chips and hard-drive connections.

"Good afternoon, Veronica Signo. System ready."

"Thank you. System, please describe the Proficiency company concern and obligations with current deep-space industry at planet Jupiter? 30 seconds or less, please."

The computer's voice was soothing and comforting. "Program Proficiency is responsible to World Nations inquiry and concern that deep-space Jupiter-program functions and efficiency and safety, have been compromised or damaged, by sources either within or external to the Jupiter company program and leadership. Secret or private reports from Jupiter-program workers and pilots, have shown a worry among the space-workers, following a ship-to-docking-platform disaster about two years ago, with the 'Ferrous-2'. Other sources and indicators or data show that Jupiter-program workers and staffs are suffering a wide variety of symptoms and illness, related to fears of off-world or extra-stellar alien life-form sentient awareness, with consequential related or un-related performance failures, including life-sustain systems, ship-navigational failures, pilot and crew aberrant behavior or rebelliousness, and other. Data-sources are unreliable and non-conclusive. Program Proficiency is intended to sort out true from false details and discover how to repair or correct failures, for the good of all concerned."

They paused. "Thank you, system," Veronica said.

Angelo glanced at a time-piece on his wrist. "That was at least 45 seconds," he said.

"The application is flexible to the request, Mr. Mendoza," Veronica answered him.

"I'd call that an error," he said curtly.

Deveroux broke in now with his own questions. "System, please display a map or schematic of the Jupiter program overall, including all bases and ships, with current locations, navigational pathways, planet position, resource locations, Earth-docking, and major bases or resources."

"One moment, please," the computer-voice said. It took a few moments, but soon the various screens and displays were popping up maps and schematics, data-lists, and all sorts of related details Daniel had asked for. Now they could see the over-view. The Earth, Mars, the Sun and inner-planets, the asteroid-belt, and then Jupiter, were indicated with overlay lines that showed the regular routes for the whale haulers, there and back again.

More specific screens showed each base or docking-orbital station, such as Benway's at Ganymede, and also the others. The computer was able to list major resources available at the smaller bases, such as ships and method-related gear, and also man-power available, even by name and rank or position. A split-screen close-up featured the Alpha-Base at the Amalthea planetoid, with the same kind of detail. Large whale-hauler transports were listed or positioned as animated figures, with more detail. Another screen showed Earth-moon orbital docking and transport un-load positions, and another showed current orbital-storage materials like the huge ice-blocks at Ganymede, and other. One split-screen display showed only planet-Jupiter positions for moons, collector bases, and materials positions on planets and moons, or on Jupiter itself, with specific substance types. There was also telemetry data such as orbit paths and orbit times, or flight times and fuel-duration, etc. Whatever else they could conclude from the request, once the computer was finished, it was clear the Jupiter-program was big.

Deverous, Angel-Face and 'V' studied the displays and maps for some time, chatting about what they felt they knew. The two researchers were rather ignorant of all the details, it was a specialized area, nothing they had been involved with in the past much at all. It seemed to Daniel that there were so many variables and factors involved, that if there was a silver-bullet or single-cause behind whatever complaints were surfacing for the Jupiter-program, they may never find it, it was a needle in a billion-mile wide haystack.

"It could take years," Daniel said. "A nut-job assassin on one of these small ships. A sabotage-mission hidden with some passengers or students, from some upset union-interests. A program failure for navigational going back 80 or 90 years no one caught. An industrial service-vendor with vast wealth dedicated to some minor part or machinery, or special fuels, or special gear. A planetary abnormality we didn't predict, gravity-waves, dark-matter, miscalculated stellar aspects. Who knows? It could take years."

"Or a religious cult," Angel-Face added. "Like mystics and spiritualists up there with their own agendas. Malcontents, or misfits, looking for something un-related to the program needs. Like colony-seeders, or long-term immortality, that sort of thing. Using the program, and defeating its purposes."

"Yes, we thought of that, and those other theories," said Veronica. "Or what about galaxy-level aliens from outside our solar-system. That's the horrifying rumor. Popular delusion anyway. But, you're right, where do you begin?"

They paused briefly, overwhelmed. It was a stunning amount of information. PPP's only job was to track down the rumors and accusations, distinguish true from false, and focus on what was really messing things up. But even at a glance, it seemed to go much deeper.

"Here's an idea," Veronica said, and again tapped the control-panel on the kiosk display computer system. "System active, please, Veronica Signo. Please respond."

"System active, go ahead please, Veronica Signo."

"System, inquiry: what is the best entry-point for PPP agents and human-resource researchers to begin an on-site investigation into PPP goals for Jupiter-program assignment? Please process, 30-seconds or less for reply, thank you, system. Waiting."

Mendoza again checked his watch. "Thirty seconds, ha!" he said aloud.

The system paused and there were more buzzing tones, but in fact it was almost totally silent, for about a minute. The robotic female voice next answered, somehow moving the voice-data from the human consciousness of assignment-committee member 'V', into an electronic data-base of almost limitless powerful processing information and analytical capacity, yet without even a speck of awareness or actual consciousness of its own.

"Best success researcher point-of-entry into current Jupiter-Program PPP assignment is hometown family and friends of deep-space transport pilot Martin Brandeis, of the transport 'Down', current resident of rural Oklahoma farmlands, including second wife, two sons, younger daughter, and grandmother, and also relationships with horse-training and horse-breeding interests in that region."

Deveroux and Angel-Face were amazed by this answer.

"Why, computer? Why his family in Oklahoma? What about the ships and the Alpha-Base?" Angel-Face now chimed in, curious to say the least. His voice tended to pipe.

"System cannot explain to satisfaction in 30-seconds or less, please," the machine answered. "Best success for PPP agents point-of-entry is a subjective value, please."

"Go ahead! You cheated before!"

They paused again. Veronica had a drink, sipping slyly. Deveroux had by now taken to gazing out the window.

"I hate these brainless drones with all their information," Deveroux said tersely.

"It's just a machine, Daniel," Veronica said.

"All right then, dammit. I'm in. I'll take the assignment. What about you, Angelo? It's a mess, it really is. Maybe I'm just stupid enough to want to figure it out."

"We call that 'curiosity'. Yeah, well, I'll go. I have a feeling about it. I'm in."

Veronica seemed pleased. "You'll have fun out there, Daniel. Way out in space like that with the planets. I'll miss you."

He laughed. "Fine," he said. "Look, let's break, and then get back to this new system of yours, and find out what the hell she meant. Even if it takes considerably longer than 30-seconds. The stupid thing obviously had some sort of processing-analysis about the pilot on the 'Down', and his relationships back home in Oklahoma. I want to know what that processing-analysis was."

"Brandeis was the transport pilot with private files and research and a secret complaint that was passed around," Veronica said. "He was the guy who thought he knew something."

"We call that a whistle-blower. Whatever he knew, your computer has added it up for us, supposedly. The answer starts at a horse-farm in Oklahoma, fine. Let's break." Deveroux said.

"I'll log you both as on the case,' Veronica said.

CHAPTER 8: Some Use For Horses

Like many people who lived in this period, about 2,400 AD (Common-Era), both Deveroux and his partner Mendoza had a distrust and distaste for computers, the computer-mind, and the general usage of interaction between real human beings and machines. Of course they really had no choice. The modern world had developed by then such that everything depended on very powerful computer systems that managed all sorts of affairs.

"No brain or heart," Deveroux quipped.

"Bob Dylan, 20th Century poet."

With the Planetary Program Proficiency, the use of computer data-base, research and information details, was so common that in almost every case, all of the PPP researchers relied on those types of sources. Transportation, air-flight, homes and accommodations, public identity and social identity, space-travel, agriculture, family, religion and worship, birth and death, all were handled at the level of computer-logs and data-bases that kept track of things. The machines even made decisions, and were also points of operational control for management. And despite the cautionary warnings of previous generations of a Big Brother-type society, full of fear and controlled people, the reality was that the culture, by then, was generally beneficent, and allowed for all sorts of comforts, freedoms, and conveniences, and was not particularly oppressive, provided you were on the up-side of any particular robotic system that managed what was going on.

But an educated person didn't tend to trust the computers, and wanted to disallow that the machines were in-charge, given the end-goal of the human factor. The computers and Intra-Net in-use at that time, were still fallible and full of mistakes. Such systems were often utilized improperly by operators, or even manipulated by corrupt interests, with various schemes to their own advantage, that the PPP sometimes hoped to remedy, for the good of all. Deveroux and Angel-Face held fast to the bottom-line for the PPP, that if the programs were not proficient, as complex and powerful as they were, failure could cause a lot of damage, cost a lot of money, or become complete disasters for many who would suffer, due to errors, mis-steps, and sometimes malice.

So, when Daniel Deveroux found that he was required to take part in this new computer-data system, or to submit to its power-structure, he made it a habit to ask himself, *'servant, or master?'*. That is, was the system serving himself or others, or had it mastered those involved, and why? And what was being accomplished, and what was being intended? The agency's requirement to falsify his own death a few months previously was a good example. True, or false? It was obviously a lie, but one intended to save his life, after all. Maybe forgivable.

From an official point-of-view, logged and entered into the identity-systems and date-banks that a normal citizen was required to complete and maintain, Daniel Deveroux did not exist, and had died in an auto accident in Seattle. This was uncommon, and public records were supposed to be accurate and truthful, or would lose all usefulness. The power-structure maintained order as it always had, with records of births-and-deaths, business, legal matters, school records.

Deveroux and Mendoza often found themselves pitted directly at-odds with very advanced computer systems and programs, similar to the California Geo-Thermal Electric Power case they had been working on. Trusting those computers and data-systems was precisely what they needed to avoid, to learn the truth about what was going on. Even though Deveroux and the other PPP agent-researchers were a loyal part of that company system family, and trusted their co-workers and their jobs and tasks, they didn't really trust the PPP computer systems either. So when called upon to make a decision to accept the Jupiter case, it ruffled Deveroux's feathers that they had to go through some newly revised automated system, to find out what they needed to know, just to choose.

A day or so following the session with 'V' in upstate New York, the two men found themselves at a hotel further south in the vast New York metropolitan area. By now they were officially assigned to the investigation into failures with the Jupiter program. The session with Veronica there at the PPP-house among the shady Fall-colored trees and long, windy roads, had gone on for a few more hours, and they reviewed more information and more details. They also moved the assignment up the ladder with the PPP hierarchy, so that it was settled that the two partners would handle the case.

As it turned out, the PPP committee was also assigning two other men to look into the Jupiter-program difficulties, who would be working strictly on Earth. According to 'V', the other two researchers would be looking at the program on Earth only, which was also huge. The industrial, the vendors, parts-suppliers, fuel-companies, parts and technology, space-support and launches, training and educational, staffing and communications, the raw-materials processing industry, and many associated groups and interests, even at the planetary scale, given the usefulness of the off-world materials, were all part of the Earthside Jupiter program. For this reason, PPP was brought in, it was important.

But Deveroux and Mendoza would not know who the other researchers were, by name, or their activities. They would work off-planet, and the other team would stay home. The information was all connected, and the PPP would later coordinate whatever results and conclusions were found, until a suitable remedy was presented.

The urban New York hotel where there were staying was large and comfortably modern. They occupied together a large flat on one of the upper decks, looking out over the yet-astounding city-scape, which still at that time appealed to visitors from all over the world to enjoy. It took some time for it to sink in, that the two of them were perhaps soon to head into the Abyss, to study the Jupiter-program, and the boring affairs of the drillers and miners and raw-materials transports, and find out why there were failures, and about the recent transport crash, and also the disquiet and rumors among workers, and complaints about mysterious goings-on.

"It doesn't please me at all that you're dead, Daniel," said Mendoza, half-joking. "But I guess, it makes sense to shoot you out into space, in a way. I mean, since you're among the dearly-departed."

"I don't understand," Deveroux said. "Why does that make sense?"

"Because, when people die, they go to heaven. Heaven is out in space. Everyone knows that." Mendoza smiled as he said this, knowing he was yanking his partner's chain just a bit.

"Only if they're good boys and girls," Daniel replied. "I'm not sure I am on the schedule, I may have been a bit naughty in my lifetime, you know, prior to my death. Something I said, or something."

Angel-Face chuckled. "And after your death as well!" And they both laughed. "I hope I'm in as good a shape!!"

They were having drinks and some food, towards the evening. It was quiet, as they preferred. They would not be wasting time enjoying the news and entertainment broadcasts, which were virtually ubiquitous in a modern metropolis like New York at that time, with video on every corner and in every shop and every hallway or bathroom or phone-computer kiosk and on busses and trains and jet-flights. Deveroux had a very strong, authentic Russian vodka, and Mendoza had come up with a strange variation on the classic Tequila-Margarita, and they had sliced sausage with cheese and dried sourdough bread. They had their own portable computer systems that connected them to the PPP sources. A lot of thinking going on.

The next job was to plan out their efforts, and map out what they intended to accomplish and the steps needed. They would finally be required to present their findings, even a year or longer ahead. Without a plan they had no road-map to work from. So, common to their form of employment, they were expecting to take a few days off, relax, review the information, and try to understand what they didn't know, that they needed to know, and how they might 'figure it out'. Where were those sources of information? What were the keystone mysteries that were blocking the discovery-process with the Jupiter-program? And it was a complex job, without a doubt. Their own education and background and research skills, and resources, would all apply. They would be planning out the research, even for a week or more. They knew that things would change, as the investigation went forward. As new information came up, they might change course, or look into other areas, and assimilate information from the other researchers.

There were two ways they looked at the work: even in the best scenario, it was a dry, boring, highly detailed and specific study, that might not even involve travel, or meeting with strangers, or to become involved in clandestine operations, or handle highly volatile facts at all. Best, because it meant they could actually 'get to the bottom of things', in an accurate, useful and un-biased way. The forms of truths required were like any other factual data forms, and were not threatening or cruel, or mysterious. Bookish, maybe, then, or perhaps in contradiction to the claims of other powerful people and groups. But still, dry and boring, and not particularly exciting.

The other view, more popular with girlfriends and outsiders, was that the work at the PPP was similar to police work, or so-called military special operations, with weapons and space-ships and thrilling dangers and conflicts. It was all very romantic, but well-they-knew, a complete falsehood. The hero-myth encouraged each man personally that at least what they did was somehow worhtwhile, and it kept them on their toes.

Daniel looked dapper in a black-and-gray pull-over jump-suit, very stylish and dressy, with his black hair and European features. Mendoza may as well have been a dock-worker or farm-laborer. He tended to be messy, truly a Latin/Hispanic male role-model, even for the times, with a longish face, and a certain strong-timber to his build, such as might handle a long hunting knife well, or an automatic weapon, as-needed.

"I don't see what the computer was suggesting to be so important about the transport-pilot's horse-training deals in Oklahoma," Deveroux was saying. "How could all that possibly be related? The computer must have been reasoning in terms of the relative travel difficulties, I mean, compared to a trip to Jupiter. But still. Who cares? Why Oklahoma? Why horse-breeding?"

Angel-Face hummed softly, his thoughts coming into place, his eyes closed for a moment. "He breeds horses, or his estate does, his family and business, back home. He's only home every year or so, maybe every two years. He makes a fortune running the transport, they all do. The pilots are paid very well. So, he has to do something with all his money. The horses, and horse-breeding, it's like a hobby. But I guess he was doing some sort of genetic stuff with the breeding and foals, and it caused a stir. Besides, Brandeis is sort of a celebrity, back home, the wealthy space-pilot no one ever sees who sends them all money, and his outstanding, superior and very rare horses. His DNA-science and genetic species modifications were thought to be unethical. But the connection to anything out there, on the ships and with the materials-transports at Jupiter, well, you got me there. They don't use horses in the space-program, despite what you may have heard."

"A lot of horse-power in those Ion-Accelerator deep-space engine-thrusters, I guess," Deveroux replied.

"Uh, yeah. A lot, ha! A lot of horse-power. You're funny."

"Well, we'll have to follow-up, but I don't trust the computer's ideas here, it just doesn't seem like a natural entry-point for the investigation at all. I mean, Brandeis pilots the transport, the one called the 'Down', right? But who was the pilot or commander operating the 'Ferrous-2', the one that crashed? That's how all this opened up. They must have known each other, the pilots are like a brotherhood or union."

"I have that information, somewhere, let me access my data-base," Mendoza said. He started to work on his own portable PC-tablet. They had already gone over the name and identity of the man, but it had been passed over quickly. The details of how the 'Ferrous-2' had crashed, concerning the technology and navigation, and errors made, were looked at more closely than details about the commander. But it did make sense that he would have known Martin Brandeis, and the other pilots, and it was Brandeis who had spent two years gathering details and accusatory-complaints, in his 'manifesto' against the program. Brandeis was a high-class, powerful whistle-blower, with all the clout he needed.

"Shall we just go ahead now and rule out sabotage, or mutiny, some sort of intentional ruin, or revenge, that the other pilot felt would serve his masters?" Deveroux asked absently, as Angel-Face was bringing up the data-base connection to secure PPP lines and sources. All classified, for what that was worth.

"Of course not," Mendoza said, quite serious. "Unless you are thinking they have some use for horses up there."

Again, they chuckled a bit. But it was a serious business, and they knew it. It looked to be a long investigation.

CHAPTER 9: Welcome To The Hotel Caligasta

So who was the pilot-commander of the 'Ferrous-2'? That year, about two Earth-cycles prior to when Deveroux and Angel-Face were brought in by the proficiency-people to investigate why the Jupiter program was experiencing odd failures? Who was he previously? Of course it was no secret, and the information was readily available, but as researchers know, a photograph, an ID-number, a name, a resume of past-service, education and training, they didn't mean much at all when things went wrong. So without further reference to the inexhaustible PPP computer data-base system, Deveroux decided they wanted to know more about this pilot-commander, as a first step.

"Whoever he is, I pity him," Deveroux said. "Those ships never crash or even bump into anything, at all, ever."

"Well, this one did," Mendoza added.

A fallen hero, a ruined career. For a man who had spent much of his life preparing to pilot the whale-hauler transports and those types of deep-space vessels safely and with complete control, such was the disgraced retirement of a deep-space commander who had seriously abrogated his duties, and caused quite a fuss. The crash was a major-incident for the program. The large transports and other ships, large enough on their own, were thought to be virtually indestructible, like the ancient ocean-liner 'Titanic', (circa 1910). They only very rarely suffered any kind of a malfunction that was quite so serious. An actual crash or collision, was thought absurdly unnecessary, it was never supposed to happen, and was far-beyond well-understood safety methods for the Jupiter-program's advanced level of deep-space motions.

There were other mess-up's in the program, but the 'Ferrous-2' was the most glaring and costly in recent memory. Some else of what they had seen included a launch-navigation/orbital exit from the Europa-moon at Jupiter, headed towards the Alpha-base half a a million miles away at Amalthea. That was mis-handled so badly, the ship spent three months off-course, and had to re-route their navigation to another Jupiter-moon at Io. This was a considerable effort for the ship's being so large, and the rocketry and trajectory for navigation being over such great distances. Ten other scheduled flights and docking procedures were cancelled or delayed. It also happened in the middle of a program-review or inspection, and was thought to be Union-related, to bring attention to complaints.

Likewise, the command-chain from Earth to Alpha-Command was rife with squabbles and disputes, such that Earth-sources felt they had lost control of the program's long-term goals, in favor of power-chasing, ladder-rung, gate-keepers seeking greater levels of position and pay. Jupiter-program goals never changed much: harvesting raw-materials from the stars, the exploration and general space-travel success for long-term Earth-support. But the players changed, and in some sense the environment on the moons and the planet Jupiter herself changed, even violently. Things got so bad, that a simple directive-request from Earth to re-direct from harvesting ice-born H20, to other materials, (specifically a valued rare-element called Helium-3) was disputed and contradicted for so long that no less than three Jupiter moon-base platform commanders were dismissed, and one of these men was later found displaying a deadly weapon at the Alpha-base in a court-yard social-area, telling others that Doctor Wu Menuda was the target of his vengeance.

He wisely relented, and things returned to almost-normal. There were other incidents, too, not really understood, yet not really unexplainable, or all-that difficult to examine or review for human-failure and technological mis-application.

"So the program in general has numerous system failures, with no single cause-and-effect relationship to anything we could point to, as a successful conclusion to our investigation, correct?" Mendoza quizzed his partner Daniel, as they continued.

"I tend to agree," Daniel said. "It's a mess."

The commander of the Ferrous-2 was a mid-life male named Hector Pillow, also known as Ralph White. He was a typical space-pilot, husky and bold, well-educated and fit-for-duty, without any serious black-marks against him for flight-duty-readiness or in his past service. All of the space-programs were quasi-military, they had to be, given the nature of the work. There was no enemy and were no battles, but routine, duty-and-honor, uniform, command-structure, rank, task-service or assignment, and weapons or defensive preparedness, were all part of the picture. And it was in this world that Hector entered, much younger, for training-education-physical fitness, and all that any of the pilots needed to do their jobs.

"Hector Pillow is not his real name," Deverouz said. "Look at this on his University records. The same person? Or no?"

"Not surprising," Mendoza said. "Considering what happened."

The US and East, the developed world powers, all maintained very advanced colleges and University programs and academies, with no other function than to staff such as the Jupiter program (there were other programs in space at the same era, as well). Like hundreds and thousands of others, Hector was groomed and glammed to the very appealing position in life of deep-space Jupiter-program raw-materials transport pilot-commander, circa 2,412AD/CE. He had no serious reservations and felt it was a suitable personal dharma, or duty, to help mankind feed itself, or provide fuels, or replenish the environment.

Two years after the collision and physical damage to the 'Ferrous-2', and other damage, Hector was a broken man. He could be found by then, if any were seeking him, in a Reno, Nevada-area long-stay hotel. It made sense, to hide his shame, out in the reddish, dusty and hot desert highlands and windy, high-sky granite mountains, where the lizards panted for water, and feral cats or coyotes also sought their prey. His spot by then was actually many miles from Reno, still an entertainment center, but by then a larger metropolis, with much work devoted to wind-generated and solar-generated energy systems, very large farms and open areas where workers kept the new-style windmill-giants and vast arrays of solar-collection panels operating smoothly, on-the-grid.

The future, then, looked much like the past, in many parts of the world, and the Western US desert region was one of those places, beyond the high-rise sky-scraper centers for tourists and high-roller business-types in the Las Vegas-style Tahoe, Reno, Atlanta, Los Angeles areas, and so on. Pillow's long-term hotel was not the biggest or best, and by no means super-modern. It was preferred to keep with an Old West style, which by year-2,400 was a mere hint of the rustic, survivalist, farm-world. A 1950's California Blue Highway traveler's dream of small cafes and wood-stove beer-halls, huddled together in the freezing cold mornings. Like pentimento, the artist's pallet peeled back, layer upon layer for lack of canvas, history and time-passage, in culture and architecture, showing as many styles and types as the land had endured, even as far back as the American Indian and the Spanish Land-Grant period.

Known as the Caligasta Hotel, Commander Pillow could be found, by any government men, space-program inquisitors, tax-debt collectors with accounts of damages to billion-dollar space-craft, news-media or film-documentary makers, lawyers, writers with book-proposals, Union-reps, or ex-wives, also called the 'Hotel Caligasta' (with some humor). The rooms were simple but very pleasant, there was a guitar-bar, pool and hot-tub, communications-entertainment, and a small telescope observatory for dark, starry nights.

A light had also gone out somewhere inside Hector's heart, that he and loved ones hoped would perhaps one day re-ignite, and that evening star-gazing might sooth or comfort, now that his career was over. The Caligasta Hotel was somewhat of an old space-pilot's retirement home. He had other friends who lived there, but his wife, sons and children in school or college, and the thrill of piloting the whale-haulers, calling out commands and figuring his navigations as the man in charge of a regular mission to Jupiter and beyond, these were gone.

He had trouble accepting it, emotionally. The small computer-operated observatory at the hotel was for hobbyists and amateurs, like all the space-workers, perhaps to spy some distant super-nova or comet, or smile back at the 'Butterfly Galaxy' (Galaxy-72aB3, by the records, recently re-named at that date).

It was just another Thursday afternoon, and Hector was there, at the coffee-shop, enjoying his favorite breakfast (from the 'breakfast all day menu'), of eggs and toast with hot-coffee. Only a few of the regulars recognized him, or had any interest in his affairs. He was tall and gaunt-looking, and had grown a shaggy-gray short-cropped beard. He tended to dress for outdoors, as if to go hiking, with a floppy hat and sun-glasses, and carried around a back-pack and walking stick, carved from wood with figures like stars and moons, in polished brown cedar-wood, with rubber-plastic tips and a brass-handle on top. A younger man, about age 30-years, was with him at the booth in the restaurant. Hector was about 55-years old.

"You're not going to make any money just hauling stuff in that tractor-trailer like you're going, Billy," Hector was saying. "There's all kinds of junk and technology scrap folks need moved around, it's true. And your robot rig is in good shape, hard-to-get out here unless you happen to have one that's working. But the thing is, the scrap and junk guys have no money either, they can't pay. You'll go broke in a week."

"Already been working it for two months, Hector," the younger man said. "I got it figured out now."

"Yeah, and you can't afford breakfast!! See?"

The younger man was trying to set up a small business operating a robot-driven highway hauler service, for the farms and industrial businesses, in the region, including the wind-power and solar-power companies. The robot rigs were hard to get, and somewhat expensive. Much like any tractor-trailer rig, they could move a lot of cargo very reliably and fast, too, on regular public highways and roads. All without a driver or operator, completely driven even on turns and stops, or in the cities, by robotics and radio-guided computer navigation. Anyone with access, and a functioning rig, could cash-in on a regular route, provided enough clients could pay his fee, to move heavy items across country to a specific destination, like a dump, recycling facility, outbound transport by sea or rail, etc. The robotic-trucks were supported by government-highway electronics and special satellite-tracked lanes, and were very successful. The younger man's name was Bill, and he didn't particularly trust Hector's judgment any more than his own. But they were friends, having spent time together there in the bars. Hector had been living there for nine months.

"I got dinner, you got breakfast. I got money, Hector," Bill was saying. "It's after six at night, and you want eggs-and-toast?"

"I want a pardon from the Global Counsel, that's what I want," Hector said grimly. Bill knew what he meant. He didn't want to get into it again with the older man, about all that. Hector sometimes took to drinking, and other substance abuse. Alone, morbidly pre-occupied with the past. He was never really abusive, but he was sullen, somewhat paranoid, moody. They sat a moment, enjoying the meal, the desert sky blistered by another blinding sunset, outside the diner, as far away as any distant world.

"Well, anyway, I can get this going. I have the rig, it works fine. I'm certified by the state and regional, my 'bots are the best, state-of-the-art. It's gonna' be fine."

"Happy robotics, then," Hector said. "The loads you're hauling aren't worth a fraction of the rigs, Bill. That's all. And the techno-scrap guys are going broke."

"What would you suggest, then? I should haul diamonds instead? Gold? Rubies? Smuggle wealthy alien-immigrants? Your pardon from the Global Counsel is identical to the space-program Summary Dismissal, after the crash, Hector. You were hauling, too. The raw helium or huge vats of hydrogen or ice-water or whatever. The Summary Dismissal applies to any of the pilots, no matter what happens, you're forgiven. No matter what. Because they know the space-work is dangerous. So, be happy about it. Screw the global people, or their counsel deal, I say."

"Helium-3 is the hot-element right now. They're only finding lots of it at Ganymede."

They had finished eating and were resting a bit. Hector was having a hot-coffee, Bill had a beer. Then there was a beeping tone-alert. "My phone, just a minute," Hector said, plucking his mobile-phone from a pocket in his backpack, at his feet under the table. "Just a second."

"Sure," Bill replied, smirking. "Maybe it's the counsel."

So he took the call.

"Ya, hello, this is Hector, who is calling please?" the former space-commander spoke casually, into the portable phone-device, turning a bit in his seat at the café booth. "Yes, yes, go ahead. With who, what agency? I've never even heard of you people. Just give me your name, what the hell do you want? Yes, I told you, my name is Hector Pillow. Yes, what? Yes. THAT Hector Pillow, what the fuck do you care? Why? When? Absolutely not, I don't care who you're with. Big deal. Then get a subpoena, or maybe show up with a SWAT-team, if you're going to be that way about it. How did you get this number, anyway? Oh sure. Oh sure you did. Oh sure. The Program Proficient what? Like hell. Sure. You contact me through proper channels if you want to talk to me. No. The answer is no, got it? No means no. I don't give a damn, I got my own problems. The incident has been totally documented and reviewed by the proper authorities and is on-file with the Jupiter program. I don't owe you even that much information. Well, then, that's fine. Fine. I'm sure that's all true. Contact my attorney about it, or the Jupiter program. I'm retired and under program authority protected dismissal, and that's the end of this phone-call, sir. Please do not contact me again. Thank you. Cid-Bixi-Mimim to you, then. Good-bye."

His friend Bill just watched him, placid. Hector had taken similar calls before, in the nine months he had been staying at the Caligasta Hotel. The other retired space-workers got different calls, for awards and public apearances, or to make speeches.

Hector shut down the mobile-phone, tucking it back in his pack. Bill looked at him somewhat askew, sipping his beer. "Never mind," Hector said, "A bunch of vultures in a tree waiting for me to die."

"I do like condors out in this part of the country," Bill joked. "Much superior bird."

"Not a carrion-eater, that's true. I guess sometimes. Or am I wrong? Anyway, what you need with your robotic rigs is a decent, high-paying client. I have a pal, with the space-program, has high-end horses in Oklahoma, real steady operation, breeding the race-horses, the Arabians, like that. You certified for your robo-truck in Okalhoma?"

"Not a big deal, you apply and show them your credentials. Hauling animals is different, though. Most of the runners have the same standards. Who is this guy?"

"Uh, oh, another one of the pilots, on the deep-space. Rich as hell, nothing to do with his money. You cut a deal with him, move the horses maybe, or other stuff for his operation, he can pay you well. It's all totally legit."

Bill wavered. "Hauling livestock is way different. The robotic-highway systems aren't always equipped for that. The animals get screwy or they need food and water. So it's not done much."

"Well, he also hauls other stuff. Just trying to help."

CHAPTER 10: Horse Farm

If there was a connection between the Jupiter Program's front-runner whistle-blower, (pilot-commander Martin Brendeis of the 'Down',) and his high-end Earth-side horse-breeding operation, back to troubles with the program, it had not been revealed to anyone. Other rumors and so-called explanations were even more absurd. The Jupiter-Program failures and systemic issues, including upset and rebellious workers, and the more spectacular accidents, were what Deveroux and Mendoza were called upon to look into. They knew it could take them a year or more to learn anything of value. So the pace for now was leisurely, not hurried or rushed, step-by-step, so they could clear away the fog and think as clearly as possible, review whatever leads they had, and compile their findings later.

The Brandeis horse-ranch was just a place to start, and there were many other strings and threads to pull together, including those at the program's deep-space facility at Amalthea. Far too many strings and threads, creating a huge mess of information. In this way, for many very large and advanced programs by year 2,400 CE/AD, when things went wrong, it was almost impossible to learn the truth. The cover-ups could go on endlessly, never truly resolved.

"That's what we do," Mendoza was saying. "Big deal."

"We're in the Truth Business. As unpopular as ever," Deveroux joked.

The two of them had by now arrived in the countryside area of Oklahoma, near the Brendeis horse-ranch. They had arranged for a simple tour, under the guise of official government buyers, supplied with credentials by the PPP. A transport was acquired, and Mendoza was at the wheel. This was a modern 'Solace'-type automobile, completely electric, with six wheels, and a globe-style overhead canopy plastic-glass. It was rather large for a private passenger truck, but it allowed them to haul some other gear as well, such as computers and data-base storage, and communications. "We find out the truth no one cares about," Mendoza continued. "It takes us years at a time, and if we hit, what we learn can save zillions of money-units and thousands of lives."

"Nice to know what we're actually trying to accomplish from time-to-time," Deveroux responded dryly. "But it gets old."

The exotic truck swept down the highways and roads into the dry, green, lushly vegetated Oklahoma hills and valleys. It was typical of the North American continent during this period. Old and new clashed and existed side-by-side, sometimes in stark contrast. Oklahoma and the rest of the US-regional authority was by this era, civilized and developed for 600 years, counting from about 1776. By comparison, Europe was a solid 3000 years older, and China even much older than that. But in the West, future-history passed in concert with an entirely new paradigm, mostly the high-tech and machine-based or computer-based and aero-space. So Oklahoma, at that time, was not much different than it might have been viewed after year 2,000. Farms, dairy-facilities, crop-fields, large mountains and green valleys, high blue skies, and the cities. The roads were not too bad, for auto-transport, and in general the land was tended in its naturally restored state. For at least 200 years, ecology and sustainable Earth-recovery was by then earning a refreshed environment.

But for a casual observer, there were clear signs that 'not in Kansas anymore, Dorothy', was an under-statement. Huge towering power-structures gleamed on the hillsides, sunlight-energy collectors developed to surpass solar-cells, row after row for miles. Communications centers, like spidery-spindles of inter-laced steel-aluminum lattice-work, with convex dishes and antenna-arrays, and control-centers below, were placed here and there in the hills, as a common sight.

Private personal aircraft were common. Many people were now using private levitation transports. They were a lot of fun, new, and yes, a lot like the 1960's-era 'Jetson's' cartoon cars'. Not very fast, smaller and operating from dubiously acquired anti-gravity assist technology, but great for recreational uses and sports. No one seemed to care about rumors of an off-world alien-conspiracy that had gifted the Earth the needed technology and science for these, (except the military, and those powers under-employed for lack of wars, by that era). So these could be seen as well, on established air-space routes, not really that many, often professionals with long-distance commutes.

Yet, down below, where Daniel and Angel-Face tooled through the windy hills in their fancy 'Solace'-type auto, the rustic old wooden barns, grain silos, hay-fields or rolls of hay, gas-powered tractors (though rare), small mom-and-pop markets with soda-pop and beer, young girls learning to ride horses and young men learning to drink whiskey, were still the way it was. What everyone loved about life here was preserved and protected as All-American and valued for unique local and historic character. On a quality highway, Medonza may have traveled in this vehicle at 160 kilometers per hour, without much worry. as they approached the farmlands, he barely managed 90 kph.

They had arrived by air in the area a day before. Their PPP connection and task-assignment manager 'V' (Veronica), set things up for them to tour the Horse-Head Nebula Arabian Breeders Ranch, now just 30 or 40 miles ahead. This was Brandeis' horse-farm. They were masquerading as government buyers, courtesy of another PPP sleight-of-hand, the same by which Deveroux had been pronounced dead, officially, back in Seattle, now six or eight months behind him.

"My instincts tell me, if there is anything to learn here at all, it would be something to do with the DNA and genetic manipulation of the animals, Angel," Deveroux said. "A horse is a horse. They can't be messing with the genes much, certainly not at the level where it would have any impact on the Jupiter program. But, for instance if the science-program boys at Amalthea had some DNA-gene work of their own going on, and wanted to try some experiments, or wanted to conceal the tests on the animals."

"Right, like talking horses, that would be a good one," Mendoza replied. "Someone had said in some report I was reading that the long-term space-programs wanted to mess with the astronaut's genetic basics, to create stronger space-workers. Change the astronaut's DNA so they worked better in space. But that's horse-shit, if you'll pardon the expression. They'd have to breed astronauts from birth. Maybe for long-term, I guess."

"Well, the other aspect is, if anything here is screwy, it constitutes an accusation against the whistle-blower, Brandeis, Commander of the transport. The other one, the 'Down', not the one that crashed."

"The guy in Reno. White."

"You mean former Commander Pillow? White is the other name he uses. Ralph White."

"Yes, poor sap," Deveroux continued. "So, if they have significant dirt on Brandeis, it is easier to dismiss his manifesto reports on problems with the Jupiter program overall, as vested interests. And if he's in any way, shall we say, cruel to animals, or crossing the line on experiments to create race-track winners from these rare Arabian thoroughbreds, then it's not really connected to the space-program, but it relates."

"Accuse the accuser who is accusing the accused, bait-and-switch, hide it all away, and run things any way they like until ships are crashing," said Mendoza, still driving through the countryside with an easy skill. The Solace-truck now had to slow and climb a long, tree-lined hillside road, up towards their destination. The sunlight was a kindness through the trees. Mendoza (Angel-Face) was a husky, dark-skinned Hispanic, but not without a light hand when on a machine like their Solace-transport. A rather cold wind surrounded them in a hush of motion.

"Yes, well, it is more than a hundred years old or so, the Jupiter materials program. To be expected," said Deveroux. "No big deal, really."

"I wouldn't expect too much down here at this place," Mendoza added. "They're not just going to spill the beans, if any. Here we come now, there's the entrance, I think."

At one side on the road, an ornate wooden sign indicated a long, dirt road into the ranch, reading 'Horse-Head Nebula

Arabian Breeders Ranch'. The six-wheeled Solace-transport paused, then tooled like a stately tour-bus onto the ranch road in, beneath the signage. There were green pastures on either side, many horse-barns and training stations and stalls, and 40 or 50 beautiful horses wandering in the fields. They could see trucks and watering-tanks, and larger houses and offices, adorned with glistening golden-colored steel or metal roofing, and towers with flags. The place must have been worth many billions of money-units. A deep-space transport pilot is a rich man, even a local celebrity.

Half a mile down the dirt road, they easily parked the auto, and then found the offices, looking for a manager named Matthew Borrick, with whom they had scheduled ahead by appointment.

"Cowboys," Deveroux quipped.

Borrick liked the cowboy horse-trader lifestyle, and was second or third-level manager at Brendeis' ranch. By their information, Commander Brandeis was at that moment in orbit at the helm-deck of the 'Down', at one of Jupiter's moons. He rarely had much of a chance to visit the horse-ranch at all. Borrick wore thick blue-jeans and high-end boots, a colored work shirt in cotton-fabric and undershirt, leather gear and of course the classic cowboy hat, a felt and straw type, all in grey. They met at first just outside the doorway to the offices and labs, where a laborer told them Borrick could be found.

"Hi there, welcome to the ranch," Borrick said, pulling off a pair of thick work-gloves. "I'm Mat Borrick, I'm assistant manager for training operations. How are you?"

"Greetings. I'm Tom Portunity, and this is Michael Rareus," Deveroux lied. "We're with the regional authority for land-management acquisitions. Did you get our certificates?"

"Yeah, yeah, I did, they're on my desk." Borrick said. "Land management acquisitions means what, exactly? So many agencies these days. A little help there, could you? None of our land here is available, obviously."

Deveroux showed him an official ID.

"Uh, we're inheriting the Earth, you might say. We are U.S. regional open space bureaucracy, basically. Remaining parks and recreation, wilderness, open forests, desert, wasteland, public lands, also disaster-area recovery and clean-up. Myself and my partner are looking at your operation in terms of how our agency can acquire or breed horses for placement in wild-lands, based on superior DNA stock. The idea is to replenish the wild horse breeds with superior breeds. Not my idea, by the way."

Borrick squinted at them in the sun. "No shit. Wow, that's truly stupid, sorry to be frank with you," he said.

Deveroux laughed and Menodza chuckled. "Well, it's government, what else would you expect?" Mendoza said.

Borrick laughed as well. "Don't get me started," he said. "Well, I have you scheduled until noon for the basic tour of the place. You can ask me any questions as we go. It's a big place, so if you have specific areas you want to see or learn about, tell me now, I can narrow it down. Too much to see, you'll only scratch the surface anyway, as far as anything you might learn. But I can tell you about most of it, off-hand."

Deveroux had a hand-held computer tablet, reviewing a file-page quickly. "Well, it might be good to go directly into your breeding labs," he said. "My bosses mostly want me to see if you guys can supply fresh semen-stock or breeding for wild-horse recovery, in the open lands."

"Truly stupid. It would never work. The Arabians and hybrid stock horses don't do well in the wild. They have special needs. You'd be taking a Cadillac and putting it on trainer wheels. No way."

They started to walk, Borrick leading the way. They left the main managerial offices and moved down short dirt-road paths, passing some workers and trucks moving hay and feed. The sun was now gaining heat in the sky. The ranch had trees everywhere for shade, fences. Just beyond were the gorgeous horses, seeming free and pastoral, tended far better than many of the Earth's poor, each worth millions of money-units, among the finest first-class Arabian thoroughbreds available anywhere.

"We'll have a look anyway," Deveroux said pleasantly.

"Sure thing, Mister Portunity, or whoever you are," Borrick said. "But like I told you, that's a somewhat ignorant theory. Our horses are among the world's finest Arabian race-horses, none better. The breeders work years, to refine the genetics from the best stallions, based on their performance, and match them with the champion mares. Every detail of the animal's lineage and genetic history is traced back four or five generations to some of the most famous sires in race-horse history. Breeding our horses to those wild disease-ridden ponies and saggy-backed donkeys out in the hills is truly a dumb idea, a total waste. We wouldn't even actually sell our seed-stock to you guys, tell you the truth. It dilutes our market value. Not even considered."

Deveroux knew how to play his lie, officially known as a 'cover-story'. He could always blame the bosses (who didn't exist), and string along Borrick with whatever opened the door for them to check things out at Brandeis' ranch, as if they were merely looking into a proposal. True enough, it was unlikely they would really find out anything of value to the Jupiter Program PPP investigation, either. But they could get a feel for things.
In a few minutes they were inside a dark-shady barn area, with sanitized stalls and inner labs and offices. A few workers in white lab-coats, and others like typical farm-workers, moved around, and a few horses could be seen in the stalls.

"Who is the owner, for this facility, then, Mister Borrick? We know about the space-pilot, but are there others? Like a group corporation?" Deveroux asked.

"Brandeis owns it, but he has an association of about ten other partners," Borrick said. "The ranch is not a publicly-held property. It's all private, if that's what you mean. No stock investors or anything like that."

"Do they partner with any university or science programs at all?" Mendoza asked.

"Sure, of course, but not as investors or for-profit, except by arrangement for services," Borrick said. "We're signatory with all the appropriate agricultural-science approval and regulatory oversight agencies, most of them linked to university. We have in-house lab-science for breeders. It's standard livestock science for many years."

Deveroux was standing at one of the stalls. A beautiful cream-colored mare hung her long face over the bar, and he patted her nose gently.

"Beautiful animal," he commented. The horse huffed and snorted a windy blow from her velvet nostrils.

"They truly are," Borrick said. "Among nature's most fantastic creatures. This one has to be put down, however. Sad."

"Put down?" Mendoza paused. "You're going to kill this animal? Why?"

Now Borrick showed them, they lifted the stall-bar gate, and could see the horse's rear left leg was heavily bandaged, with thick white medical tape and hosiery, and special plastic splints. The horse moved heavily from side-to-side, huffing her breath.

"Bad leg," he said. "They don't improve, once they get a break like that. For her to live with a bad leg like this the rest of her life would be torture, cruel. They're runners, it's their nature. So we put them down. It's standard."

They admired the horse a bit more, then moved away and closed her inside the sterilized pristine manger. "Too bad," Deveroux said. Then they continued the tour for another hour or so.

CHAPTER 11: The Crash of the Ferrous-2: Part One

More than two years previously, the Ferrous-2 was in a parking orbit around the Earth's moon, and then later departed on a routine transport run for the Jupiter regions. At the helm or flight-deck was Commander Ralph H. White (also known as Hector Pillow), the same man who was later disgraced and dismissed, ending up at the Caligasta Hotel outside of Reno, in Nevada, at a sort retirement home for space-pilots and other deep-space workers. Prior to any trouble with his ship, the man was admired as a responsible, mature and reliable commander of these large ships.

After more than 100 years of space-travel and work in space, Commanders at White's level were part of an elite group, and quite respected. There were numerous organizations and clubs, the labor unions and also associated link-ups, like medical needs groups, male and female orientation groups, technical groups, levels of professional or military achievement, with names like the Golden Eagles of the Abyss, the Space Dockers, Radio and Infared Communications Workers Local Amalthea-Alpha Base, The Gentry Commander's Pilot Unification, and so on. White had a Level-8 authority as a whale-hauler. This meant he had more-or-less absolute power when the ship was in flight (and it always was, one might say), at least with himself on-deck. His knowledge was general, he knew all the ship's systems well-enough, down to the low-flow toilets, that in any situation or emergency, he could make reasoned choices for the well-being of them all. Other than this, the Ferrous-2 had a large crew, as many as 80 individuals and specialists, all working at unique tasks that were essential, though often seeming very minor. The commander's job could be quite dull.

The ship was later in transit to planet Jupiter, a distance of some 700-million miles. White and his team again plotted course to leave orbit from Earth's moon, as the seasonal position of the planets and asteroids was favorable. The Huschcroft Ion

Accelerator engines kicked in after about two weeks, moving the ship to a fractional light-speed of 4-million miles per hour (roughly), and then gradually increasing to 6-million miles per hour. For weeks and months, the Ferrous-2 tracked a parallel path towards Jupiter, much only a distant reddish star, far, far away.

White often felt the helm or flight deck was claustrophobic, with no real view of the surrounding environment, deep-space or the vast nothingness, indigo, empty, bitterly cold. There wasn't anything to see, and certainly no man had ever seen anything in space un-assisted by telescopes or radio-video cameras and internal screens. But nearing Jupiter finally, and then setting up to orbit the large planet itself, he and his Second, a command-level flight-transit large-vessel space pilot named Tamwa (at that time), could begin to make out the shape and form of the huge gas giant, Jove, or Jupiter, truly an incredible thing.

"There she is, we will get a better view in a few weeks," Tamwa was saying. He was a tall African ethnic man, with the best education. They had been on three runs together. White had commanded his ship on as many as 15 runs.

"I've seen it before," he answered. "Beautiful, big, and extremely dangerous. Every time is like the first time, isn't it?"

"Yes, sir," Second Commander Tamwa said pleasantly.

Which was certainly true. Jupiter could not have been less hospitable to the human life-form. Giant hardly described her, the planet was all gas, swirling as dense poisonous mists and streaming ether, layered in forms and colors, and then much denser beneath, compressed into a thickness as liquid, and then even solid, far beneath, from intense pressure as the weight and mass of the gravity well of the globe of it all, pushing it all into a huge mass with more efficiency than any trash-compactor or crushing machine.

The planet is second only to the solar-System sun in size, for our system. About 300 times the size of home at Earth, and yet spinning at a mad pace to every completed orbital cycle, once around, in only ten hours. She had more than 60 small moons and planetoids that spun around her, some fast, some slow, and some even in opposite direction to all the others, or in large elliptical orbits. The Ferrous-2 must place herself, eventually, first into orbit around Jupiter herself, in the very largest circle one might imagine, and this took 10 days at least, after they 'arrived'. The journey was just shy of a twenty-eight weeks, for them by then, and problematic to their flight-path was the passage between Mars and through the asteroids, beyond which was Jupiter.

The maneuver was much as a space-ship would be placed into orbit around Earth's moon. Commander White and his staff manipulated the Ferrous-2 into Jupiter's stunning orbital circle. By now the view included more or less only a solid wall or colors and gasses associated with the misty surface of this strange and bizarre world.

"What is so important about this Helium-3 substance they want now?" White said, just making conversation. "I guess some of the science guys found a mother-lode, and now it's the flavor-of-the-month. I assume you have the loading schedule ready, and Europa-base is confirmed?"

"Yes, sir," said Tamwa, his Second. "Of course. We have helium-3 in all tank-vats forward of in-line split-sections three and four."

"All of them? That's a butt load of helium."

"Yes, Commander. A butt-load, sir."

The objective now was to take the big ship down to the orbiting base at Europa, dock, and load the raw substances. This meant more plotting for navigation, from their current orbit, to the Europa moon. Everything was in motion all at once, so Europa would be plotted, the motion of the small docking-loading orbital base, and the position of the Ferrous-2, in her orbit. The big transport would then leave Jupiter's orbit, and maneuver down to Europa, and enter orbit around the moon itself first. This process took another few days.

Once in orbit at the Europa moon, the path was mapped again, to move the whale-hauler near the docking-loading base, also in orbit, and then with the help of pilot-ships or 'space tugs', after a long slowing down, Ferrous-2 would hook up connected to the base, for loading. Another few days of work, with various points along the way for course-changes, thruster action, engine applications and movements, all precisely planned, with Commander White reviewing every moment, and every smallest or larger move of his ship.

"Planet Europa is minor degrees forward of her position mapped previously from Earth-base tracking," Second Commander Tamwa was saying, both men on the flight-deck. Other flight Deck Helms staffers also attended. All around them, the monitors and controls, computers, mapping screens, communications, command-controls for engines, life-sustain detail logs, and a host of other information, with the staff and technicians needed at each station, buzzed and hummed. Yet, for almost all of them, it was a routine run, nothing out-of-the-ordinary at all, and no one was expecting any problems.

"We don't know why, it's common, three degrees is very small out here, but it means our local tracking will be more accurate as we approach," Tamwa added. "For your information, sir."

White worked at his tasks like any of the Commander-level men and women nder his command. He was methodical and deliberate. He left no room for costly mistakes, but if anything was dis-organized or in error, he left the staff mostly blameless, check it himself and work together with his people until suitable remedies were in place and all was running smoothly. Some of the other pilots felt White was 'weak'. He wasn't inclined to yell and scream, or dismiss staff over minor matters. But the job was intense at times, tedious, and for many weeks and months, simply dull.

The ship was so large, all sense of proportion could easily be lost in relation to other real physical objects in space, and for the most part, tracking the Ferrous-2 around the solar-system from point-to-point was rather like a very demanding game of mathematics. He also had to make sure all the loaded goods and minerals or gasses and rare-exotic fluids, were handled properly.

But part of ship's crew aboard the Ferrous-2 had no other job at all, they only worked the materials and handling.

"Submit all details on that to navigations-plotting right away, it can't mean much, I guess one never knows. Go ahead with formal approach communications to Europa base," Commander White said. "They know we're here, it's on-schedule, and we've had contact since leaving the asteroids. Just tell them we're on the way, and what our position is and arrival details, by radio-common. Thank you, Second Commander."

They continued about their work, White relaxed. Three degrees of error to the position of an in-motion planet-moon like Europa might have all sorts of consequences. The flight deck was like a rare tin-can of stale air and technology and people, with beautiful displays and emotionally comforting furnishings. Like many of the program's ships and stations, only partial gravity was available for workers inside (artificial gravity). This meant that in any repose or position, men and women workers would often move about as if floating in the air, pulling themselves along by railings. Frequently the appearance of things was much like normal gravity, as far as work routines. So White could enjoy food and snacks or drinks, review reports, read, check his maps and other activities, during travel time.

The Ferrous-2 was now moving from perogee orbit around Europa, towards the orbiting dock-loader base. Speeds were greatly reduced, the tracking would move the ship once around Europa in the opposite direction as the base was moving, she would then drop down toward the base like a huge, floating whale of dark steel, almost a mile long and a quarter-mile wide. As this took place, pilot-craft 'space tugs' would lock onto her hull at three points, using large buttressed clip-like features on both ships, very small by comparison.

These space-tugs had their own powerful engines, and with no gravity, it took only small amounts of action-reaction energy to nudge the ship this way or that. But with a ship this size, once in motion forward, like a large ship-at-sea back on Earth, there really was no stopping. Instead, Ferrous-2 only would match-orbit to the base, then gently move towards her and settle to an aspect against a series of long, straight-edge ramps and pumping positions, and then lock the hull down to those structures, held in an embrace, while other work was done. It was an amazing ballet in deep space, as gentle as a flower petal in the rain hitting a concrete sidewalk, unappreciated other than by nature herself, and the thankful space-workers.

White was looking over what he could find out about helium-3, and what it was used for. Apparently the stuff was an excellent source of energy, for many applications, especially high-end technical, such as new-era generators that supplied electricity to entire cities. Very efficient, too, most pre-burner high-level starter blows.

Then there was a loud tone at the door of his command office, it was the Second, Tamwa, the tall African-ethnic.

"Europa-base sends her welcome and greetings, commander," he announced. "They are looking forward to whatever goodies stock-and-supply has for their crews. Someone said a shipment of high-quality beef from Montana? And I guess also the usual media-news, films, new music, and special food and drink. Supposedly the radio-operator told someone they had shortages out here, for some reason."

White laughed. "Sounds fine. What hour for the pilot tugs to connect? How far out?"

"They have our range and position, sir. I am guessing, roughly five hours. Maybe a rest-period would be appropriate?"

"All right," White responded. He was now on a ten-hour shift anyway, an hour of rest would help his alertness when they were docking to the orbital base later. "I will retire to quarters for two hours. Take the helm, Second Commander. Alert me when we are in maneuvers. And have stock-and-supply set aside a portion of those goods for ship's staff as well, no reason not be so generous, I guess."

"Certainly," Tamwa responded. So for a few hours, White retired to his temporary sleeping quarters at the helms-deck, as it all was proceeding.

Forty-minutes later, there was an incoming call from the Europa-base, which the common-radio operator could identify from the alert-tone that announced to him the communication-link was active. The Ferrous-2 was in constant contact with the same kind of navigators and telemetry workers at the Europa base, as aboard the big transport, comparing their maps and positions. The radio waves moved almost instantly through the empty space between the two objects, millions of miles from home in the cold and dark nothingness.

Second Commander Tamwa found himself yawning, from the flight-deck.

CHAPTER 12: The Crash of the Ferrous-2: Part Two

The tone that alerted the radio-link operator on the command deck of the 'Ferrous-2', was signaling from the base that orbited the moon called Europa. It was not an urgent message, or particularly witnessed by the command crew at that time, as of any real interest. The radio operator's job, here, was to take the message, log the information, and route the details to the proper stations aboard the ship. Many other functions for the communications stations that were spreckled all over Commander White's ship, it was an essential role. The Europa loading dock orbited the small Jovian moon, and had gear and equipment, ships, living quarters, suitable for many years of service, bringing back the raw-materials on this small, desolate, chemical and mineral rich moon.

The Europa base was an orbiting station, meaning the program founders had constructed a facility, much like a large, floating ocean-bouy, about the size in this case of a ten-story hotel or skyscraper. On either side were dis-embarking points, or hangar-doors, releasing ships, and also a longish, straight-edged cat-walk ramp that extended outward in a series of platforms, along with energy panels and pump-machines, and then angled configurations and arrangements at various cross-points, producing what was used for loading and unloading materials to the large transport ships. This, and also the small 'space tugs', and other more transient flying machines, that would handle rigs for large raw materials, loaded to the ships like the 'Ferrous-2', or the 'Down'.

The message by voice-operator on the Europa base simply indicated that the Europa base telemetry that tracked the big ship's motions were observing a very slight off-skew angle in the Ferrous-2's approach. In modern terms, this might be compared to an air-traffic control flight monitor, signaling the pilot of a jet-liner still miles away, that radar was showing his landing path required a correction, such as to veer left a few degrees. The protocol assumption here was that the navigation and helm at the Ferrous-2 was aware of this as well, and the observation would be confirmed by the Europa teams, for double accuracy. Then whatever minor thruster-movement needed would be applied, well within the correction-opportunity window. So it wasn't thought abnormal, and for any pilot, it was expected when docking or during a close-maneuver.

From within the base in orbit at Europa, the Ferrous-2 could now be seen by visual, meaning with unaided observation. There was a high tower-type center, and inside was the telemetry staff, other workers, and the maze of tech-gear in the air-sealed environment, also mostly weightless. The orbiting station here had about 45 regular workers, on-site in shifts for many months. Sometimes the stations were almost completely abandoned or empty of any staff, until the long voyages of the big transports grew closer, and the staffs would transport to the stations at the various moons, from more comfortable assignments for R-n-R, such as at Amalthea Base, the program HQ, far, far away. From a main view-port, two of the staffers could see the big ship lumbering slowly towards their position.

It was quite a sight, and there was often a video record. The big ship was elongated, with Jupiter like a wall of colors far beyond or behind her, and the stars and other moons, along with the much-smaller 'tug boat' ships, now attached by railings at three points on the main hull of White's ship. All in total silence, with the exception of on-going tinny-can sounding radio chatter.

"Commander is Ralph White, good-old boy, knows his stuff I guess," said one of the operators.

"Sure, I suppose he'd better. Right now if he doesn't correct a half-point degree towards planet, he may prove you wrong," said the other man.

They spent twenty minutes more, tending their own work, and just watching White's ship crawling slowly towards them.

"He still has an hour, the pilot ships will handle it," said the other man as they continued their vigil. By now the big ship was almost ready to hook-up to the station.

There were alerts, and outside crews on-hand to deal with the event, involving lining up a small portion of the Ferrous-2, designed specifically for docking, and then locking down essential mechanics, huge clip-like buttresses. Once complete, the Europa base and the transport ship would move in orbit together as one, and crew and staff could easily move back-and-forth, until loading of the raw-materials could start.

As we know, at some point, the docking procedure started to malfunction. The whistle-blower research from Brandeis' was much more specific. A radio-communication from one of the 'space tugs' seem to indicate an unusual level of concern.

"Europa base, please recognize, this is your number three pilot craft. I think we have a problem here. Come in, please, on this frequency," the tug flight-pilot said on the radio. It could also be heard on the deck of the big transport (White's ship).

"Go ahead, number three, Europa base tower. How big a problem?"

There was an unusual noise on the other end of the call. "Oh, about fifty billion tons of transport deep space heavy-weight coming down your throat at 300 miles per hour if he can't use his Western front nose thruster in the next two or three minutes, sir," the tug-pilot seemed to be saying. "No bullshit, base. The tugs pulled her East, the damn thing made a tiny jolt West towards you, 10 minutes ago. We're trying to correct, but she has the momentum on it. Please advise, recommend emergency lock-down on all your externals right away."

There was a long pause. "Okay, pilot craft, yeah, that qualifies as big enough, I guess," the base-station operator replied.

So, as one might say, 'all hell now breaks loose', as the word spread from one point to another, 700-million miles from Earth, near the Jupiter moon Europa. The Europa base crews were now in a fight for their lives, with almost no warning. If the big transport was coming in wrong, at a significant error-level, it could mean disaster, and indeed, it was a disaster. Things might only have been worse, as someone commented later, it the Ferrous-2 had burst into flames, which it did not.

No one knew what had caused the error. Commander White's flight-deck was frantic in attempts to shoot a blast of thruster-energy from the Western nose of the ship, at significant enough levels to change her position prior to the next two or perhaps five to ten minutes of forward motion. Sabotage was later the recommended explanation, even two years later, but it hardly mattered. All three of the pilot-craft 'tug ships' also tried to heave the beast the right way, with thruster blasts of their own. But those small-craft pilots knew their own ships were connected by powerful bolting mechanisms, directly to the Ferrous-2's hull at various points. If she crashed, so would they. Radio-chatter was full of venom and cursing.

"What the hell Ferrous? You guys taking a nap!! Come on, dammit, move her East five points degree now! You're off!" one of the pilot craft men broadcast into the communications system they all shared.

"Uh, Ferrous-2, this is Europa base, this is now an official level-four emergency, please comply. Come in, please, Ferrous."

Of course White knew what was going on. As near as he could tell, the ship was emitting the small maneuvering thruster blasts that were very rarely used at all, in conflict with each other, one working the opposite against the other. He had no idea why, they were all set on a pre-programmed cycle operated by a second team, who now were in the hot-seat to figure it out. The tiny thrusters were supposed to work together in perfect unison automatically, but now they were fighting each other, sending the whole approach off whack.

Computer analysis was fruitless, as seconds ticked by with the ship in motion. So White was not too quick to answer Europa base without something to tell them, based on actual knowledge.

"Second officer, is it going to hit with no correction? Yes or no? And what will it hit? Report, please," said White.

His Second Commander had also been quick to sum things up. "Yes sir, I'm sorry. Apparently the ship will swipe a portion of the first-rung docking ramp, with no correction, in another two minutes, sir."

"What? Dammit, you can't be serious!! What the fuck happened to those robot-thrusters?? All right, brace for any collision, notify the base, and continue to work manuel thruster power as needed to the Western nose immediately. This is without precedent. Now, Second, please."

White knew the Ferrous-2 would not suffer any damage. The base structures were little more to her than a pelican floating near a battleship on the ocean back on Earth. But for Europa-base and her crews, and the space-tugs, it could mean more damage and death than any of them had seen in their entire careers.

Two of the pilot crafts were able to disconnect from where they were locked to the big ship's hull to push and pull her in. The giant bolted-clip connections exploded with pressure-gasses, as they were designed to do in an emergency, sending a colored spray of yellow mist into empty space from various points. Then the tug-ships veered off, tumbling down and down in the weightless dark abyss, then to regain their own flight-control.

"That's it, we're off, it's over, she's done!" one of the pilot craft commanders shouted into his ship's radio.

The two small space-boat tugs veered off with their own engines firing, the pilots were saving their own lives, no one would blame them. But the third pilot craft was not able to disconnect. "Europa base, I can't drop mine off from this position, I'm right on top of your dock-structure ramp-wingset, if I pull out now I'll hit them both," he said, on the radio. His voice was a painful cry for help.

White passed off the duty of informing Europa-base to a radio-operator by hand-written note. The radio-man turned pale as he looked it over. "Now," White told him.

"Europa, this is Ferrous-2. Please brace for impact in 90-seconds at your first-rung docking ramp and wing-set, this is confirmed from ours for 90 seconds. Emergency warning please Europa, Ferrous-2 is out-of-control to normal docking. Brace for impact at first rung docking, it looks like she'll rub her hard. And god help us all."

A reply or confirmation from the Europa-base was not necessary. What happened then was perhaps an instance of space-travel history. The first-rung docking ramp and wing-set is a large assembly of solid steel structures, sticking out like a straight-edge in one direction from the man base, along a series of such structures extended as far as half-a-mile. Due to the weightless environment in space, these didn't need to be big or huge, but they were very strong, sufficient to essentially hold the mated parts at the underside of the hull of the big transport, in a firm embrace that would sustain throughout the loading. Normally, the space-tugs would match-motion of the big ship to the base, or, full-stop. But this was deceptive, they were all still actually in-motion.

Ninety-seconds passed. By this time, only a few of the work-crews at the dock-positions on the wingset by the first-rung ramp, had not been hastily evacuated. The wingset structures were portable motion-detector operated personel-movers, that easily went from point-to-point along the entire platform, directed by vibrations in the steel railings and wheel-bearings alone, in perfect air-tight safety. Most of the workers had found a way off the ramps within the past ten minutes after the initial alert, but five or six could not, for whatever reasons. These astronauts found themselves powerless within their small air-shells, as the Ferrous-2 impacted, something perhaps resembling a mountain suddenly dropped upon their heads at high speed.

Oddly enough, there was no corresponding noise or crashing sound, now could there be. But from within, the entire Europa base shook like a bowl of cold steel jelly. The gigantic Ferrous-2 was only off by the slightest degree, but it was certainly enough, the angle against the ramps only fractionally in error. The ships' giant dark hull underside swept past, with all the mechanized-tits and bump-points and connection ports, window-ops, antenna-gear, giant barricaded ramp-doors, each passing through the final position of the numerous base towers and floating-bell shaped mains, like a retarded and very unwelcome guest, honored and welcomed only hours before. Within another two minutes, five men had died.

And for still longer minutes, every living soul on-board the Europa base from the highest tower, to the very bottom departure pods, was treated to a terrifying experience. Inside, the walls, doorways, halls, computers, everything they knew as their temporary deep-space home, only shook and shook, stronger at first, then more, then less and finally less, with lights going out and systems suddenly down on all sides.

The big transport vessel continued her swath of destruction another three long minutes. Many of the ramp-docks were lobbed off, falling away weightless like pick-up sticks of stainless steel and technology. External lighted areas went suddenly dark. Dozens of tube-connecting loaders and pumps were crushed or ruined. The pitiable pilot-commander of the tug-ship that was not able to disembark of disconnect safely from the big ship, was locked to the larger ship's hull. He now found that he was merely along for the ride, and a wild ride it was for his small tug-ship and two crewmen.

The tug-ship was now locked at the front end of the giant Ferrous-2, for a cutting edge view of the entirety of the wreck and ruin, bashing into anything in her way. Whatever the Ferrous-2 would hit or scrape or bust up, his small tug-ship would hit or scrape or bust it up first. As if any miracle would find them or any of the survivors, but he and his crew were eventually found alive and unhurt, with no deaths. The radios still chattered.

"There goes my favorite fuel-loader," the tug-pilot captain could be heard as almost a joke, by radio. "Pretty wild show out here, base-command."

"Very funny, pilot craft three," someone still at thier post, at the base, replied by radio. One very useful function of their humor was to restore calm and keep with the idea that some had survived, so it was allowed and encouraged. The other radios buzzed and popped, as the survival communications was fast and furious.

After three or four minutes like this, at about a half mile past the first-rung loading ramp or docking position and structures, somehow the helm of the Ferrous-2 and White's teams had managed a sustained fractional push-thrust in the Eastern direction. This would shove the giant, weightless but massive. frictionless space-craft off from the various surfaces of the base itself, where all the destruction was happening.

"Maintain the same thruster array, we'll fall off beyond toward planet, don't release those forward engine thrusts until my command, understood?" White was yelling at Tamwa, his Second-in-Command, but also on ship's radio. The engine navigations-staff were at another control position, but understood perfectly well his urgency. By finally hitting the right buttons, even the relatively small thrust-push in the right action-reaction position, would nose the whale-hauler off the base-surfaces (the ramps and loaders), until she simply fell away toward the planet, saving the base further harm.

"Big fish rolling over away from you now, Europa base," came the word.

Deep space was plenty big for them to then regain control and move the Ferrous-2 into an entirely different path. None of the other ship's system's were harmed, as for White's Command. But of course it was all totally without warning for them all.

"We have a trajectory course on navigation-plotting now for the movement, commander," Tamwa said. "Once we pull off, she'll nose down planet-side. We'll move the course to orbit around Europa itself, from vertical instead, sir. It is the easiest way for this mistake, for this ship, at this time, sir. Plotting now for thrusters to position."

Commander White looked at his Second, stunned. The command deck was unharmed, there was no visible damage, but he knew, without a doubt, his career was over, and he would never pilot such a ship again. None of the large transports had ever suffered such a physical collision with actual base structures in his entire career, or in much of the Jupiter program history. And of course he knew the Europa base was an incredible mess. From the radio-chatter, he also knew some had died by now. All this had happened in about ten minutes. Inwardly, Commander Ralphie White had never been so angry in his entire lifetime.

"Relay radio to whoever has survived at the Europa loader base, operator," he now spoke coldly. "Offer them my apologies. Also establish long-distance emergency beacons."

"Yes, sir," said the radio-operator.

CHAPTER 13: Bunch of Junk

The fallout and consequences of the collision of the Ferrous-2, and the miner's base at the Europa moon, was significant. Even after two years of repairs, the base in orbit there was not quite ever the same, there were factors that made on-going operations 'shakey', and some work was still on-going, and would be for some time.

White's vessel, at that time, about 2012AD/CE, had performed as his Second, Commander Tamwa had said, pushing away from the orbiting structures, and then moving into the depths of the nothing in a vertical dive or roll, even as large as the ship was, finally entering a second successful orbit around the gravity-well of Europa.

Survivors at Europa-base were in deep trouble. Breaches to air-supply and life-sustaining systems were the first order or emergency-protocol, obviously to remain among the living, they had to immediately ascertain if there had been ruptures to the main base hull, or dense protective outer-shell. Even under normal conditions, this was always a concern, and was checked and maintained meticulously, due to things like small meteors, or flying debri at high-speed, the dangerous flotsam of space itself.

After the collision, there was every sort of danger, many of the outer structures along the docking rungs and wingsets, had fallen away, broken off like twigs, floating away weightless. Power systems associated with many functions had failed, some of them essential to their well-being. And there were life-sustain failures within the base, with great fear of suffocation and death in the icy cold and airless vacuum of space.

These were well-trained and courageous astronauts, but it was a considerable threat and highly unusual, none of them had really been through anything like it, because it was so rare. Many other things were to be dealt with urgently, as Ferrous-2 rolled off like behemoth, dark and huge, a graceful Godzilla they had anticipated that day, for no difficulties at all.

Later, the event became legend among the Jupiter Program workers and planners. After a time, with other failures among them, Daniel Deveroux and his partner Mendoza, and investigators for the Planet Program Proficiency group, were brought in to look at this event, and other circumstances.

The approach was to calmly and patiently review many aspects, and 'get to the bottom of things', to reveal what was going wrong, however complex it might be. This was reasonable, given that technology advances on Terra (Earth), by then, included all sorts of programs that extended successful life; power grids, water-supplies, food supplies and distribution, identity and population, labor, money-management and money-systems, computer grids and data-base, university and science exploration or experiments, travel and air-traffic, satellite, oceanic forms like under-sea development and deep-ocean foods, housing, weapons, medical hospitals, all of these were within the scope of the PPP.

Deveroux and Angel-Face were by then back in New York, doing more research at the hidden or off-the-map campus-like facilities of the PPP, talking about the case, following what little they had been able to learn at the high-bred horse farm owned by Martin Brandeis, the Commander of the Down, and chief whistle-blower and researcher on Jupiter Program failures.

"Nothing about horses," Mendoza observed. "I've looked over the logs and records. No horses at all."

"That figures," Deveroux replied.

The two men now had more time to spend with the PPP data-base links and other research resources. They used the offices and rooms, technology, and expert-helpers, freely. Both of them understood, it would be a difficult case, and could take a long time. But neither of them anticipated any trouble as far as serious resistance or violent opposition to their efforts to learn the truth, despite the high-profile of the Jupiter Program and the gigantic qualities of related finances and Earth-side wealth and power. Yet it was never far from them that they may run into groups or powerful people, who didn't care for what they might learn, it was part of the job. And of course they understood, the horse-ranch in Oklahoma was only a place to start, a tiny first-step, with no real belief in any valuable leads from anything happening there. Mendoza was joking.

"So, something screwed up and they bashed up the Europa base a bit," Deveroux said.

"More than just a bit, Dan, it was really rough. Ten people had died there."

"Ten? Or five? Right, may they rest in peace and may their memory be honored," he went on. "But to reach the truth we want, the Ferrous-2 crash, is not the whole story, is it? If you look at the overall system for the Jupiter Program, since about 10 years ago, there had been a variety of complaints and failures. We have them listed, and surely some we don't know."

"Yes, but they're un-related" said Mendoza, seated at a large computer display with a cup of hot coffee. "You just have a hodge-podge of failures, some worse than other, some technical, some human-error, some seeming like sabotage or even conspiracy, I guess the labor-unions, or the vendors who sell them supplies. You have suspicious organizations that have taken root in the program, spiritual-religions who feel they need special insights for the space-workers. The management doesn't like them, they are rather strange, but essentially harmless. The only possible connection to Brandeis' horse-ranch, which is a long-shot, are the DNA-genetic work they are doing with the horses, thought to be extreme, breeding for physical qualities to win horse-races. And then the idea that someone, somewhere, would apply the same science to a new generation of astronauts, to breed super-astronauts who can work longer hours. Brave New World stuff, no one likes that either, and it doesn't even really function, and there's no connection to Brandeis because he's not management, he is the plaintiff, with his whistle-blower research on it all. So he has no motivation, no motive. It's a mess. There's no one, single answer."

"No one single problem, you mean," Deveroux quipped.

"Except the whole damn mess of it," his partner replied dryly. "On the other hand, operations are going ahead every day, they seem basically to be handling everything business-as-usual. Ferrous-2 was two years ago. Nothing has changed."

Mendoza shrugged. The PPP worked for other global powers, he and his partner were little more than efficiency experts, and the broadest goal for all they did, was to save time and money. But the difference between five dead, and ten dead, it hurt his heart and Daniel's in ways they would never show outwardly.

Following this event, however, the Jupiter Program management realized they had to review things at a higher level. Menuda Wu, the sleazy program head-honcho who ran things from the distant Amalthe Alpha-Base, built into that large, floating space-rock like a crustacean or barnacle beneath the waves of the deepest blue ocean imaginable, mostly then squirmed and bloviated his way through the crisis, to appease all the various and highly-biased stake-holders.

"The Jupiter Program mourns losses at Europa. We deeply regret this situation, and will seek understanding and correction with all diligence," he said in a unevenly reviewed statement.

Space-travel and labor at that time was an elitist and very coveted privilege. Only the very wealthy and powerful, or the very well-educated and well-trained, could take part. It was something much-desired, thought to be a splendid adventure. But only if it was totally safe. So all the programs in space (to Earth's moon, to Mars, to the asteroids, at Jupiter, and any general local system exploration elsewhere), were built for absolute safety, and could pretty much boast of that record.

In 200 years, as things developed for space-science, many things were standard, and life-sustaining safety was the first-order at every point. So, with all that, the Ferrous-2 had now humiliated them all. Few if any in leadership were willing to forgive and forget, if only with fearful thoughts about their own next trip to 'out there', and what it's like to die in space.

Doctor Wu's first response was to shut down or restrict all the current deep-space large-vessel transports, and scheduled transports, following the collision at Europa-base. At least he had done something. Emergency rescue and repair teams were immediately dispatched to assist at Europa-base, even within an hour of the event. These were sent not from Alpha-Base, but from the other Jupiter-moon bases, which were closer.

The disposition of the dead and their families, the screams and protests from Earth-side management, and those anticipating supplies and raw-materials shipments, the screams and protests of the labor-people for the space-workers, the long-term repairs and ship-recovery and maintenance (which could be very laborious and detailed for the smaller ships), dealings and negotiations with Commander White, and his ship and crew, at the time of the event, and what their orders were supposed to be, since any docking or loading at Europa was now impossible, it only continued, on and on.

"Bunch of junk," Wu told himself privately. He felt he was running a chess-board in which all the team-pieces were at odds and not co-operating, but in competition, out-of-control. They were like mini-empires, a Balkanized program divided by vast distances and widely variable circumstances or interests. He tended to spend a lot of time at the Alpha-Base gym, or taking the hot-tubs, to relax and ease his stress.

A year later, the Europa-base Commander, summed things up at a dinner with his staff: "Europa moon-base has handled 247 whale-hauler loads without incident, since year 2,322 Common Era.. Ferrous-2 was 248, and you all know what happened. The Commander, Ralph White, denies any wrong-doing on his part. Each of his actions and choices for the docking procedure were logged and reviewed, he seemed to make no real errors, prior to the mishap. He has been dismissed from service, of course."

Cheers and hoots from the Europa base staff, banging on their dish-ware with forks and spoons, and loud, rude jokes.

"No one is to blame, but we lost ten men, a very sad thing. Some of you knew them, or were friends or related by blood. Five were lost on the loading dock rungs in the wingsets, ruptured for air-supply. It was over in minutes. Two men and a woman died trying to recover an essential power-supply that broke loose on the lower-level life-sustain stations. Space-walkers Sinaloa and Troy. The heavy equipment got out-of-control during the rush-work, one was crushed and two others lost air. Another man died as the Ferrous-2 passed, and was thrown off one of the cat-walks, and another man died trying to save him."

He paused. It wasn't an easy thing for him. Charlie Benway had not been Europa-base Commander at the time of the event, and wanted the staff to feel confident and assured about the future. "Today marks one year since the crash. I know you all have worked very hard to restore order and safety measures. But there is still fear, fear that it might happen again. Europa-base brings up raw-materials from this moon, some of it very valuable back home, some of it more routine. Our work is not meaningless. The ships bringing materials back to Earth are highly valued, and so is our work, because the substances at Europa moon are rare and there are large quantities, and the uses on Earth serve many millions of people, and improve their lives."

Another pause. "Commander Benway?" said one of the staff, a surface-materials pilot. "Question, please, sir?"

"Yes, go ahead."

"The Quantuum Mind Unitarian Fellowship is telling everyone the collision was the work of sabotage by off-world alien beings from other worlds. Can you comment?"

"That's horse-shit," Benway replied quickly. "Are you a part of that group? I've heard those rumors and so has everyone else. They think anything that goes wrong in the program is related to some sort of galactic-level mind-projection with negative goals. That kind of thinking is pointless, it creates fear. Their only proof is one of their spiritual gurus hears voices in his head, he's crazy, I don't know why it's even permitted."

"Their recent newsletter suggests there are some kind of intelligent robotic stations or long-term structures deep inside Jupiter itself, sir," the staffer said. "They claim to have evidence."

"Then that's their claim," Benway responded. "I highly recommend you or any of your staff ignore such runors."

CHAPTER 14: Circulating Super Human

"The music of the spheres, unbearably beautiful among men for ages, are less than minor shop-talk among the gods, often far more interested in the sustaining gifts of property-laws and legal matters pertaining to insurance fraud, albeit at a much grander scale," an anonymous enlightened individual associated with the Quantuum Mind Unitarian Group, 2,410 AD/CE.

Disappointed with their work so far, Deveroux and Mendoza found the PPP had a surprise waiting for them when they had finally given up entirely on the program's automated lead generator, as far as the Brandeis' horse ranch, and anything learned about that. Commander Brandeis, they now learned, was at that time, now back on Earth, and ready to meet with both of the investigators, in secret, to present his research and 'manifesto' concerning Jupiter Program failures, informally.

The only clue they thought of any value from the connection to the horse-farm, was a vaguely possible conspiracy among planners for the space-program, to genetically alter the space-workers of tomorrow, for super-human endurance in the unique environment of the Abyss.

"Inhuman, don't you think?" Deveroux said to his pal, at one point.

"Well, hay is for horses," Mendoza said. "Fresh, clean breathable oxygen-air is for space-workers."

Deveroux and Mendoza were the natural choice to handle the meeting, which would also include case-manager (Vanessa, or 'V'), and other bosses associated with the PPP heirarchy. No one was closer-to-the-flame, or deeper inside the current situation within the Jupiter program, than Martin Brandeis, commander of the 'Down'. His ship was almost identical to Hector 'Ralph' White's ship. And as it was often noted, 'in space no one can hear you blow your whistle', no matter how loudly or urgently one might blow. Brandeis was now a whistle-blower celebrity. Yeah, the whole thing sucked, as Brandeis was more than ready to set forth, given his considerable power and respect, as a materials transport commander.

Deveroux and Mendoza had no idea the commander would be meeting them, until 'V' let them know. Daniel took her call from where they had finally settled as their personal base-of-operations, a small mountain cabin in upstate, a snowy wilderness that was hard to reach. Yes, Daniel and Vanessa were lovers. Both of them led lonely lives, even for months at a time. They were very hard workers and devoted to their jobs and the well-being of others involved. After Dev's faked death, he felt that a mourning period or some kind of celebration might be appropriate, and so did she ('V' was of course privvy to the fraud, set up to protect him).

Thus the legendary 'highly effective sexual-bliss couples stimulation device' (Version 3.2.2, otherwise known as The Circulator), was joyously explored while he was in San Francisco with her. No one was much the wiser about it, since he was dead, except perhaps Al Mendoza, who got an earful about the affair as they travelled about doing their work. Conversation about the Jupiter program had grown dull. She reached him by phone with the news about Brandeis.

"He wants to tell-all, as you know," Vanessa was saying in their private conversation. "His cargo-transport is on routine stand-over at the moon, so it seemed like a good time. It's been two years or more from the Ferrous-2. It may as well be us, the Global Space Authority is not ready for him. He claims that much of what he has learned is, uh, very provocative, or controversial, and he says he can prove it."

"I don't know why program proficiency doesn't just go with all of his research, and call it a completed investigation. Instead of waiting five or ten years, which is how long it seems like it will take me and Mendoza to finish at our current pace" Deveroux replied. "We'll just be repeating his research eventually. Hell, I should live so long."

"Well, he's biased, for one thing," Vanessa said. "As a transport pilot, his conclusions would be suspect, and challenged, that's why proficiency is more effective, we're outsiders. Also, some powerful people don't like him, for whatever reason. And Daniel, please don't forget you're officially dead, so you can't really 'live so long', can you? Okay? Your grand-kids will complete the investigation into the Jupiter scam, or whatever it actually is."

"Maybe you can help me with that," he joked. "Grand-kids, I mean."

She laughed. Vanessa found him very pleasant and sexy, and sweet, too, with his dark, European features and runner's build. Suave? *"Keep circulating, oh heart of mine."*

The meeting was arranged for the next morning, about 11 a.m., at the PPP's hidden facility above the New York Regional. She gave Dev' a little more to go on, as far as what to expect. There would be other PPP bosses, and a legal-rep for Brandeis. The whole thing was very secret, which Deveroux had to admit he always enjoyed, it made him feel important.

They decided that Mendoza would not attend, to keep their work as investigators protected, although Dev' knew that Mendoza would find out in other ways, even in great detail. The evening passed, the small mountain cabin was cold and dimly lit, and the two men played cards, slowly drinking old brown whiskey alcohol. By 8 a.m. the next morning they had prepared and found their way 120 miles south to the meeting-location, where the commander of the deep-space transport 'Down' would tell-all, travelling in the same type of electric 'Solace' auto-truck they had driven to Oklahoma.

For a man who spent many months and years in deep space, in a weightless environment, Martin Brandeis was a very healthy and robust-looking man. This was at least somewhat deceptive, those close to him knew he had minor health problems, something a professional pilot wanted to remain private, just between himself and the program medicals. Of course the program had a complete health-and-exercise regimine, to maintain top health, even in deep space, or, especially in deep-space. By the time Brandeis and his legal man appeared there at the ivy-covered hidden campus for the PPP meeting, he looked every inch the majestic tower of manhood they all thought of him, or some did.

Without a doubt, he had his enemies, so real health-problems were ammunition for various types of career blackmail. It was something one would only have observed if you were looking for it, the way he walked, when standing or walking on the Mother Earth, on his own two strong legs, after returning from eight months of weightlessness and stressful service in space.

"Who are the researchers?" Brandeis asked his legal, a mature man of smaller stature, as they entered the building from beneath a shady awning of Sycamore trees older than either of them.

"I don't have their names," the legal said. "PPP never really gets personal about things. Their whole schtick is objectivity. But they're experienced, they've been assigned to look over all aspects of the program, top-to-bottom. But Martin, it could takes years."

"Sound like a couple of academic Nazi's," Brandeis said. "Probably taking bribes."

Inside they went. Somewhere in Brandeis' sub-conscious mind, he adjusted to the idea that a few weeks before, he would have moved through similar doorways and halls, floating four-feet above the bottom-floor, pulling himself by rails and netting in the weightlessness of space, aboard the 'Down'.

It was a long meeting. Brandeis had a complete over-view of all his various allegations, in detail and with some very basic attribution as far as sources and confirmed truths. At that time, this type of meeting was all accomplished on inter-linked hand-pads, or computers, where forms of information could be shared and move from hand-to-hand electronically, along with all sorts of tools, and then also move into the center of the room on a large vid-screen, which as it happened was being projected as well in other places, including off-planet, undisclosed except to a few.

They began officially (on the record), at about 6 pm, and didn't adjourn until about 2 or 3 a.m. Brandeis was deeply loyal to the program, and felt his personal review and research would help, or even save lives. He was very emotional about it all, and had spent at least 18-months on the reports, including work by private companies and individuals he had hired.

There were five main components and areas of inquiry he addressed: an overview or laundry list of Jupiter-Program disputes, failures, dangerous conditions, serious and fearful complaints, rumors, technical short-comings, leadership avarice and mistakes, painful and very laborious working conditions, and worse. From this it was clear, it was not only the crash of the Ferrous-2, that moved his complaints. This was only the most spectacular event in recent years. Brandeis' work for his report went back as far as 30 or 40 years.

Detailed speculation and suspicions about what actually went wrong with the Ferrous-2 docking, more than two years previously. Global Space Authority, Amalthea Alpha-Base, and other agencies had their own reports and investigations about the crash. But as a transport ship commander, Brandeis' view carried a certain weight. He went on at length about why and how the small-thruster auto-programming and outer-hull directional technology was absolutely fool-proof. He knew its design and operations very well. Much of the rest of this part was mere guess-work, and related to the rest. The general assumption was that the thruster-directionals and robot-programming had failed or were tampered with.

Brandeis very boldly recommended that the entire Jupiter leadership at Amalthea Alpha-base be relieved of duty immediately. He then provided a detailed report on the actions and commands of about 30 of the program's main decision-makers, over the same time-period. Poor choices and bad decisions, that is, and their many consequences. He also provided program regulations indicating very specific officer and leadership dismissal requirements for neglected reviews, including Commander Ralph White, for whom he had little sympathy, or any real personal relationship other than membership in the Golden Eagles of the Abyss pilot's union.

In another double-whammy, Brandeis then detailed his own very hard opinions about deep-seated elements among the staff and workers and even leadership of the Jupiter-Program, that held destructive and counter-productive, secretive views, something the PPP and other Earth-bound authorities did not take very seriously. Some of the groups had traditions that were more than 100 years-old, by 2,412. Brandeis felt these were more like conspiracies. The first whammy was his allegations against the cultic and disturbingly popular space-religions or science-spiritualism, among the program off-planet program population. Brandeis held the view that much of this was tearing apart the program from within, creating fear and division.

The second whammy was related: the persistent rumor that galactic-level aliens were 'messing around in the vicinity of Jupiter and the Earth', causing various disturbances. Again, the global authorities that guided the program did not take much of this seriously, but from whatever sources such rumors and half-truths arose among them, like good astronauts would, the views and conspiracy ideas were backed up by sightings, science, photographs or video, radio-telescope, experts and much more.

Typically, Commander Brandeis cared little for any proofs or revelations concerning UFO's. His main concern was very real, that his crews and staff, other workers and leaders, were obsessed with these ideas and that program efficiency and peaceful unity of purpose was suffering greatly, psychologically.

Each of his overview-areas of these reports was discussed for hours. This was only the short version, he had much more to tell, and many other attributed sources of information. It all seemed very well-done and complete, it was hard to argue with his findings. Deverouz kept notes and recordings, drawing back on any argument or strong views of his own. He didn't want to be florid about his role, he was only working for the PPP agencies as an investigator, it was just another case for him, maybe higher profile.

Twice, while they all sat around a large table, late into the night, with their monitors and communications, and the ideas and information flowed back-and-forth, Daniel noticed that one of the food workers who brought them meals or snacks-and-drinks, looked a lot like his partner Mendoza, probably with some kind of a recording device under his white server's smock. Vanessa seemed not to notice, also drawing back. During a break, he cornered 'V' and they could speak in private about what they were getting from Brandeis.

"Vanessa, if the Jupiter Program is screwed up with technology or pissed-off workers or lazy officials, and all that, I can deal with it," he said, his voice low, as they rested in a food-area. "But if PPP wants me to try and figure out if there are actual real off-planet extra-terrestrials or whatever, with light-ships and some bizarre interest in Jupiter, no chance. No way. That's way beyond my scope. I don't think so, it can't be done. Know what I mean? No chance. Seriously, I may as well try to prove the existence of god, as an efficiency matter. Understand? You're the case-manager on this one. Please don't send me down that road, that's all."

"Of course not, Dev'," she said. "That's Brandeis' thing, but he's right, he had to include it. It doesn't matter if its true. What matters is that his people believe in the fear of it, and it has an effect. So any of your work would be on the effect, not the cause. Problem solved. Want a dough-nut?"

Nearby, his partner Angelo made good time cleaning up dishes and plates the group had been eating from, disguised as a food-service worker. Their eyes met only once.

CHAPTER 15: Chevalier Chevelle Carolinas

Brandeis was seldom enough back home on Earth, that his testimony to the PPP at the secret meeting was not his only activity, during his vacation. Among Jupiter-program staffers, Brandeis was known well-enough that his private and personal (professional) inquiry about the program and its various troubles, was not really any big news to anyone who might have a legitimate interest. At least to the extent that they were not among the accused or indicted, in his data-base of wrongs and errors. As his schedule was written for a planned hold-over period for regular up-keep and maintenance for the 'Down', PPP insiders applied their leverage with other agencies to invite (or summon) his testimony, and he was happy to comply.

"Bunch of shit-heads," he was heard to comment later. His legal rep only laughed, and agreed.

But once the meeting was thankfully behind him, there were other things Brandeis wanted to tend to, including personal family stuff, business, and recreation. So he left it all elsewhere in his thoughts. He didn't feel well about cultivating an obsession with systemic corruption and failure among his own peers, after all. Work in space was rather suffocating and not really all that adventurous and thrilling.

Commanding his ship was not like the old TV shows and science-fiction films. Brandeis never saw any other worlds, or made any epic discoveries. He never fired any photon-torpedoes at enemy ships. He had no exotic and enticing contact or uniqe encounters, or audience with alien beauties, or met with strange emissaries in grand palaces to consult about galactic wars, as a rare and urgently required representative from Earth.

To make matters worse, moving the 'Down' to and from Jupiter, was not even nearly as much fun as piloting a simple aircraft like a small plane or jet. One did not much compare the experience to flight at all. It was very claustrophobic, the transport was so large it never really seemed to actually be in flight, and the perception was only that of various types of deeper or higher-energy vibrating and rumbling, not high-speed motion. The distances were so vast, the navigations mere mathematics (and those computerized).

Any excitement or power-trip thrill of command was eventually reduced to a matter of giving his subordinates the proper order to push the proper button at a certain very predictable moment during the journey. Joyless, trapped with his childhood dreams of such a career, within a giant sterile machine, sent to gather materials like hydrogen, ammonia, or methane, and then home.

"All space-pilot and no Earth-man play makes the Captain a dull person," Brandeis would sometimes remind himself, and any time back home was such a time, for almost all of them in the program. In the past, during earlier periods of the space-program, (or really, many different space-programs), workers in space would return home more frequently, as a requirement for their sanity. It was a no-brainer, the weightlessness, the cold and danger and dark, the isolation and pressure-cooker technology of ship-to-ship docking and safety-first space-walks, the distance navigations, far from normal and far from loved ones. Men had been known to lose their wits.

Not very commanding for the sake of humiliation or respect, deep-space work could bring strong men to the brink of mindless, drooling psychosis, given certain avoidable circumstances. To prevent or heal this, they were supposed to take regular breaks from it all back home, and treated very well indeed. But sometimes the 'shit-heads' forgot those rules and reasons, in favor of whatever else was going on or needed, as if somehow what was known and learned of space-travel, and its dangers and demands, or very unhappy space-men, might sometimes not apply at all, as if merely for convenience.

Thus, a few days later, Brandeis could be found at a popular horse-racing track in North Carolina's newly restored entertainment-recreational regional open-lands, along with PPP investigators Deveroux and Mendoza. It now had boiled down to some man-to-man chit-chat about the Jupiter program, off the record, and as researchers assigned to the complex and difficult case (formally known as PPP Information Account 971-Jupiter/Deveroux), they were thankful Brandeis felt comfortable enough to accommodate them in pleasant privatude.

True enough, the so-called sport of kings, (that being horse-racing), had not changed very much by that period in the future. Though thought to be antique and maybe obsolete, much the three of them viewed at the Chevalier-Chevelle Carolinas horse-track, near the city of Raliegh, was identical to such as in the past. It was intended that way, by 2,412AD/CE, for the Old World charms, especially in architecture, furnishings, dress and attire, food, music and the arts, transportation, and almost anything at all to do with Mother Nature and the outdoors, was carefully preserved and restored.

The Chevalier-Chevelle Carolinas were such facilities, catering to complete nostalgia for earlier times, as far as a horse-track and stables, casino and betting, etc. It was late in the day, a twilight, the weatherful Atlantic coastal mist of an early summertime sky above, dotted with colorful clouds and light from the West. The greens were well-kept, but perhaps not exactly manicured, or trimmed by hand with shears. Commander Brandeis, Mendoza and Deveroux were walking along rock-gravel-dirt roads and small paths, with horse-feed and hay-bales and watering, also stables and barns, and beautiful trees. At times for Brandeis, the space-man, something as simple as a tree, natural to Mother Earth, could bring tears to his eyes. And here so many, so green and fading yellow or orange, the falling leaves.

He reached out his strong right arm and lovingly handled the neck and brown mane of a deep brown thoroughbred horse, tethered to a wooden fence for grooming. A horse-husband worked a tine-brush over the animal's shiny coat by the shoulder on the other side, closer to the fence, beyond which jockeys and handlers were walking other horses towards a starting gate, for some race or other.

"He won't be running today," Brandeis was talking. Mendoza had a bucketful of grain, and held out palm-fulls beneath the creature's velvet nozzle. The horse's bright eyes shifted and blinked, his ears perking up, and then back against the rear-corners of his forehead, delighted with the feed-grain, but annoyed by the attention, and more time apart from the mares.

"This one's yours?" Deveroux asked.

"Yes, yes it is," Brandeis said. "Beautiful, isn't he?"

Mendoza brightened and blinked then, too, just like the horse. He had his personal speculative theories about almost anything in life, and never failed to share his opinion, no matter how unpopular.

"A horse, an animal like this, they are telling me now that the rarity of the manifestation of such a complex and amazing life-form, and its repeated DNA-forms or duplicating life-forms from historic times, is evidence of galactic-level truths about our own world and human life, meaning we are in fact of a sort of animal paradise, rare among the far more numerous barren and dead worlds, devoid of any life at all, like Jupiter, by comparison, and elsewhere in the galaxy and the Universe," he said. "What do you think? Do you agree?"

Brandeis chuckled, still patting the horse. "Well, yeah, beats the heck out of some dead moon or ammonia gas-world of unbearable cold and instant death, for sure," he said. "I get you there."

"Edgar Rice Burroughs called a horse with six legs a 'Thoat'," Deveroux added. "Science-fiction writer from the 1900's. Tarzan."

"Tarzan? Tarzan wasn't science-fiction," Brandeis said. "You mean his fantasy books about Mars."

"Sure he was," Deveroux said. "A displaced British royal who chooses to live among the apes? Futurism back then didn't always visualize machines and space-ships, you know. They worked the future-vision backwards, too, to more natural and Bohemian or even jungle-laws, men living like animals by choice to escape themselves, to escape the modern mess. Like a wild, sexual or even violent Walden Pond. Henry David Thoreau with gorillas.

Burroughs also wrote about Mars, long before they could even see it very clearly through telescopes."

They paused a bit, easy about it all, as far as any crisis for miners and raw-materials on Jupiter's moons. "No animal like a horse could ever have six legs, though," Mendoza said. "That Tarzan writer was ape-shit or something. It would never work."

"Nice horse," Deveroux said. "Beautiful creature."

Brandeis addressed the horse groomer. "Have him exercised, okay?"

"Of course, Mister Brandeis," the groom said.

"Thank you. Let's walk a bit, gentlemen. I'd like to stretch my space-legs." He meant Angel-Face and Daniel, so they started to stroll along the wooden fence, towards the casino, which looked like a large, open circus-tent, a huge canvas with tall wooden poles and underneath many tables and activities and people, a quarter mile from the main track.

Deveroux wanted very much at this point to make some kind of progress on the work he had been called to, with Jupiter. Brandeis was a significant source. He couldn't afford to alienate him with any back-handed notions of blame or complicity or negligence, including the questions or line of inquiry he might bring to the chat. He also strongly sensed that Brandeis was sincere. They were on the same wave, similar in their goals and general outlook. And probably in their ignorance as well. When getting to the bottom of things, in work for the PPP, Deveroux had found that in most cases, the bottom of things didn't really exist. Just as there was no one single-bullet solution, there was often no single problem or blame. Usually the real truths were far more complex and layered.

"Have you ever travelled into the outer-space, Mister proficiency investigator?" Brandeis was saying.

"Me? Uh, no, no, the opportunity was never there," said Deveroux. "A few orbital experiences, that's about it, like everyone."

"What about you, Jose?"

"My name is Mendoza. Folks call me 'Al'. No, never into deep-space at all, sorry. But yeah, orbital, like Dan. I went to a resort on the moon for a month a few years ago. It was fun."

"I did the space-wheel orbital hotel-resort," Deveroux boasted, "Fun, true, really. Hardly anything like travel to Jupiter. I guess once you leave Earth orbit, or beyond the moon, it's all pretty much the same, isn't it?"

Brandeis seemed to reflect. "Hard to describe," he said finally. "Same, yeah, all the same."

They were walking on the soft gravel, horses and trainers passed by on the other side of the fence, birds above speeding past. "Let me ask you," Deveroux said. "You comprehend my role here, Commander. I am a professional nobody, an elite, high-authorization blank page, and I'm sent to fill the blanks up with useful truths, accurately rendered to the cause of understanding something. Something you are much closer to than I will ever be, personally. So, let me ask you then, in your opinion, where do I look, where do I go, what is the rabbit hole and what are the rabbits in the rabbit holes? You get my meaning?"

Brandeis grew tense. "It's all in my report. You heard what I had to say, at the deposition. It's all there."

He didn't like to be pushed, clearly.

"Only the effects, Commander. Excellent work, by the way, probably earn you a number of powerful enemies, you may even get fired. Whistle-blowers, traditional lambs-to-slaughter anyway, right? The effects, the fallout, the consequences of everything you were able to identify, I need to think of those as effects and not causes. Not the underlaying source of what's really going on up there. See? What you told us at the meeting, the crash of the Ferrous-2, the lazy or bored leadership, the space-religions, the frightening rumors. I don't want to mistake those matters as other than symptoms'of what I really need, for what I need to do. We are looking for a cause, or a general motivation. Where do I look?"

"You're wrong, pal. All those things are causing a whole bunch of other problems, that's the whole point. Something causes something, then that causes something else. The space-religions cause some worker to falsify some data, make them look closer to some obscure spiritual truth or something. That false data, maybe even minor, later, maybe a habit, or a long dark thread going back years. Then some calculation is screwed up on a navigation, no one even notices. It's not sabotage, but it adds up until there's a failure. People, human error, I guess. The Jupiter program is old, real old. Somehow we've forgotten the essential disciplines or science or basic laws of nature, that cost us more failures. So, I think you're wrong there. My report is a list of the sources, or causes of problems. It's basically just my own grievances after years with the program."

Now they stopped for a moment, as the current race, back at the main track behind them at a quarter-mile away, finally started with a loud bell, and the crowd in the stands cheered for their favorites. Brandeis' horse, the one they had stopped to see, was not in the race as he had said. The shaded their eyes and watched, the bell rang, the gate dropped, the horses ran their paces and the jockeys did what they could to win-win-win.

It wasn't the money. They didn't even use money, by that time in America, as we may use it now. It was the notion, the concept, something about it. After a long while a leader out-paced the others, then finally it was over, with a shout and a thrill, a cloud of dust. Only Brandeis seemed to actually know any of the horses or riders. The three men traded comments, turning away again towards the tent-casino, only a few hundred yards away now.

"All right, investigator," Brandeis said. "You want a plum, I understand, you want to win your race. Here's something that wasn't in my report, or any report that I know of. You want to know what rabbit-hole to look down for your cause-and-effect?"

Deveroux paused, paying more attention. "Of course, sir."

Brandeis said, "Listen, about four years ago, way before the Ferrous-2, special teams were sent to do measurements of the planet Jupiter itself, okay? The planners wanted to measure various substances for potential harvesting directly from Jupiter, not just the moons. So there were all kinds of measurements, a spectrum of analysis and research. They X-rayed that entire damn giant colored-ball, in fact I guess they did, actually, from data I saw. All the top-end science and tech on this, at that time, okay?"

Deveroux said, "I believe you. Go ahead."

"So, one of the science-guys on that team told me about a classified discovery they felt was true," Brandeis said. "And he is not a bull-shitter, not a crazy guy, this is a reliable source, right? He said they definitely found un-natural solid formations way down deep under all the surface gasses and substances and compressed fluids, deep inside Jupiter. Understand me?"

A long pause."Un-natural structures like what? Un-natural in what way?" Deveroux said.

"That's your rabbit-hole, not mine," Brandeis said, laughing a bit to himself. "No one believes the story anyway. You find out about that, I bet you'll get somewhere with your process. Because I don't know, truthfully, I don't know, and I don't know anyone who does."

"But why is that significant?" Mendoza asked. "Surely there's something inside the giant planet? It obvisouly has an inner substance, and if it's solid, so what?"

Brandeis said, "Within a year, the entire project to harvest raw-materials from Jupiter herself was dropped, and all analysis and measurements, and then after that I never heard any more about what my friend had told me. It had obviously been silenced, information black-out, but that's common. You find out, find out what those structures are, or what they thought they had discovered. Because it crosses all the lines and connects all the dots, but we just don't know. And the matter is forbidden as far as the necessary tools and technology and time or labor. There. Happy now?"

"I still don't understand," Mendoza said.

Brandeis said, "My friend, Al. If they did all that high-end research, and then shut it down, information blackout, and the idea suddenly dropped, what does that tell you?"

"It tells me they were wrong when they started out. All right, what was his name? Your researcher friend who told you this. What department?"

"It doesn't matter, Mister PPP," Brandeis said. "He's dead. Died in a car crash in Seattle, Washington regional, about a year or so ago, somewhat suspicious."

"Wait. A car crash in Seattle, a year ago?"

"Yeah. All his records and research and whatever work he had done, gone, nada, zip. It's as if he never existed. Odd thing, isn't it? Besides, I wouldn't tell you his name anyway, he has surviving family, I'm not stupid. He was a University researcher, in Seattle, and he was declared dead, but it was all faked. He may have been murdered. All the more reason to conceive that someone wanted him silenced about the same information, don't you think?"

Mendoza and Daniel exchanged knowing glances. After a tense moment or two, they entered the casino area and enjoyed some of the entertainment and games of chance, but Brandeis didn't offer anything else of interest.

CHAPTER 16: Children With Toys

"When you were once one again, and once was oneness like an only friend, one, not two, but the two is you, and the otherness never ends,"—Marciel Janus Penieur, poet and space-traveller, frequently seen doing guitar-shows at the Jupiter-program Alpha-base on Amalthea, from his song, 2,412AD/CE

At this point, agents Deveroux and Angel-Face Mendoza, his large-economy sized Hispanic partner, now began to sense the depth of their otherwise harmless review of Jupiter-program malfunctions. Brandeis' pointlessly impossible black hole of related information, added insult to injury by insinuating some sort of background knowledge about Deveroux's fraudulent death, in Seattle, almost a year previously. Very sly of him, they were under-cover for good reason, not merely to protect themselves from harm or hurt, but to guarantee an un-tainted investigation, critical to their success. Of course, Brandeis was sympatico to the investigation's goals, but this space-ship Commander didn't enjoy wasting time or games, that much was clear.

"Maybe you're not quite so dead as we figured," Mendoza joked, later.

"That one threw me, true," Deveroux said. "It's not important, I suppose, he is on our side of the matter, after all."

The Solace Electron auto truck they had commandeered for the time-being, tooled sweetly with other ground traffic, again into the stunning metropolis. At this period in US Continental land-mass history, municipal regions, or cities, like New York or Atlanta, had expanded into very large areas that covered even thousands of miles in area-perimeter. They were thought of as regional authorities, not cities, and essentially ran themselves as far as politics or government.   The Urbano Fantasma. And they were beautiful, stunning, Oz-like in towering structures and lighted displays, an endless array of residential, business, corporate, industrial, and population. To keep it all going, was one reason the Jupiter program was so valuable; a limitless supply of high-quality raw-hydrogen and helium, re-inventing life on Earth, by about 2,300AD/CE, with an almost truly endless clean energy source to power every imaginable modern tool and convenience. But as usual, humanity in her greed and cruelty, was busy finding new and inventive ways to glut her appetite for every desire and avarice. *Children with toys*, someone had said.

The highway into the regional cities around New York were filled with fast-sleek electric cars like their six-wheeled Solace Electron and its large, scoop-shaped plexi-glass canopy-style passenger area. Traffic moved forward at about 120-miles an hour, and would slow, and then speed up again, with the trams and people-mover trains to either side as well, and air-traffic overhead on different paths, as anti-gravity assisted small air-vehicles.

Mendoza drove, and they continued in silence a while, watching the passing scenes wistfully. Deveroux tuned the dash-system for some music, choosing earlier-era jazz tunes, his personal favorite, though he liked all music and most arts. He reclined into his thoughts. It was important in his work to fully understand how unimportant his own life was, in terms of any grandious anticipated outcome from his research with his co-workers at the PPP.

Proficient? He thought, what about us? *Are we proficient ourselves? Don't we also lie and cheat to get information?* Spy versus spy, agencies sneaking around getting into each other's stew-pots for what they could learn to an advantage here or there. If Planetary Program Proficiency was not flawed as well, then how did Brandeis know anything about his own short-term stay in Seattle? His 'I know that you know' comment was slathered with irony, and the stuff about deep-Jupiter measurements and inner structures was, perhaps, only hopeless.

Brandeis was trying to tell Deveroux something, his own informed opinion, and it looked a lot like a red flag, stay away from this one, too hot, too big, too deep, don't get hurt. So for his own sake, DD tried to consistently cultivate a self-referenced meaninglessness, given that he certainly could be hurt, and that all that hydrogen and helium and so on could wait, or was simply not his problem. *Big deal*, he thought. 100 years from now it won't amount to much, at least not for himself. But a free trip to the inner-depths of gaseous Jupiter in some horrid pressurized diving bell, well, why? Why should he be so bold? For fun? He had to laugh.

"Al," he said after a while, "let's take some time off from all this with Jupiter. I've been talking with our management, and it's starting to look like we're eventually going for a ride to the Amalthea Colony and the base they have there to run the Jupiter hauler-transports. Seriously, it's been planned out, so, I want to take time off for about a month, before anything like that is needed."

"God, I'm not going to enjoy that part much," Mendoza complained. "Space travel to a small Jupiter moon base? Oh god."

"It's not so bad, we'll probably travel as passengers on one of the big transports, maybe even aboard with Martin Brandeis on his, if the schedules are compatable," Deveroux said. "But for now, spend some time with your wife, you know, just enjoy things, rest up."

"Which one? I got the wife in California, but she had an illness. The wife in Belize is nice, I like her. What about my wife in Great Britain? That one is hot, let me tell you. Please be specific."

They laughed, it was a thing between them, as mature men. Mendoza had many women and wives, but Deveroux had never married and enjoyed his romancing only as a single. Mendoza also had a lot of children.

He settled on travel to South-America, or Central America, formerly known as Belize, and the woman he kept there, a delightful brown-skinned girl in a wooden house on stilts by the beaches and Caribbean waves in the sun. Two weeks later, the Urbano Fantasma visions of mega-urban Long Island and New York, the sky-towers and lights and craziness, faded, and faded more, in Al's memory. It only took him two days by personal aircraft, (a fast one such as they had taken from Seattle a year ago), to arrive in Belize, a tropical wilderness by comparison, hot and humid, full of tangled weeds and vines, tall trees of an aged sort, palms, small villages and dirt-roads, beaches, Latin food and people and languages. Welcome home.

Meanwhile, Daniel had other things to do, it hardly mattered. He was right. A serious journey into the Abyss to check out planets and space-ships and space-stations, that far out, for two novices who had hardly been into orbit around Earth, they might not even ever return. *So this is good-bye,* he thought.

Mendoza's woman, Marta, was no girl at age 36-years and very attractive, though short, about five-foot, four-inches, with cute tits and healthy flesh on her belly and thighs, and a joyous smile and way about her. Yes, he loved her, and she loved him, more than maybe they would ever really appreciate. Al was not like Deveroux so much, he was a simpler type of man. Mendoza had begun life as a mostly under-educated military servant, operating heavy equipment, and then technical gear, and more into his education for the purposes of the government, which as it worked out was to serve in the PPP as a 'researcher'.

But this did not mean academia, that sort was more Deveroux's territory. The books and intelligence, the math, or science and theories, the bigger-picture and knowledge about the way things worked, and the way they did not. They made a good team. There was no way to find out anything about the high-tech programs they were investigating, unless they understood enough about them to recognize all that was going on.

At the same time, a hands-on approach, such as Mendoza could provide, was often much more useful in the field. When he finally reached Marta's side, there on the coastal Central-America, she was expecting him. Their three children, Vera, Chuck and Dave, were ready for a party, complete with his favorite foods and music, and some local friends.

Marta met him at the car-port area, he had arrived in a sort of jeep-transport designed for off-road. *"Mi marido, mi amante! ¿Dónde has estado, mi hombre pobre niña! Entrar, los niños son waiting. Oh!" (My husband, my lover! Where have you been, my poor baby man! Come inside, the children are waiting. Oh!),* her high-decible love-chatter swooned over him and she made a fuss, as of course she must have. It was mid-day, warm, palm-trees swaying, he always enjoyed the tropical mid-Americas and people, Guatemala, Nicaragua.

"Marta, Marta! Sí, sí, ¡Te quiero!", he called to her, travel bags in hand *(Marta, Marta! Yes, yes, I love you!).* They hugged and kissed, laughing and chatting in Spanish, then went into the home there. He hadn't seen her in about six months at least, the home-comings were always the best part, he felt very loved and respected, he found he could smile again, even forever, somehow, from within.

It was small home, but large enough, old-style architecture, similar to many in the region. Marta kept house-and-home together, and he knew she had her ways when he was gone. It was a marriage of convenience. There were storms, in the region, off and on, and the children were in school. Poor little Vera had an autistic palsy, at age ten years old, the boys Dave and Chuck, age 16 and 22 years, seemed only interested in surfing, girls and parties.

As they day wore on and night enclosed them softly, Mendoza was very satsified, warm and cozy, happy, content. She spoke at him in rapid Spanish, later, over beer and the traditional food she had ready, over the soft light and darkness beyond, quiet and still, like wind in the trees, so unlike New York.

She said, *"Al, por favor, debe ser una broma, esto es serio. Usted no puede ir al espacio en una nave espacial. No, no. Usted no debe ir en absoluto, es muy peligroso, pueden encontrar a otro hombre. Que se preocupa por su planeta estúpido Júpiter o Saturno o Marte. Te necesito, los niños, que nos envíe dinero, pero yo realmente te amo. Sé que hay otras mujeres, para usted, es de roble. Pero, ¿cuándo le sirvo? ¿Cuándo la mía? ¿No lo entiendes? ¿Por qué te vas mi mundo? (Al, please, you must be joking, this is serious. You cannot go out into space in a space-ship. No, no. You must not go at all, it is very dangerous. They can find another man. Who cares about their stupid planet Jupiter or Saturn or Mars. I need you, the children, you send us money but I truly love you. I know there is other women, for you, it's oaky. But when do I get you? When are you mine? Don't you understand? Why are you leaving my world?)*

If he only had an answer, he wouldn't need to leave her, he thought. True enough, Mendoza didn't wish to head out to Jupiter at all. Maybe the shcedules were wrong, or would change. It was a lot like the military, for him, and Marta looked at his work that way. The wife in California was more educated, an artist, and saw things differently.

The UK wife also was fairly savvy, an attorney or legal-politico, very passionate. So of course he communicated with them all, it was fairly simple, and the women knew about each other, he had no secrets. In a good year, he would rotate husband duties, travel from home to home, this was the modern way, and he was a well-paid man, so all the needs were met. Mendoza could boast of as many as 12 children he knew about.

Little Marta, though, she was the simple one, a farm-girl, not much to her life but religion and household matters, farming and gardens and small livestock, and she loved to watch the current soap-operas on TV. Even if he could explain, she really would never understand. In many ways, he enjoyed talking with her more than he did the intelligent ones. And then of course they shared a Latin heritage and language, it was a special thing for him, in his heart, so dark and mysterious.

"No te preocupes, mujer," he replied. *"¿Cómo es poco Vera y los chicos? ¿Qué han estado haciendo? No importa todo eso, si me voy, me voy."* (*Never mind, woman. How is little Vera and the boys? What have they been doing? Never mind all that, if I go, I go.*)

So, she told him all about it, and they shared family matters. The house needed repairs because of a storm, and that had taken weeks, and she admitted openly she was having sex with the construction man, a younger, husky Mexican. Mendoza laughed, but there was a pang of hurt, they were married, after all, at least in some sense. Vera, the one with autism, was in school and mostly doing as well as she could.

There were special arrangements for her, she had trouble talking, but could read, and at times became confused and wandered off, but Marta had help to watch her. Chuck was working on a local farm and learning crop-science, but Marta suspected it was a coca plantation, for the cocaine drug, and didn't approve, he had only been working there a month or so, but was well-paid, maybe too well paid. David, the younger one at age 16-years, was still in local schools, which were poor-quality, and was interested in sports, especially track-and-field events, he had won some awards and competitions.

All three of their children were named for an old Beatles song. The children loved their dad, and missed him, and he spent time with the boys a day later, much the fatherly type as he could be, full of advice and some sorrow. But they respected him, he was very wealthy by local standards, and always laughing and teaching and smiling whenever they saw him. But Vera, she hardly recognized him at all, just the way it was, with that condition.

For a week or so, he had Marta, and she had him, it was always too short a time together, which perhaps made their loving more intense. She was such a sweet thing to him, like an animal, and himself as well, devouring her passionately, in their hidden bed together, sweating moistly in the tropical heat. Just before he was set to leave again, Dave asked him a painful question.

"Papa, I know you work for the governments and the world and all that," the boy said, in English, as they walked on the sand, the distant blue waves like an unawareness that beckoned without end. "But, papa, the governments are bad, you know? They are corrupt and they do bad things. The prisons and the military and wars, the new drugs for poor people, all the mental-powers on TV and the news, the money is no good, they take it all, like Communists, sometimes the people have nothing, that's what they say here. What I mean is, whose side are you on? Are you really for the people, or for the world government and space program? Which one, for you, I'm just saying, you know?"

In a way, he was pleased, his younger son was a socially responsible citizen, and wanted to choose the right future, and know what was happening in the world. So he was challenging his father's loyalties, given the work he did. The trouble was, Mendoza really didn't have an answer. He felt like a man without a country, his true loyalties were much harder to explain or express.

They spent many hours like this, trying to express what they really knew they never would, a family. *"Hoy el teatro vacio,"* Mendoza finally said. *(It is an empty theater.)*

CHAPTER 17: Before You Go, Deveroux, Deveroux

"Yeah, god," he laughed, as they gazed up at the silvery pin-points of stars in the deep indigo stillness and bliss, late at night in the coastal Northwestern hills."That's her. Nice outfit." --Marrciel Janas Peniuer, travelling poet-musican, Jupiter program Amalthea Alpha-base Gin Fizz Club, 2413AD/CE.

The coastal Northwest landscape-backdrop and natural beauty, was also a live HoloCast projection from home, that they could enjoy at the Amalthea Colony, 700 million miles away. Off the coast of Washington in some of that regions roughest and most difficult coastal terrain, Earth's terrestrial glory was never so gray. The live-streaming image was placed in a data-stream by Elsewhere Eye Holocast company, and was available at the Amalthea Alpha-base, there on the rim-of-the-worlds asteroid belt orbital path.

"I wonder if there's fish?" someone said, at one of the base clubs, relaxing and enjoying the view.

"Artificial fish," his companion said. Just another day in deep space.

It was a comfort, even a bliss, and about the best that technology-communications could offer at that time, 420 years following the so-called computer revolution. For Penieur, it was the best gig he'd landed since times when he found himself playing guitar tunes in a public park, in a gang-controlled local population back home. Urban, dank and dreary, with some killer or other often known to lurk, and him a music-lover, of course.

700 million miles away from that mess, playing his heart out on a very advanced space-station, and good money-pay, too!! What a deal!!

For some reason everyone expected him to dress as some kind of homeless bum or 'hobo', like an indigent or penniless wanderer, mostly of urban sort. Thus, no mistake, the bliss of star-gazing at this distant outpost, with the contribution of the Endless

Eye Holographic projection of night or evening natural scenes, it was a dram job. The video images from home played out like natural scenery in real time onto a wrapped-circular foil-type screen-density surface. A transporting delight, they would say.`

None of which held much attraction for the PPP's lead man on the Jupiter Program investigation, Daniel Deveroux, considering such enjoyments needed for dangerous travel. But in his view, long-term re-location from the considerable superiority of the real thing back home, and some really awful conditions and terms that made space-travel at the level of the Jupiter program highly stressful at times, were not so great a deal. *Ain't nothing like the real thing, baby,* Deveroux would self-regard the topic. He had his own musical talents, after all.

As Mendoza was spending time in Belize with Marta, Dev' had travelled by now, back to the Seattle, in the Washington area, to the same area as where he had died. In this case he went there by under-ground (and transel-overhead bullet tram), coast-to-coast in about four hours. For a distance of 3,000-miles, transit time per unit-mile, a traditional ground speed of ten miles per minute (60 seconds), or about 600 miles per hours.

This was a higher velocity than a Mach-One jet airplane. Much faster, but the average passenger would be quite comfortable. North America was interlaced by then with thousands of miles of high-speed rails like this, super-modern glass-and-steel high-speed pneumatic people movers. A passenger was restfully seated, and could view the passing scenes in a blur. The extremely high speeds were attained through the use of magnetic-lift super-conductor levitation rails and pneumatic tube-ways. From within his berth, Daniel was able to speak briefly with Angel-Face, by private mobile connect.

"How do you like the fast train? I've taken it many times," Mendoza said, still in Central America's Belize region with his beloved. Then he laughed. *600 miles per hour? Why not just send me as a holographic projection?*

The natural image of America, from where Deveroux lay prone, cuddled snugly by a perfect-fit body-berth, with a private view-port, was a blur. That was all that could be said. "Yes, well, fine, Mendoza, you horny dog, god I wish you'd lay off with telling me about your women!! This bullet train vibrates like an earthquake!! I will re-connect in a few hours. Out."

"All right, Dev'," Mendoza said. The call was a high-hander, it was helpful at times to readi-check the gear anyway, such as communications. "Adios. God speed, my friend. Off."

Seattle only a short distance ahead, the super-train speeding along unseen, from beyond the Rockies, a downhill run North into the regions above Nevada, then North along California, like a long, long mechanical snake. The passengers were not athletes (as space-travel would partly require), and included single moms, children, older travellers, and those with disabilities. But, it was fast, for ground travel, only some very specialize actual surface road transit types could approach that speed at all. Why Seattle?

Deveroux had died there, for one thing, and there was an old friend he wanted to consult with, prior to any departure for Jupiter. He had property he wanted to check on, too, and needed to settle some personal matters, as any responsible person would.

"Sophistry, anyone?" Deveroux joked with himself, much later, 20 hours later, actually, when pausing to reflect, by then somewhere in the general urban vicinity, more or less laying low. His bones were still rattling from the Bullet-Tram journey. The thing may as well have been made of wooden beams and rusty iron. Air-flight was much preferred.

He had an awful feeling of being lost, like a disorienting, wavering dread. The Seatlle area and the coast were a lonely and wild part of the US, (all the best parts were, even by then). There was a house here, with no one living in it. There were cars no one drove. There were completely-equipped privately available storage lockers packed with every known technological item he might need in his investigation, including some weapons, and wealth-units, that sort of thing. Records and personal data, or life-passage evidence, useful at least for various formal passages; no one kept them, that is, they were official proofs that did not officially exist. Similar records indicated his death in the car crash, a year before.

Same as Commander Brandeis' friend or associate in the Jupiter program, with his original research on deep-interior Jupiter structures. That is, the unofficial truth, the solid rock of the unknown, unknowable, and usually unwelcome facts.

'Don't try this at home. Jump right in and find out.' Such was one of Dev's favorite mantras, which he repeated often as an affirmation. Indigo the night, deep blue, her immobile firmament, Shiva they say, or an immaterial non-life. Soul? In the same way, his own image, or that of others, could appear in those deftly exaggerated interference-overlap resonating form-shapers, HoloCast, perhaps to smile or wave at such as Marciel Penieur, the poet they kept at the Alpha Base. This also didn't appeal, it was an invasion of privacy. The Beatific beating, beaten, the bleating sheep eats meat.

A suit of clothes no one ever wore, and entire wardrobe, and outfits for women, such as his 'V', the elusive Vanessa Signo, PPP case-manager, part-time lover. Other garb and get-ups, he was not much into disguises. An airplane no one ever flew. A mailbox no one ever opened, and where very little new mail ever appeared, an actual paper mail-box, somewhat unusual at that time, and known as Care-Boxes, they were common enough.

And more. He found here in a data-base, a certain set of social contacts, who would never see or hear from the one contacted, again. A set of private, personal memories, photographs and books, discs with images and voice-recordings, a dry-pressed flower, a lover's ring, some oddball souvenirs, like the high-vibration LED levitating signal style image of his mantra, *'Don't try this at home.'*

And, for those who truly knew Daniel Deveroux, such as his pal Mendoza, a very special set of security-clearance proofs about Dan's great-grandfather. Associated accusation against Deveroux included that his great-grandfather, Alex, was part of an off-world program at that time, many years prior to that, to re-locate humanoid aliens, very much if not exactly identical to the human form, from other worlds, that is 'extra-terrestrial'. In other words, that Deveroux, this Daniel Deveroux, was not entirely human, but fractionally from a planet other than Earth, by blood. No one really minded, but it was part of his personal life-story owned-items.

Money no one ever spent or used. Lots of it.

Tickets to yearly-annual events and shows, entertainments, big theater spectaculars, music events, passes to annual sporting events, hotel passes, restaurant and bar passes. No one ever used or enjoyed them.

Club memberships no one ever participated in. Retail and shopping passes for purchases never made.

Authenticated Universal Magnetic Passcode items to many of the West's most sensitive resource facilities and military functions, though not nearly all. Never used or applied, not a single one. These provided him an entry-pass clearance to secure facilities if needed.

An array of weapons. Never fired, no one killed or wounded, in pristine, un-use, well-oiled, fully loaded, idolatrous and murderous in perfect readiness, with more ammunition than he could possibly use.

Un-used prophylactics, still sealed and hygenicly clean.

Stacks of survival foods, un-eaten. No one hungry for them.

Medical kits, no one needed them. No one (who was dead, anyway) was sick or injured.

A few women around the Seattle area, gal-friends, lovers, now behind him, in the past, in terms of relationships, just pictures and contact numbers, memories of sweet embraces. No one called, and he didn't call on them, except in his own subconscious animal mind. Deveroux wondered, did they attend his funeral when he was announced dead? Did they even shed a tear? But wait, there was no funeral.

So, the women he knew while in the area, they were only lovers who didn't call any more.

And a son, too, a 30-year old man, his birth-child by blood, who didn't know his father.

So it was a lost and lonely feeling, sad, a certain emptiness, distressing and sometimes leading him where he would not go, if only for now-and-then companionship. The Northwest Corridor by year-2,413 was also a very populated region, very attractive for Americans for its plentiful natural water-supplies, lakes, rivers, streams, and ocean too. A beautiful city, and a beautiful life. No one shared it with him. So, maybe good-bye wouldn't hurt so much, to leave the planet, heading outward for a year or 18 months or even two years, to the hell they all knew Jupiter really was, without a doubt. Jupiter, the Abyss, good-bye, for a while, maybe even forever.

Dev' found himself at a hotel again, deep within the urban boundaries of a part of the metropolis known as Lake Aspirin. He puzzled more over all the Jupiter program research, enjoying a meal and some chat with folks he met at a sports bar. The research was pointless, not so much a wild goose chase, but a performance of the same style, without a goose at all. And that was frustrating. Seeking what wasn't even there. Even after all they had learned and planned to investigate, it was still a hodge-podge of information that lacked a center, and the work he was doing would probably eventually simply resemble what Brandeis had already presented to the PPP advisory council on the case. Commander Brandeis had done all his work for him already before he started.

He resolved to spend time with an old friend who was a long-term Ph.D. astronomer at the Washington State University Science Faculty. Knowing that he did not know something was Deveroux's best bet almost every time out-to-bat for PPP. The man was also somewhat a friend, sympatico with the work, trusted.

They spoke by mobile link, from Deveroux at the bar, to his friend.

"Yes, Doctor Podliakov, my friend, I will be in Seattle-area for as long as a month," Deveroux was saying. He could also see an image of his friend, a youngish-looking Caucasian man with curly-wavy hair, a pudgy baby-face, and a big laugh. The video image seemed to move past his doubts, a familiar face, collegiate.

"Good to hear from you," Podliakov said, on the other end of the call. "How about in the morning, 6 a.m. for breakfast at the joint across from the sports arena, you know the one? Little yellow building? We used to have breakfast there a few times. Don't worry, no one will recognize you but myself. I mean, as far as your death is concerned."

They both laughed. "I wish some or a few would recognize me, here and there," Deveroux replied. "Anyone I knew when I was here is not in contact any more. Lonely feeling."

So, the next day, Deveroux found the breakfast café, in the yellow-building, where he knew it was from memory. So often in his travels, he did not enjoy the deja-vu of a familiar place, simply because he moved around so much, like a fugitive. Inside at a wooden table enjoying a freshly hot egg-omelete breakfast, with his friend, one of the University Astronimcal Science professors of long-standing, Richard Podliakov, Ph.D.Astronomy, and other titles. They melted the ice between them with jokes about current headlines and media-news affairs, politics, and so on. Then the talk started back and back and back in their thoughts, to discuss the Jupiter program, and what Deveroux was hoping he could accomplish.

"Space travel is even less fun than the Inter-Continental High Speed Bullet Tram, " Podliakov said. "You won't like it. You're talking about as long as three years, maybe with a short stop-over back home, in-between. In three years you could end up with serious anorexic-anemia, you understand this? Muscle loss."

"Well, I heard about that, yeah," Deveroux answered him. "But, well, it will be an adventure. My job as lead on this for my bunch, includes simply directing the thesis towards some sort of useful new information. And there seems to be no useful information."

"From a science-view, new information is impossible. Useful, as you know, depends on the eye-of-the-beholder," his friend said. "But trust me, you'll hate a year in deep space, seriously."

They enjoyed their meal and more chat, leaving heavier topics to fall away soundlessly.

"What the hell is the Circulator, Daniel?" Rich asked him at one point.

"Hmmm?" Deveroux looked up from his toast, slathering some sweet butter and jam, so purple, so modern, so fake. "Uh, a rather large sex toy for two, for couples."

Podliakov grinned silently.

"It's new," Daniel added dryly. "Don't give me that, you've heard of them!! They use epidermal absorbant stimulant electronics, full-coverage, for two."

Rich wiped hin chin and beard with a yellow cotton napkin, coughing lightly. "Look, my friend, blessings and peace to you," then he paused, taking a breath. "Look, on the Jupiter program, it's obvious what's not working, is not obvious. It's an old program, okay, more than 100 years old, given the research and construction era. So, it is going through corruption, at many levels, too many for any one single aspect to break the case and clear things up with your proficiency group."

"It's always that way, Rich," Deveroux answered. "No big deal, we mostly go for matters that open things up so higher powers can get their work done to make needed changes, without a war on their hands."

"Let me state my assistance plainly," Richard said. "Planet Jupiter is a mystery and always will be, period. The reasons are many, but from what I've seen over a lifetime, the planet is, it's impenetrable to our attempts to really understand it. The thing is vast, unbelievably hard to connect to normal sources of analysis. So, they gave up, years ago, in fact. Yes, some said it was a cover-up and they found out something spectacular, or frightening, but it wasn't true. So, it's somewhat true, you're on a fool's mission. The sort of truths you are talking about, a sort of Unified Paranoid Field Theory of high-tech gremlins that don't exist. It's a cosmic joke on us all, and anyone who looks into things of that nature. Those types of truths can't be understood or expressed and so are inevitably mis-understood by people who believe in them. Do you feel bad about it, like a loss, emotionally? It's like a joke, you learn to laugh. We have limits. That's my two cents. I have other thoughts about it all, maybe talk in more detail later."

They started to prepare to leave the table and end their meal together.

"They have us scheduled to ship out in six weeks," Deveroux said forlornly. "I didn't have much choice here."

"You're number one, Dev'," his friend said, folding his yellow cotton napkin on the wooden table. "Understand? You."

CHAPTER 18: The Second Commander

"No, it's not forever, that would never do. We only need the truth to bend a little, just for now, in order to survive some terrible, terrible mistakes, just a while longer. And it does bend, you know, it does."—Second Commander Menima Pearl, Jupiter transport 'Down' second officer, under Commander Martin Brandeis, speaking to the Quantuum Mind Fellowship. at a luncheon to honor the dead and fallen, following the wreck of the 'Ferrous-2', about a week after the disaster.

She was one of these remarkable women, a rare one, and you'd see them all the time by this period on Earth, circa 2,413-14AD/CE. Amazonian, perhaps, though not in that way, athletic to be sure, stronger than a man in many ways, different, but with the work to be done, a maleness or certainty, excellence. The Second Commander of the transport whale-hauler 'Down', was somewhat dimunitive, or short in stature, about five-foot five inches tall, with short-trimmed dark hair, slender overall build and light-freckled skin, but she seemed like a strong little human tree or curious animal, clean and perky, bright, very intelligent.

Dangerous would not come to mind for anyone she knew or worked with, as far as a personality trait, but the female of the species is known for such a different hardness, given the needs of any children and intimacy of child-birth, or sisterhood, and misogony against women in general, even over hundreds of years, that to form a relationship, under her considerable authority aboard the 'Down', anyone would approach with caution, and think twice. A powerful woman, an unusual fate or role in life. And then there was her mind, as well.

"Second Commander Pearl on-duty at flight-deck helm Pilot's Desk, logging for the record at 10-30 hours and 14 minutes. Please confirm for command-desk authorization to control-command for this duty shift. Thank you."

Leadership and pilot-duties in the Jupiter program were an elite club or small clique, an insider deal. The positions were limited, the technical and training and education were demanding. Stress, personality or leadership values, decision-making, willingness, readiness and career-path, all came into play long before anyone found themselves calling the shots on a gigantic deep-space transport route to Jupiter 700-million miles from Earth to grab 50-million tons of pure helium. And calling the shots for those runs, was boring, dry, dull, somewhat senseless and automated.

A cold person? A hard heart? A Space-Nazi with no real romantic sense or loving ways, no desire for other aspects of life, no babies, no gardens, no home-cooked meals, no secret unicorns or dreams, no poetry. Many people loved and admired Second Commander Pearl, and she was a sweet person as a child, in many ways wonderful and alluring of her private mysteries. And then there was her mind.

Consciousness or awareness in space-travel was certainly considered a specifically human activity. Levels of awareness could be very important, and the dead machines, the dead nothing of the dark Abyss, the dead computers invested with artifical life-imitating chatter and data, they wre nothing without the human factor. Much as a form or highly technical device, with some important function or other, a bomb to blow up the world, or an instant cure for billions of cancer patients on Earth, beamed towards home from some wonderful orbiting medical healing miracle platform, and the joy of a way to save or extend life; much as, let's say, a wonderful new invention to rid the world of all diseases, or even remove death itself, some miracle like this, would await the push of a certain button, a certain code, a certain authority who knew which button, which code. And even waiting a very, very long time.

Until finally the right button or combination was initiated, and self-activating machinery started to hum and buzz, and all that might follow, according to its nature. But it was always and ever only a human hand, a human mind, a physical animal with two legs, arms and hands, and hairy top and sexual parts, male and female, that would ultimately push the button. And awareness, actual animal-human knowingness or certainty, in space-travel at that era, was maybe what was needed the most at every turn for things to work well, when buttons were being pushed, etc. Second Commander Menima Pearl's mind was 'aware', and this was a good thing, in her work, cultivated, preserved, developed. *Big deal, she thought, I was born this way.*

Details about Pearl's career were typical. She was long enough in space-labor and space-flight, with many awards and accomplishments, and so on. Her astrological sign was Aquarius, she was un-married at age 45-years but had six children back home, fathered by two different men in other lines of work. Hobbies included robotics, collecting ancient calculating devices and obsolete technology, and religion and spirituaity.

Thus, unknown to many, the Amazonian Second Commander was a priestess or spiritual diva-guru, among the Quantuum Mind Fellowship, and other Religious Orders. She was long-term in that priestly role for many years, and her work was serious and positive, thought to encourage loney space-travellers, and to strengthen lovingly in the face of so many unusual hardships and the complexities of the Abyss.

Others pointed out that the Jupiter program and other space-programs at that time, only tolerated the space-religions, and that there were down-sides, among them a loyalty and fanaticism that was dissipating and incestuous, somewhat claustrophobic, and sometimes scary.

The Abyss was not home, and never would be. After even a few months in deep space, life on Earth was forgotten to a degree, along with its beatific Natural Wonders, its normal gravity, warm sunshine, beaches and cities. The dark, cold, infinite Shiva-Mother, upside down, lightless and sightless between points, other features, created a sort of caccon or shell of life, like a living sea-shell of human thought and ideas, in each person, drawn in close with each other together, protecting from the infinite and eternal, the death, the unknown after. The Quantuum

Mind Fellowship nurtured and developed or protected the traditions and practices that sustained the multiple living organism of groups of human beings travelling great distances in space. They had no dark design or sinister intentions. But they were human beings, with all that could mean, both the light and the dark, which people have as their parts.

When the Ferrous-2 crashed and the deaths, Menima, the Quantuum Mind Priestess, did not leap into action with glee or pounce on the chance event as a moment to express her religion. Not at all. Deaths in space, or as part of the Jupiter program, were rare, but did happen at long intervals between. So, there were the funerals, or course, last rites, mourning. And for those workers in the program, death in space was unique enough, to recommend to their thoughts and dreams a hopeless, dismal, sad and nihilistic view of reality and spiritual truth, that those who died would make no voyage home, and had no spiritual home at all, no afterlife or heavenly reward, but merely blinked out, life ended, and the vast infinity of emptiness and cold dark death all around, space, only seeming more heartless, less a home for a soul or mind, than had they not courageously ventured there at all.

And yet they must, compelled by circumstance, and the choices and decisions and claims of a very powerful industry, like children who dreamed of adventure, and later found it a ruin to their souls.

Menima and the other ladies of the Quantuum Mind Fellowship had prepared a short ceremony, to honor the dead, about a week after the crash. At this point in the history of the Jupiter program, such groups were many years old, with orders and rituals and meetings, like any church. So the women wore their white, thick-cotton robes, and colorful sashes and signs or symbols. There were other devotional items, as for a funeral, not like a wedding or prayer-meeting.

The women entered, carrying a large stone. This was a devotional Oneness idol that Second Commander Menima Pearl would use, aboard the transport 'Down' at that season. It was merely a large, jet-black stone, formed as an irregular and jagged circle, framed in the same very strong titanium-metals that the ships were made, in an ornate fashion. The black stone was said to have been collected from among the asteroids, and the art-work was supposedly created by the family of one of the Jupiter-program's earliest deaths-in-space, as a memorial.

It was an idol, of course, but it made the funeral tones appropriate, it seemed to her and others. More cheerful matters also attended, eulogy, memories, photographs, they gathered with an audienced of those concerned. A funeral or memorial in deep-space, something different, something the same.

And then there was her mind, or awareness, that was also the Second Commander. Of course she was sharp, educated, intelligent. But in her dreams, and private life, she was hearing of something other than herself, that she alone was only to know or hear, there in the dark, within. Ideas about herself, the program and the men and workers, and other ideas, such as oracles, visionary, immortality, destiny. Sacred Learnings that the Quantuum Mind Fellowship maintained and brought forward, even specifically for the space program, and also connected to regular churches back home, and to University, theology, ancient and traditional. The reason was that as human history had allowed it, new ventures into space, had also opened into an arena of new spirituality and new types of spiritual truth.

She was only a woman, a human being, not strong, not a big, muscular person, or with stock piles of exotic weapons, or armed guards, or a military cadre or marching men, none of this was her way. So no one feared her, personally, nor should. Menima was a decent, harmless person. But as witnessed of Martin Brandeis' report to the PPP, and among the Menuda leadership at Alpha-base, the fear and to a certain extent the power, lay in the notion that the space-religions were a corrupting influence with an overall negative effect on the program itself and program goals. What created Menima Pearl as a threat, was her intelligence or awareness, and her position in life.

The fellowship also included men. The meetings were like any other. Menima would retire privately to chambers, when she could, between stints on deck or at the navigation helm as second commander. It was a luxuriously furnished, a private woman's quarters, not so large, but very pleasant, which by this time in space-travel was the standard, rather than the cold-stark-sterile-militaristic space-ships of the past.

Among her possessions or personal items, was a decorative robotic toy, which she called a 'minook'. This was a self-activating machine-like self or beingness, looking a lot like a little Buddha or pleasantly fat, cherubic angel. The thing had its own mouth and speaking-capacity, and was oriented by computer-programming to an almost limitless supply of stories, songs, poems, new combinations of wise-sayings, predictions, analysis, ancient books, and more. The minook didn't travel or move about, like the other robots they used.

A minook was somewhat of a collector's item, old-fashioned, from as long as 200 years previously, faddish then, no longer very popular at all. This one was ornate, rather beautiful in its way, the original series were intended as sort of robotic advise-companions or amusements, based on new computer technology and advances, for use in space to relieve stress. So their programming was intentionally harmless, and always positive or helpful. A quality minook could even advise on health remedies and space-sickness, romance, gifts and purchases. dream-interpretation, or cooking meals. Menima had this one on a counter-top, with beads, candles, some dried flowers, necklaces, and oils or polished stones.

"A high-demand falsehood reveals the way," the little minook said. The voice was like a music, and could change, a holographic form would also create the illusion of real life.

Menima found this to be a dismal and up-setting random selection. The woman was essentially rather lonely, in her work. It never really left her thoughts that something like this toy idol was a mere machine, dead and not alive at all, with no real thoughts of its own, and no eyes or heart. It was not a life. The wise-sayings would repeat a while at intervals, as she worked around her apartment on other things.

A high-demand falsehood? Leading the way? The way to where? It didn't sound very promising, but on the other hand this old-school minook could come up with all kinds of poetic words like this. It could even be programmed to interact directly with her as one would speak to another living person.

Suicide? No, of course not, no.

She continued about her rest hours and private time.

CHAPTER 19: Diving Board to Jupiter

It was no small thing. The two research associates were now setting foot on a path that, even at that date, only a tiny fraction of the Earth's citizenry, could boast to have either enjoyed, or not. Al Mendoza, and Deveroux were now to be prepped and informed by 'V' at Planetary Program Proficiency.

"You'll do fine," she kept telling them. "Just act naturally."

Vanessa may have had a cold heart in terms of Deveroux's future absence, they were not serious about a relationship beyond a few romps in various hot-tubs. Vanessa was not a person seeking to be rescued by a man. Though perhaps true, Deveroux was a man wishing for a woman to rescue him, mostly from loneliness. Daniel Deveroux' life was one of somewhat empty and socially lacking obscurity (*'Given my recent death,'* he thought bitterly to himself), a year-long journey into space was maybe the worst choice for his mental health. Likewise Mendoza was loathe to make the journey, for other reasons, mostly the mechanical nature of the machines and his revulsion at leaving Mother Earth behind, in general.

There were two travel choices they were given by the PPP planners, to decide on, or at least partcipate in choosing. Why did they have to travel outward from Earth some 700-million miles, to parley words and accusations with the princes of the Jupiter Program? Because the program is not proficient, might have been an easy answer. More than this, PPP wanted them to seriously document, infiltrate, prove or disprove, and archive data-base, any true or false notions of an actual conspiracy, working from within the Jupiter Program. This alone was rather like sending a sheep to prove the wolves are hungry. Thus, they would have anonymous-status undercover persona roles, that would not be revealed whatsoever, to any of the actual players in the investigation scenario.

Brandeis, the commander of the 'Down' had met with them both and knew who they worked for, and even more, and so did others. The general rule was, that the PPP was a global public-transparent service, so researchers would use false names and identities only as a 'fig leaf lie'. The computer-world they now lived and worked in, and the essential authenticity and trust needed by the agency, really meant nothing if anyone involved found out who they really were. So they didn't hide it, but the false identities were applied until it became painfully obvious.

This was similar to the 'ruse' they had employed posing as DNA scouts for a local University program interested in the horse-breeding operations at the Horse-Head Nebula Arabian Breeders in Oklahoma. Computer-society by then was such that a major split, public population philosophy had moved all information into the realm of transparency or public availability and discussion, including personal identity. It was a quirk of Western history from about 200 years previously, at that time. They had no secrets, but only few could understand the complexities and technology and huge portions of data, or even bothered. So such things were like popular falsehoods or open lies, and no one really cared.

"The other principle reason to send you up, is to locate and interview the four remaining planetary-research specialists involved in the early exploration of the deep-interior of Jupiter itself," Vanessa told Angel-Face and Deveroux, about four weeks prior to their departure. "It's understood that we don't know what's wrong with the Jupiter program, of if there's any one specific thing. A conspiracy, the unions, sabotage, the space-religions, and the technical leadership failures, all that from Brandeis. But the visionary deal here is whether or not there is something down there deep inside Jupiter, that shut down their research, years ago, or freaked everyone out, or ended the transparent science exploration, for some reason. That's your rabbit. That's your rabbit hole."

The three of them were now together again, in up-state New York, beyond the regional urban metropolis, nearby the hidden PPP facility (one of many around the world, for this minor global agency, circa year-2,400). Vanessa, the case-manager and PPP council-member on assignments, was once again cunningly beautiful, her dark hair and robust hips only more appealing as the trip into deep space grew closer.

"So we find the old men, and interview them, on-site, with no bias or intimidation, no punishment for telling the truth," Deveroux repeated back to her.

"Well, unless they LIKE it, Dev'," she answered. "Maybe it's just their thing!"

There was a chuckle. They had salad-plates and hot-tea on a patio-area outdoors, beneath rows of sycamore trees and hedge-bushes, in view of the low, dirty-looking hills beyond, and the campus-like PPP rooms and offices. Al had enjoyed time with his family by now, or at least one of his families, and Daniel was still pondering what he had learned from his high-science pal Rich Podliakov in Seattle.

In truth, they knew very little about Vanessa's life, either, where she lived or any family. They weren't supposed to know, but they still cared. They had become an extended family.

"Well, nothing living could exist inside that hell," Angel-Face commented, between bites of macaroni salad. "Inside Jupiter."

"Maybe not," Vanessa continued. "We all talk to the same people, the same sources. I don't know either, of course. But the deeper flow is the dangerous talk and rumors, the ideas related to extra-terrestrial or so-called alien minds with some vague interest in planet Earth affairs, maybe even millions of years in effect, some cosmic ticking clock or astral-plane schedule-of-events. That stuff, we need to speak privately and with complete assurance to these four men, too. These are ground-floor program scientists. Two are at the Alpha-base, in minor roles, at Amalthea. The other two are stationed at various bases, and aboard one of the other large ships. Secrets, you guys? Understand? Big secrets, little secrets. And other sources, but they hide it all away. We want to know what they know. End-run. Mission-accomplished. Sort of. Anyway, it could tie together many parts of the investigation."

They sat silently a few moments, listening to the birds in the trees, flitting about lively. Other PPP workers came by or went into the buildings. The Jupiter-Program investigation was not the only major PPP activity, at that time, by any means. Only one of many. Other work, with other researchers, included investigations into militarization of deep-ocean Earth fish-farms intended to feed the world's billions of hungry mouths with low-cost, high-protien fish, raised as a new aqua-culture industry in the deep oceans as vast fish-farms.

But this was now subverted by military interests, despite the lack of any real wars. There was also an investigation into entertainment and gaming interests, casinos and pleasure-resorts, on the Earth's moon, which had all but over-run any real new exploration, with a spectrum of troubles that had also resulted in safety failures. Deveroux may have wished he and his partner had taken on that job. Much shorter travel time to the moon, than to Jupiter.

"We have all the briefs and data-base for the investigation," he said. "All that will travel with us as stored data, so the names, the groups, the bases or orbit-stations, the ships and commanders, easily available. And we'll be linked by secure communication back here. But our cover-identities, you know, the genetic-modification thing, isn't that a hot-potato, a little controversial? Maybe we should be more conservative, just to avoid attention."

"You're genetic-design engineers for Astro-Health Group, a large and well-known medical science corporation that serves world space programs, all sorts of related health issues for any space-travel or space-workers," Vanessa said. "Al's cover is a work-efficiency analyst, or human resource specialist, reviewing space-labor procedures and related medical needs, for Astro-Health. Dev's cover is as a deep-space medical programs specialist, looking at space-worker health in general."

"So I'm a doctor? Yes, Vanessa, I know," Deveroux replied sourly. "But AHG is known to many as among the first to develop human genetic experiments, to create super-healthy workers for the space-industry, or workers with unique, deep-space genetic physical features, by dinking the DNA and genes, etc. That was the connection at the Brandeis horse-farm, for whatever it was worth.

It's a very unpopular idea right now. So AHG is a poor choice for us, don't you think?"

Vanessa paused, humming to herself, a sort of purr there beneath her ample breasts. "Well, we had some food-service identities, but management didn't like them as well. You know, food suppliers and basic-essential life-sustain products, like water, oxygen, or barbecued pork-ribs. You know. Food."

"What's wrong with that?" Angel-Face replied. "Let's go with those. Low-profile, we can get more done. Go with the food company."

" I agree," Daniel added.

"Well, sure, but management was looking at the AHG roles, it gives you way more mobility, within the Jupiter program. If your covers are the AHG roles, your laison at the Alpha-base is directed to show you anything and everything, including small-ship transport from point-to-point, even various moons and planetoids, like the bases at Io, Europa or Ganymede."

"No, no," Daniel was muttering to himself.

Vanessa ignored him. "With AHG, the labor-conditions, worker health-needs, and worker health-programs, extend everywhere the program goes, with various conditions you would naturally want to know about."

"Like the external space-suit hydrogen miners on Europa need for health and survival," Mendoza said. "But they all need food, too, Ms. Signo."

"But that's different from what they need at Amalthea Alpha-base, 40 million miles away in an air-tight high-tech space-hotel. Way different. Food-service jobs may not provide that. AHG is a main supplier up there."

"Well, I don't like it as much," Deveroux ended his objection stiffly. "Too controversial, sets us up as adversaries. They'll think we're breeding their children as genetic monsters. You don't work in the field, Vanessa. If they hate your guts, they don't cooperate."

"Even astronauts like something good to eat," Mendoza concluded, taking Dev's side.

Another pause, by now they had finished their salads. With only a pulled wax-type string, the effort of a finger and thumb, each of the fabricated plates, spoons, dishes and cups they used to eat their meal, suddenly evaporated into a harmless, clean mist, then gone, like a harmless, pine-fragrance vapor. It was a new anti-litter product application they all used, preventing disposal problems.

"All right, what about transport and our own prep? We will have medical screening and travel schedule, and protocol for transport, of course?"

"Medical prep and tests, and your basic space-travel training-review, three days next week on Long Island," Vanessa said. "The travel options look like this. You can travel to Jupiter with Brandeis on the next outward bound run for his big whale-hauler. They move guests and important people back and forth all the time. He would connect to smaller ships, then move you to the Alpha-base. This would be fine, maybe not what we want, he's too close to the fire with his whistle-blower status."

"He also knows we're PPP," Angel-Face added. "By now."

"That's no big deal anyway," Deveroux said. "They can find out if they want. Like the Alpha-base people, or the director, Dr. Menuda Wu or whoever he is. We mostly have to just assume the cover-identities are no good."

"Well, there is also a high-speed deep-space people-mover headed out, about the same schedule, three weeks from now. It's one of the needle-ships, or they call them needle-crafts. Very high-speed, only carry about ten people maximum. Direct to the Amalthea-base in only four months travel-time, for deep-space."

"How long does the transport take, like the 'Down'?"

"At least nine months, more like a year, depending on the position of the planets and departure times."

"Let's just take the needle, then," Al said. "Why not?"

Within a few days, both men were almost ready for the voyage. Medical tests went on for an entire day, as the doctors were looking for signs that Deveroux or Medoza might not survive deep-space travel demands. If they had any chronic disease conditons, like high-blood pressure, psychosis, viral infections such as the flu or even the common cold, it might change things.

They also needed some basic-training review, so they would fully understand how to handle their roles during space-flight. Both of them were novices, and had traveled in space only very briefly, in Earth orbit or to the moon.

So they needed a lot of information. *What if I can't sleep during the third-month? What about bowel-movements or body-fluids? What type of clothing for the launch departure/acceleration? Same as during flight? Can we talk or chat about anything at all, or do the pilots find it up-setting if we are trepidatious about conditions or fearful in our talk, or just complaining? How bad are conditions on the flights, or on the 'needle-craft' as compared with a large transport like the 'Down'? How do we handle the anti-gravity, or weightless effect? Which ships or bases have artifical gravity, and which did not? What about the robots? Can anyone use them? What types of emergency conditions may arrise, and how do we respond as passengers? Is personal or private photography allowed? What about romance or sexual activity? Yes, or no? How do we use the exercise rooms?*

It was no small thing, the voyage they were taking.

Things took shape, and only dread and self-strengthening resolve remained for each of them as the final days bled into the past, prior to their departure times. Once they had figured things out, working with PPP and global space-schedules, it was decided that Mendoza would travel to the Amalthea Alpha-base for the Jupiter Program abourd one of the needle-ships, a vessel called the 'Aerotica'; Deveroux was booked to travel to the same point near Jupiter, aboard Brandies' transport, the 'Down', for other reasons (among them to respectfully get to know Second Commander Menima Pearl, and learn about her activities with the Quantuuum Mind Fellowship). So they would split up, and not meet again for nine months, at the Jupiter Program's deep-space HQ.

"What a waste of time, " Mendoza said. "Four months aboard a high-speed transport? I may as well take up knitting. Read a novel, I guess."

"No, no," Deveroux said. "Look at your data-files. One of the other passengers aboard the 'Aerotica' will have plenty to tell you. Doctor Menuda Wu, the program director, will be on the same flight. You can spend time with him, he thinks you are a human resource engineer for the Astro-Health Group. Wu will be heading back to the Alpha-base from other meetings on Earth. So it makes sense."

"He can't possibly be that stupid," Mendoza said. "He knows who I am, with the PPP."

"Well, it matters not for our purposes," Deveroux said. "If we all tell the same lies, it's similar enough to the truth to get our work done, isn't it?"

Mendoza laughed. "Yeah, but some idiot is always telling the truth, along the way, right?" he said. "And that messes things up horribly for everyone else, huh?"

Laughter, then Deveroux paused, he was reading one of the protocol handbooks, there in a lounge-area where they were getting tests and training. "I'm not sure I understand your meaning at all, my friend," he said. Mendoza laughed again.

CHAPTER 20: Departure

*"I hate to travel. But sometimes it's the only way to get from
one place to another," ('The Accidental Space-Traveller: Your
Guide to Deep-Space Motion Sickness and Other Un-Ending Joys',
by Stephanie Mie Stawp, published 2345AD/CE, Tor Books)*

More like a death than perhaps Deveroux or Mendoza were
willing to confess, buy this time-era in the Western global Earth
scenario, legitimate long-voyage space-travel to local planets had
two main components for any passengers or travellers. One was
the luxury handling and courted elite power group or leadership
class, and how they were treated on such trips. The deep-space
travel was not for the uninitiated, and for any VIP's, they were
rare guests, with red carpet personal assistants, almost as if for
some strange athletic event.

Additionally, their acommodations and treatment, included
the finest of everything, and far more hours 'awake on deck' than
others. They got to use the view-ports, the rest-areas or food-
courts, the music and entertainment features, exercise areas, and
even enjoyed tours. So did the crew; however, this was still deep-
space travel, and there were both dangers and tedious medical
changes that made life 'out there' stressful and debilitating for
almost anyone.

Thus, Door Number Two: 'Not Luxury Class'. Only the
professional astronauts and workers were really acclimated to
long-term space-travel conditions, and knew how to cope with its
side-effects. For anyone else on these voyages, there was a
learning curve that was either met, or not. The staff and crew, the
medical and health people and basic life-sustain operators, would
deal with any living thing the same way, they had no choice. But
from the moment Daniel Deveroux, and Al 'Angel-Face' Mendoza,
docked and boarded with their transports to the Alpha-Base
Jupiter Program Amalthea home, until they actually came home,
back to good old Earth (a period they anticipated to last as long as

a year or more), the side-effects were the main attraction as far as their comfort, mobility, and ability to simply survive.

"Maybe we can fake it, and not go?" Mendoza said, to Deveroux. "Just say we did, let it go at that?"

"Might work. I'll consider it." Dev' answered sarcastically.

Not to mention the challenges of getting their work done for the Planetary Program Proficiency group who had sent them. More like a cold, lonely death, than a paradise holiday. More like a bad dream than even the coldest, most wet or dusty piece of ground back home, where the foot-steps of a child might yet remind god sometimes of Gaia's holdings and love, the treasures of human life.

After further negotiations with the agency and global space-launch authorities, it was decided and confirmed that Mendoza would speed ahead of his pal Dev', aboard the high-speed carrier, 'Aerotica'. Deveroux would indeed be travelling as a passenger aboard the 'Down'. This sort of choice was typical of what they were now facing. At any time in the journey, the space-flight authorities could command their living bodies to go here, or over there, to this ship, then onto that ship, live at this base or that base, or even onto one of the so-called 'life-raft' vessels, in an emergency.

They were also liable to be sedated into total unconsciousness, by word of the ship-flight staff, and hauled around like cattle. This was standard for non-professionals, due to the occasion at times in the past of passenger fears, due to the nature of deep-space. But it wasn't done much anymore, and even the lowest-level passengers heading out to Jupiter (a student on a class-session to observe methods, or a worker replacement for a very simple clerical job), were still among the very exceptionally gifted of the Earth, just to be invited. So things would balance out as far as pride, ego, physical fitness, luxury-access and the powerful (who might include world-leaders, back home, or mission-specialists with high-end science-credentials).

"I got the better deal," Al said to Deveroux, at some point during their medical and emergancy life-support suit-up sessions and training. "The 'Aerotica' is only four or five months out to the Amalthea base. That's incredibly fast, less than half the time it will take the 'Down'."

Deveroux gazed back at him. Mendoza was un-dressed, in a pre-suit under-wear article. As a Hispanic man, it was an odd appearance, rather a 'Don Quixote in Space' type of appeal. Dan was placing a glass-dome air-helmet on his own head and neck for size. *Just like real space-men, he thought.* Mendoza was also checking out the personal suit or essential life-support emergency system that would be assigned to him. "Someone told me, you don't start really living until the day you die, Al," Deveroux replied. "What does the pace of things or any speed or velocity really have to do with it?"

Mendoza smiled. "Mine is the luxury ship. You get the cow-train. Ha-ha-ha!!"

"Mooo!" Deveroux said from within the helmet's sphere of clear-clean oxygen and dim-blue shade. Then he lifted the thing up and outward, moving it off its hooks with a twist and an unlocking turn. *Child-proof, he thought again to himself. Don't try this at home.*

And so it went. A hardy, positive attitude was essential, everyone knew about it and was expected to participate at the same level. No saboteur, no one person or individual holding all the rest hostage to some insane fear or cowardice, no special-cases, E-Ticket, you can't get off this merry-go-round once the ride starts. It was for this sort of thing that the PPP was held in high-regard as a serious Earth-systems investigative force. These two, a tall, buff Mexican and a suave-stylish European intellectual, were military grade specialists anyway. Part of the job.

Within ten days, Mendoza was up-loaded via orbital launch, to dock with the 'Aerotica', come aboard, and find his spot. The Earth orbital launches were very common. At this time, various launch-pads and up-loaders may have lifted ten or even 20 objects into orbit per week, especially at certain points in the planetary seasons. It was much like a passenger jet-aircraft, maybe a Virgin Airlines Orbital Pathways flight of dreams.

These simply entered Earth orbit, and safely carried passengers, and had docking capacity and other features. But they could not travel into space even as deeply or far as the Earth's moon. Al's flight lifted gently into the Mother Night of Earth orbit, and after a time the passengers being up-loaded could see the 'Aerotica', some 300 miles off on the blue-blue horizon, like falling into blackness and stars. The ferry ship drifted slowly towards her and changed course-and-speed to match.

Maybe like when a kid got his first car, back in the day, Medoza observed. Or maybe like your first date, your first real sexual relationship, your first wife, your first home that you owned outright, your first journey to France. Such was the sense of things when gazing out through one of the non-window windows (video-carrier port-viewer), and eye-balling the 'Aerotica'.

Angel-Face had never really travelled in space much at all. He knew, he could either fall in love with the ship and its people and ways, and learn all he could find out about it, and how it all worked and what its essential nature was, for his own sake. Or he could fear. And fear is a no-no in space-travel, pretty much any time of day or night, or any day of the week or not. As in life.

The orbital ferry-docker drifted closer. Now the long-distance silver-bird was in view, much clearer. The thing had to reach speeds of almost ten-percent the speed of light to reach Jupiter in about four months, or 127 days, as they now calcuated things, for this journey. It had the appearance of a very large Earth-born farm potato, in silvery panels and large tiles, all metallic, of course, to withstand the stress and vacuum. It was not sleek and needle-like at all, as the slang-words portrayed. It was rounded, and had a strange dip in the middle, that seemed to make no sense, so that the entire form was like a large shoe, half-bent like a folded piece of origami-paper artwork. Beautiful? One might have said the design predicted the uses and goals of the machine, and those were strange and odd as well to any normal Earth-man.

For Mendoza, it was 'love at first sight'. *Quality ride, fewer than ten people on-board, total occupancy. Amazing. Ten-percent light speed. Might even lose a few wrinkles, at that speed, he thought. Marta would like that.*

The shuttle moved cautiously into range, there was basic communication back-and-forth, signals and codes and measurements, all standard, all double-checked. Within a few short hours the two ships would embrace ('dock'), and folks could move from one ship to the other in the weightless wonder of it all. For any frequent-flyer, it was understood that the orbital view of Mother Earth was perhaps the most visually spectacular part of the trip. Much of the rest was in complete darkness.

Mendoza would be introduced as Alberto Mortissimo Gonzales, a field-analyst for Astro-Health Group, sent to examine and review Jupiter Program human resource efficiency related to worker health needs and demands, especially as-related to genetic-design and long-term worker improvements. This of course made him a Nazi to almost anyone he encountered. They all hated the idea of genetic modification for space-workers. But he was accepted and pleasantly disposed towards all, there above the whole world, stepping between two floating stones, without so much as a stumbling, ready to 'find out'.

With so few passengers, the 'Aerotica' was leaving orbit on on her way within only another 24-hours. The commander-pilot was Rolf DeNeuri, a pleasant and very athletic-looking Frenchman, who loved his ship even more than those who might be temporarily enfatuated with its simple beauty and facility for its purpose.

"Welcome, Mister Gonzales, please enjoy the journey with me to Jupiter and her asteroids," DeNeuri said, as Angel-Face soon found himself inside the second ship, slowly getting familiar with things.

The gates closed, the shuttle-ship dropped away.

Deveroux's departure took more time. Commander Martin Brandeis was in charge, who he had met and spoke with both at the PPP disclosure-briefing where the Commander's complaints against the program were reviewed by a committee, and at the North Carolina horse-racing track, and at other minor points-of-contact as well, over the past year or so. He certainly knew a lot about the guy's opinions of all that was wrong and screwy about current affairs with the Jupiter Program.

Deveroux again wondered why he and Angel-Face were even needed, if Brandeis' was to be a trusted whistle-blower eventually anyway. The Commander had made it his personal mission to document everything he could possibly call the truth, about Jupiter Program failures, since the crash of the 'Ferrous-2'.

Was Brandeis like some kind an obsessed Captain Ahab, where the white whale was already his own pet fish, and the fish-tank deeper than any human mind could possibly fathom? Deveroux had his own ideas.

Daniel's departure schedule was some time after Angelo's on the needle-craft, the high-speed 'Aerotica'. The 'Down' was again in parked orbital stationary position around the Earth's moon, along with five other similar ships, mostly the giant haulers. Viewed externally, the giant ships didn't seem that big at first, from a distance, space itself being so vast. The 'Aerotica' was about as big as two or three football fields, The 'Down' was the size of Long Island, or appeared to be anyway. It was almost a mile long. Designed to move billions of tons of liquid, gases and frozen materials, over many millions of miles of deep-space, they were not very attractive. Not exactly 'sexy'.

Most of them were a deep-blue in color, with all sorts of shades on various parts. Prior to Deveroux's docking and departure process as a passenger, the 'Down' was piloted away from the moon, into a short eliptical orbit of both the Earth and moon (an elegant loop for navigation). Boarding for any passengers would take place in-between. These ships did not usually have any passengers. So this also took time, and Deveroux's shuttle was not like Angel's. He had not to only leave Earth's ground-floor gravity and surface, and his shuttle would not park in Earth orbit to dock. This one had to venture half-way to the moon, and back, not by going backwards, but by going forwards and looping back from moon-orbit to Earth once again.

But by then Deveroux would be transferred to the bigger ship, for a much longer journey.

"Somewhat like a death," Deveroux entertained in his mind and thoughts, as all this was happening. He also was humbled by the idea that there was essentially no way to conceal from Commander Brandeis', and others, who he really was and what his real job or goal was. Not after the race-track in the Carolina's. It was too obvious. The only way was to bring Brandeis' in on things, and Dev' could think of no better ally for the PPP investigation. But the Second Commander, and other top staff at the helm of the 'Down', may never know. To everyone else, he was Doctor Peter Finches, a medical programmer with Astro-Health Group.

"What, no horses?" Deveroux would later joke with Brandeis, once 'up-loaded' to his new home.

"None to speak of. Welcome aboard, Doctor Finches." Brandeis said.

Within about five days, the 'Down' also begin her basic departure maneuvers and naviagtion movements, for her 28th run to Jupiter, in the lifetime of the ship.

But for Deveroux, it was to seriously face his own mortality, for himself, a sort of umbilical cord connecting him to Mother Earth, a sad feeling.

How many times have I died? How many more times will I die? He guessed maybe some people had perhaps died a few times, somehow, or made the journey to Jupiter. But not himself. He sensed he would not like space-travel much.

CHAPTER 21: Aboard the Aerotica

"Jupiter is hell,", *Dr. Wu Menuda, General-Manager, Earth-Jupiter Harvesting Program Alpha-Base, prior to his dismissal (about that time).*

Alberto Mortissimo-Gonzales boarded the 'needle-craft' high-speed transport 'Aerotica', headed for the Jupiter-program Alpha-base, known to his associates during the voyage (which would probably last a year or longer in deep-space, given his work and time needed at the program's asteroid-built main-operations base), as a systems-analyst for the Astro-Health company back home. Anyone else knew him as Al 'Angel-Face' Mendoza, and mostly his world was made up of family and friends, and co-workers for the Planetary Program Proficiency group. Once the ship-to-ship connection was completed, the passengers could get comfortable, waiting for the Frenchman pilot (Ship's Pilot Rolf DeNueri) to arrange his navigations and dis-embark from Earth-orbit, for the long haul.

The 'Aerotica', or other ships like it, was certainly the best-bet for any passenger heading out to the distant planets in Earth's solar-system. Non-military, non-industrial, smaller, and designed to be much, much faster, it was a people-mover, with creature-comforts and a more pleasant overall inner-design. Only about ten passengers could make any trip out, and DeNueri had a crew of another ten men and women (navigation, communications, engines and rocketry, life-sustain, various other). So, as Angel-Face had said, "I got the best deal." A luxury-ship among those heading to Jupiter, by comparison to the worker-vessels and industrial types.

So, of course, Mendoza would keep his secret, unlike his partner Dan Deveroux, also heading to the Jupiter program Alpha-base, about the huge materials-hauler ('whale-hauler'), the 'Down'. And that secret wasn't very secret, in many ways. The two men were investigating troubles and mishaps within the Jupiter program, for the PPP. Some authorities and powers on Earth would every now and then refer to the PPP as a conspiracy, usually when those powers and authorities were challenged or questioned. But the agency Mendoza and Deveroux worked for made a lot of sense, during that era, about 2,412AD/CE. The systems, computers, science-technology, unions, industry, military, prisons, food-supply, weather-control, energy-suppliers, they could get incredibly complex, and out-of-control, often as a result of graft or sleazy cheats. But, at those levels of global power, the results of errors and mistakes, could be truly disasterous, with effects that lasted a very long time, if no one bothered to check up on them.

From within the 'Aerotica', Mendoza settled into his room or 'berth'. This ship had a partial artificial-gravity, not totally effective at about 60-percent normal, but it allowed for ordinary ambulatory movement for human beings. The ship's passageways were a maze of colorful hallways, doorways, lifts or elevators, special-service rooms, viewing stations, rest-areas, a small gymnasium, and so on. Angel-Face anticipated that he would probably not be invited to the flight-deck or ship's helm, with the pilot and crew, during the entire five months on the way to Jupiter. So he spent most of his time with the other passengers, which happened to include the Amalthea-Base program manager, Doctor Wu Menuda, heading back to his kingdom following his visit back home.

They weren't actually going to Jupiter, but to the Alpha-base HQ for the program, which was built into a large floating rock ('asteroid'), known as Amalthea. This object orbited Jupiter as part of the planetary-system's wide swath of both large and small rocks, also in motion. The planners had felt the Alpha-Base could over-see many Jupiter-local operations well from this position, although it was of course in-motion, based on calculations from point-to-point in the annual paths of both celestial objects, creating an easier navigation (closer), more frequently, than other potential places they may have placed the base, then almost 90 years in-service.

Like a colorful hallway one might view to the sideways glance of a rosewood staircase, hardly seen for walking upwards only slightly, not so demanding as would challenge a non-athletic person, the movement of people within a highly-designed future-world space-ship, such as the 'Aerotica', was something that could take many weeks to master.

Byzantine, maze-like, and dark, not so brightly lit, passengers grew weary over many months. All the more reason for a quick-trip, yet the distances so great, some 400-million miles to the asteroid belt, that it simply took about that long as it did (five months), at that time. And Jupiter ws 300 million miles beyond that, although again, the position of objects were in-motion. Thus, the journey was very unlike sitting in a bus-seat and looking out the window at stars and worlds or galaxies and asteroids, as the ship speeds along, for such a period as about 127 days, very fast, all the way to Jupiter and the asteroids, from Earth.

Somehow, as only perhaps a planetary-ship pilot like Rolf Deneuri understood, the 'Aerotica' was laid-out and designed for the 10 or 12 passengers, to move around freely, if only to avoid medical problems. Anyone in a wheel-chair, or a person with no legs, or a person with a palsy or who was very weak (like Earth's famous wheel-chair bound astro-physics teacher Stephen Hawkings, from a previous era), might not do well.

So it was true, space travel in this era of 2,415-16AD/CE, was still something for athletes and strong or very healthy people. And in a strange way, Doctor Wu Menuda, the program's Alpha-base leader or Grand-Poobah, was not a very athletic and strong person, but it was mostly because he had spent so much of his life in deep-space. For whatever reason, he mostly comforted himself of his career-choices and lack of normal Earth-bound exercise, with wild parties, orgies, pleasure-chemicals, exotic foods and spicy-drinks, way-out music, and bizarre machine-like sensations that the technology could provide. Al Mendoza was a much stronger-healthier specimen.

"The Circulator?" Menuda was commenting to Mendoza (Mortissimo). "Yeah, it's pretty good, for sexual release with two or more partners. But it does require a learning-curve for proper use, which is great fun anyway. They are very user-friendly."

Mendoza knew who Menuda was, it was no secret. The same was not exactly true, the other way back. Angel-Face was masquerading as a genetic-design engineer for Astro-Health (in the Mortissimo persona). So he couldn't exactly question the space-program leader directly, and he wouldn't have done so anyway, given the nature of his work for Planetary Program Proficiency. The two agents really didn't want to compromise their research by exposing their identities.

Menuda was a powerful figure, and could find out eventually, especially as Angel-Face made the voyage to the Amalthea, with him, and would be there as long as a year or more. Al recalled his partner Deveroux's comment about the new sex toy called the Circulator, that Dan had bragged about sharing with Vanessa ('V'), their PPP task-assignment contact, herself a very healthy and active woman.

"Oh yes," Al replied. "Well, I have a friend who uses one with his lover, and I never understood how it worked, or what the big thrill was about, that's all. You mentioned about the Amalthea-base as being friendly with all types of sexuality, or healthy relations between sexes, and I guess there is one or two of these machines up there, is that right?"

"Doesn't Astro-Health Corporation keep track of these things for the Jupiter-program, Mister Gonzales? Or is it Mortissimo?"

"Call me Al," Angel-Face said. "I'm sure Astro-Health does, in every way, but my department is not that one. That's a matter of staff-health, and Alpha-base policy. As you know, I'm working on this whole thing with genetic-design for the space-workers."

After a few weeks in-flight, things settled down aboard the 'Aerotica'. Six of the passengers were Alpha-base workers; two clerical-computer operators, both women, with specialized technical-training in communications; four of them were Jupiter-moon base mining-platform workers, like the men killed when the 'Ferrous-2' collided with the orbiting dock at Europa; one Black man named Joseph Larouche, and his wife Josephine (a

European), were aboard the ship on a personal journey to Mars, to spend time at bases there as part of a cultural-exchange with people supposedly in touch with aliens from other worlds, at that time, though no one really believed in it, (the 'Aerotica' would connect half-way to Jupiter for them to disembark to another ship to Mars, they were a delightful couple, emissaries of a global artist-enclave); Mendoza and Alpha-base leader Doctor Wu made the count ten. Two other passengers were not identified or were traveling in secret, unknow to anyone. And then the crew and pilot, other staff, and maybe a gremlin or microbe hidden in some water-system within the ship.

The "Aerotica' was large. It had to be for the engines and life-sustain and so on. But for the passengers it was an odd journey, with nothing much to do but move about from hallway to hallway, room to room, and interact with other passengers, pleasant enough. For some passengers after a month or so, it was fairly easy for the ship's staff and crew to voluntarily place them in a state of deep-suspended long-form 'sleep' or semi-consciousness, with various chemicals and devices.

Al Mendoza naturally gravitated towards Menuda, in a sly way, hoping to learn something of value in the PPP investigation. Menuda was no fool, he sensed Mendoza had goals he was not sharing, like most of the people he dealt with. This only increased the sleaze-factor where Menuda was concerned.

The two of them had taken seats on an upper-deck area where they could watch dancers on a video-screen. "Well, of course the Alpha-base, the Amalthea base, and all the Jupiter-program operations have both male and female workers and staff," Menuda said. "It's an established fact for many years, that sexual health for space-workers is very important, at least as important as other health matters, for workers suffering the rigors of deep-space. So, we encourage all that, like maybe a small city back home. You know, it's not that unusual at all. So this thing, the Circulator, I guess some inventor came up with the thing, and yes, we have a couple, I guess. The base is large enough that I can't personally keep up with every little thing, you understand. Why are you interested? You have a friend using one?"

"Well, he was boasting about it, as men will do," Mendoza replied. "Are they safe?"

"Of course," Menuda said.

"How do they work? Just interested."

Menuda was a darker-skinned man than Mendoza. He was lanky, tall, and his belly stuck out, like his chest was curving inward. His skin was very smooth, hairless, and the muscles on his arms and legs were taught but not bulky. His face was not unusual for any features. He tended to wear robes or togas when working on space-ships or at the base. Yet, for whatever else he was, Doctor Wu was in charge of decisions for the entire Jupiter-program that Deveroux and Mendoza were investigating. Sex-chat among men getting to know each other in this way, on a voyage, made for a tension that Mendoza felt he could use later, a bit of a leverage, and Menuda hated him for it. Mendoza was stronger physically, better-looking. But Menuda was a far more powerful man.

"Study up on it, sir," Doctor Wu said blandly. "The device operates in a water-environment, like a shower, or hot-tub. Lovers wrap themselves together in their love-making with a sort of easy-comfortable rubberized mesh-material, that is connected to the circulator itself, a secondary unit, you can place it to one side easily. The mesh-netting is flushed with a combination of chemicals and energy-types, that permeates the flesh, including the brain-organ and sex-organs, creating heightened orgasmic response. Most people wrap them around the waist and loins. The thing works on certain vibrations deep within the human nervous-system, the energy-type is magnetic-electric link to brain-waves, and body-meridians, but there is no helmet or head-piece, just a comfortable cloth-mesh, almost like a towel or one of those old feather-boa's. They have different colors, purple, pink, rainbow. It's very stimulating, gentle, but in the sex-act, I guess people using it claim it is an ecstacy of considerable power for both partners. Like, they say, your blood is on fire. So, best wishes with your own experiences, I guess."

The dance-show they were watching on the video-monitor showed women doing gymnastic routines, nothing unusual. It was a long trip to Jupiter and the asteroid-base. They relaxed with food and drinks, in a sort of lounge-area. But it was a dull view, obviously, on a long space-voyage there was only nothing to see, and plenty of it.

"What will they think of next?" Angel Face replied. "Sounds like my buddy, though. Kinky. Fetish. I never tried one. I like hot-tubs, though."

"The base has all kinds of health-programs," Menuda said. He was in a brown toga, and worked on a lap-top command-control computer, connected to his work, known only to himself. The music with the dancers was not that great. Mendoza was enjoying his food-drink, reading a bit. There were long moments of silence, the deep-plush velvety couches arranged for the 60-percent gravity. Like any professional men, they understood, each had work to do and small talk was often a pointless gratuity. They would never be friends, it was not in the cards.

"Astro-Health wants to study my workers, and DNA-alterations for better movement and strength working in deep-space, is that correct, Mister Mortissimo?" Menudo finally asked him at one point. The dancers on the video screen were now twirling madly with costumes and music. One of the other passengers drifted through the area, with a friend. A smile and greeting, then passing.

"Yes, Commander Wu," Angel-Face said. "A briefing about AHG tasks for my work on this trip has arrived on your system-immigrations data-log. It's all there."

Wu worked more on his tablet-computer. "It would take a long time, yes?" he asked him. "To do any genetic-changes to the workers?"

"Yes, of course," Al replied. "I am only doing some basic ground-work research and interviews. It's controversial. So they want me to look at the goals and methods first-hand at the Alpha-base and at some of the orbiting harvest-stations."

Doctor Wu paused. "I don't think its very wise, myself."

The dancers dipped and spun 'round and 'round. The 'Aerotica' was humming along.

CHAPTER 22: Whale Hauler

"Base dispatch, this is AB Hesidom, with Reservoir 21 monitoring, reporting, please confirm. Waiting."

There was a pause. Astro-Biologist Hesidom, a youngish-looking worker at Alpha-base, had done his duties, yet again, at one of the vast H20 Reservoirs hidden deep within the rock that was home to the Alpha-Base at Amalthea. There were 30 water reservoir tanks in all, connected by passageways and tunnels and shafts, each equal to about a football-stadium's worth of quality water, even drinkable. Following his hours-long routine water-systems review, he would call-in the logs, for analysis higher-up.

"Hesidom, this is Alpha dispatch, go ahead please."

"Please receive logged system's monitoring report on Reservoir 21, for this date. Cue me when ready to receive as voice-record. Waiting.

Another pause. There was a 'boop'-sound, then a Computer-Robot Voice: "Official Logged Base Data for Reservoir 21 Monitor AB Hesidom for this date. Proceed."

Another 'boop' sound. Hesidom began his report.

Mendoza's partner, Daniel Deveroux, by now had already boarded the 'Down', on his own journey to the Jupiter environment, yet on the same mission. The choice to place the two investigators for the PPP on different ships made a certain kind of sense. It wasn't for lack of transports, it was to maintain some kind of secrecy or to conceal their work. All too often, when the transparency the Planetary Program Proficiency organization was known to bring to complex Earth-sourced problems, was compromised, it was because of the bias and hot interests of those involved. Bribes, threats, accusations, deliberate obsfucification of evidence or details, violence, seduction-by-power, and every sort of rotten, selfish, or flat-out greedy motivation and scheme, were applied, and Deveroux and Angel-Face were familiar with this approach.

So, for any of their work with the Jupiter program to be even somewhat effective, they needed to appear to be doing something else; in this case, Daniel was in the role of Peter Finches, a second Astro-Health Group program-application designer looking at health-and-welfare within the Jupiter program. And even this would meet with resistance, the Jupiter people all felt they knew best, and mostly they did. The other problem, for Daniel, aboard the 'Down', was that the ship's commander, Martin Brandeis, had already been contacted by the PPP, and Deveroux and Mendoza both, so there was no real value to any secret-indentity for Daniel at this point. But, for propriety's sake, he was 'Peter Finches'. Brandeis after all was the Jupiter program's best-known and most powerful whistle-blower, with exhaustive detail on program flaws, following the wreck of the 'Ferrous-2'. It seemed best to Deveroux that they simply work together.

Brandeis' ship, one of the huge 'whale-haulers', had looped-de-looped Earth's moon, docked with Deveroux's shuttle at mid-point, and then prepped her navigation towards Jupiter, something Brandeis had done many times. The ships were large, dark-looking, bulky and bulbous, not attractive or sleek. The engines were in the back, with other engines and devices for loading and unloading materials harvested from Jupiter's moons, and bringing them home. Not a passenger-ship at all, and slower than pilot Rolf Deneuri's 'Aerotica'. Deveroux would not be at the Alpha-base until at least four or even five months behind that ship.

The advantage was that he would be able to spend time with Brandeis and the crew and staff of the 'Down', and perhaps learn more.

Within a few weeks, Deveroux had acclimated suitably to the 'Down's' ship-board routine and environment. He hadn't travelled much in space at all before this, but despite his earlier fears, he found the experience acceptable overall, as it was meant to be. There were living-quarters and passenger-quarters, and much like the 'Aerotica', a slate of enjoyments and rest-areas and viewpoints; but the Down was so much larger, that Second Command Menima Pearl later offered Deveroux a private tour, with herself as guide. Few knew the ship better than she did, and she was a very pleasant person, familiar with dealing with outsiders who had no experience there in her kingdom, shared with Brandeis. They scheduled the tour, it would occupy them both for more than 12 hours.

Commander Pearl, stocky, buffed-up, a true cosmonaut and leader, finally met Deveroux in a foyer or guest area off the entrway of the ship's helm and navigation-maneuvering decks. As big as the ship was, the command-center was proportionally large, and included controls for communications, astronomy, life-sustain, docking-and-loading, and many other aspects of the ship's functions. There were 12 or 15 people working in this area at all times, with Commander Brandeis, or Menima Pearl, in charge, trading off command in shifts. Many officers on a ship like this worked daily dressed in pull-over cover-all suits that included ensignia and certain personal gear they needed. Deveroux had been fitted with a similar 'ship suit', for passengers or guests. They were silver and green, and had communication's links and small data-PC links for specific types of uses.

"Hello, Mister Finches," Pearl said warmly, entering at the waiting-area, or entryway below the flight-deck. "I'm Menima Pearl, Second Commander. Happy to show you around our space-ship for a few hours today. How are you?"

Deveroux had been resting his eyes, seated to one side by a large display of maps and diagrams, that showed the layout of the whole ship. "Hello, Commander Pearl," he said. "Pleased to meet you. Peter Finches, with Astro-Health Group." He extended his hand pleasantly and she allowed a kindly grasp and welcomed his greeting.

"Yes, I know who you are," Pearl replied, somewhat coyly. "Astro-Health has been a Jupiter-program health-and-wellness vendor company for many years. The connection is very strong."

"Thank you, it's a good company, considering the demands of work in deep-space," Dereroux answered. "How is it I rate a tour of your ship here today, the 'Down'? Why is it called that?"

"The ship was created about 20 years ago, along with four others," Pearl answered him. "These giant materials-cargo ships are very unique, and are built in space. When they name each ship, its at the discretion of a committee. For whatever reason, this ship was called the 'Down', after a mid Twentieth-Century fantasy novel called 'Watership Down', by an author named Richard Adams, from Great Britain, published in 1972, previous era. A 'down' in that book, refers to a network of underground holes and tunnels in the side of a hill, populated by rabbits. There is an original paper-copy of the hardback novel and some souvenirs in the Commander's personal office."

"Oh, I see, how odd," Deveroux said. "And here I thought it meant something about astro-physics and space-navigtion and the direction back towards Earth. You know, up and down, like that. Rabbits, huh? I never read it. 'Watership Down'. Why water?"

"It was just the name of a small town in old England. Once we leave Earth, up and down have no real meaning, Mister Finches," Commander Pearl said. She moved closer to him by the electronic map display that showed diagrams of the ship, their current location, and cut-away images so a layman could see how the ship was arranged. "Let's get started, do you mind? It will take ten hours of your time today to see it all, at least. Or we can break it up into two days, although I have my own duties to attend as well."

"Sure, sure," Daniel said. He was now standing by her side, in his green-and-silver pull-on outfit, looking much like everyone else on board. *Just like a real space-man.* He also had a small carry-bag with computer stuff for his own use. "I was looking at the map here earlier while I waited. You must be familiar with every detail, as Second Commander."

She smiled. "Well, it's not a shopping mall," she said. "I've been Second Commander under Brandeis for several years, we work well together. The ships are automatic, mostly, for a hundred million miles at very high speed, there is almost nothing to see or do, except monitor engine functions and navigation pathways. But as we change course, or docking, or enter orbit, or the loading and unloading, intermediate ships that need to connect or hook-up, and any unforseen objects like asteroids or large blocks of ice, then it gets interesting very quickly. As you can see, here on this map, it's a huge piece of equipment."

Deveroux was running his finger along the edges of the lighted, colored digital-electronic real-time map of the 'Down'. "So we're right here, just beneath the main control room deck where you run the thing?" he said.

"Yes, that's just above, over this way, there's an elevator and also a stairway," she turned and pointed, this was the entryway to where she spent most of her on-duty hours, the flight-deck. "In the forward motion of the ship, we are at the front, but not the prow or tip, the command room is set back some 500 yards or so, and then upwards. Even below us here, several levels below the command deck, are only more storage and service-areas. But the command-deck is the front-piece, you might say, it's what you see when you view the ship externally and try to figure out which way its pointed."

Deveroux chuckled. "Well, I assume you must know that," he said. "I mean, which way we're going."

Pearl smiled. "You know, Mister Finches, without my maps and navigations, it wouldn't be truthful to say that I did. We're sealed inside for many months, there is hardly even a sense of motion at all."

He paused. "Which way is the planet Jupiter, from here?"

She understood he was being fecitious. She held out her arm and pointed down past the entry-way to the command deck. "It's that-a-way, Mister Finches, I'm fairly certain," she said. "You are a funny one. No one guides a ship like this by sight or bodily sense. But it's that way, really it is."

After a while, they studied the map more, and than started on the cook's tour. They would use a small electric cart for many of the long passageways. From inside the ship, different sections were connected by long tunnels, or external ramps protected by steel-and-glass. They spent an hour or so in the people parts of the ship; the food-courts, living areas, recreation rooms, various technical stations that needed attendants, such as the communications platforms, the real navigational centers (which included long-distance radio-telemetry and plotting-methods that could figure out how to move the ship from point-to-point, and what commands were needed). The other work-centers, too, such as life-sustain control rooms, staff-scheduling and human-resources, 'suit-rooms' and astronaut departure prep rooms. She also showed him at least one or two of the external ship-to-ship docking ports, such as the one he had boarded the 'Down' on himself. Other areas as well.

After a break, the two of them maneuvered towards the mid-part of the 'Down', which housed the complicated system of vats and containers, loading-tubes and vacuum-locks and pumps, where materials from the moons of Jupiter were held for the voyage home. Of course these were huge, and there was really no way to look inside them. There were several types, arranged in long rows of giant tubes or giant cylinders, each about as tall as a 10-story building.

Second Commander Pearl tried to explain how these were used, but to view this part of the ship, Deveroux found he had to move along with her in the electric-cart, on long tunnel-ramps high above seemingly empty cavernous depths of the main cargo-holds, dimly lighted from below. It was un-nerving enough to him as a novice that he hardly cared about the volumetric capacity of the tanks.

"There is nothing to be afraid of, Mister Finches," Pearl said to him, watching as he leaned over the side of the cart, where she had paused at a platform area-ramp to view.

"Oh, of course not," he said. "It really is big, isn't it?"

"If it wasn't, we'd never get enough of the raw-materials from Io or Ganymede to justify the trips," she replied. "A few hundred pounds of raw helium is one thing. A few hundred million pounds is another."

Also along this part of the tour, Deveroux got a good look at the ship-to-shore docking and loading pumps and portals, and how these were arranged. It was fascinating, as a layman. The genius of it all was to somehow move all that helium (or hydrogen, or H20, or ammonia-methane, etc.), from the small planet-moons, into the ship's tanks, without a safety breach or airlock vent to the external vacuum of space. Or of course any internal break. So the pumps, the portals, the connecting tubes and latches, the in-between pumping, it was all sealed and double sealed, with enormous O-ring metallic gates and secondary pipeline plugs and switches.

It was the same feeling for him, a few hours later, looking at what could be seen from within, of the ship's main engines. Along this part, there were also hangers and places where other small ships from outside the 'Down', could berth or dock safely. This was the very back of the ship, about half a mile from where they started, where it all ended in a flattened, buttressed wall, laced with a descending series of ramps and stairs or ladders, then opening up to larger rooms and maintanance stations, and finally the backside of the engines themselves.

They may as well have been grain silos or radar-telescopes, they didn't seem to be any kind of engines at all, because of course the thruster-cowls were outside. Second Commander Pearl and Deveroux (or, Peter Finches) stopped at a place at the bottom of it all, in the electric cart, where they could gaze up at the walls and levels of machinery. Pearl and everoux sat there a moment in the electric cart, relatively about the size of a single pine-cone, fallen at the foot of one of California's giant redwood trees, the Earth's largest living things. Also, at this part of the tour, Deveroux now could hear a deep, smooth hum and buzz, the sound of the engines, loud in his ears.

"Big," Deveroux finally said, listening. "Big. Big ship."

"Yes, it is," Pearl said.

CHAPTER 23: The Large Form Globe Object Music Society

"Where there is no road at all, the road is only longer, and less of a view."—Daniel Deveroux, aboard the deep-space transport cargo vessel, 'Down', with Commander Martin Brandeis, on a voyage to the asteroid-belt, about 2,413AD/CE.

AB Hesidom: "Reservoir-21,Tank Safety Integrity, A-plus, no runs, no drips, no errors. Water-quality for potable, 60-percent, purification required. Cause for impurity unclear. Water-level for Reservoir 21, also down 40-percent, cause is also unclear."

He went on with his routine report another few minutes. The robot-recording-machine again went 'boop', and the transmission was officially completed. But the Task Dispatcher, a woman (they tended to prefer the female for radio-interlink communications, it was more appealing psychologically), had a personal question.

"That will do, AB Hesidom. Hit you back next cycle. Personal question for you?"

"Sure," the young space-worker said.

"Those tanks don't leak. Where's the 40-percent? That's a million gallons of priceless water, or something like that."

Another pause. "Off the record? We think there's something strange sucking it out to something else even strangre for strange reasons, like some water-sucking bacteria from some space-rock or like that. Not sure, actually, but it is distressing. That's the fifth month I've had to report losses."

"Hmmm. Wow, I'll keep it to myself, I guess. Hope Alpha-Base doesn't get thirsty. Dispatch out. Thank you, AB Hesidom, Reservoir 21."

The line went dead.

Yes, the 'Down' was quite a ship, quite a tour. In many ways Daniel was more interested in Second Commander Pearl, who she was, what she was all about, and why her position in the command structure of the Jupiter program placed her at odds with her leaders, and her personal spiritual journey, involved in various traditional space-religions. She was a pleasant enough person, large-size for a woman, bulky or 'butch', which was common enough. Also athletic and healthy, a strong person, as any space-ship taxi-driver must be. So, he easily dismissed her, as any sort of target for his work for the Planetary Proficiency Program, for the time-being, knowing that the so-called 'space-religions' were a bane to the overall program with Jupiter raw-materials.

"Probably a real sexy beast-woman in bed," Deverous self-reflected absently.

Some of the other people he might be spending time with included Brandeis, the main-player whistle-blower with his detailed reports, and then various passengers and crew. Fascinating people in any case.

One group that had found their way onto the ship for that journey, were a musical-study foundation team, doing arts-work out in the deep Abyss, having to do with the way they could make some kinds of interesting music 'among the spheres'. Classical composer Gustav Holst's 'The Planets' was a favorite. They were known as the Large-Form Globe-Object Music Society. They created music and recordings, by linking electronic scans to actual planets or moons, and connecting their movements and paths to modern electronic-music instruments. The effect could be tremendously beautiful and harmonious.

Deveroux chatted with one of the group, whose name was Beautiful-Truth Amakmid-Cornerstone, a Latin-looking, thin man with a professional demeanor, that is, an artist. It was a long journey yet, why be bored?

"You mean, your team of technicians and musical composers, are traveling to the Jupiter Alpha-base, to spend a year creating electronic-music based on scans of the moons, and of Jupiter itself?" Deveroux inquired politely.

"Yes," the man answered. They were walking together down one of the pathways from where they all had private rooms, to a common-area or 'quad', which had comforts and provisions and services. It was a long voyage yet. "It is a stunning new approach, if you are a musician."

"I can play a harmonica," Deveroux said. "I like jazz music."

"Oh yes, I like all music, the music in me, the music in you as well, my friend, Mister Finches," Beautiful-Truth said. "The Society has been working on the technique for years, this is our first shot at Jupiter. We've already done recordings of Mars and Earth's moon, and the Earth, of course."

"So this is a recording of some kind of signal or noise from the planet?"

"No, no, that is a common mistake," the other man said. "A radio-telescope device can pick up a signal from most of these large cosmic objects, it's not difficult. If you process the signal, there is a 'sound' or 'noise' that comes back. That's not what we do. We are truly creating the 'music of the spheres'. The Globe-Object Music Society will spend many weeks, creating a scan-web matrix telemetry, basically a constant overall signal scan, combining many elements of the planet Jupiter, as it really is."

"Like a real-time multi-level scan?"

"Yes. The motion, the layers, the substances and materials and gasses, the density, the other motion in orbit around the Sun, and even the slow orbital position from the center of the Galaxy, and the moons, the light and shadow from the Sun, the temperatures, like this. There is a great deal of magnetics involved. Once this is done, all the signals are united as one thread of electronic activity, that the music-composers link to their electronic music synthesizers, and reproduction, and recording. This allows the same combined data-stream to be applied to common musical notation, you understand? So in this way, it's really true, we are playing with the planets like they were musical instruments themselves. It is very popular back home, but we are new."

"Astonishing," Deveroux said. "Must be amazing work."

"Yes, it is amazing. I am very pleased to be a part of it."

For another week or so, Dev' was free to enjoy the leisure-time he needed to re-consider how he and Angel-Face might be able to handle things, once they arrived and reunited at the Alpha-base. The investigation wasn't going well, they were a year into it, and the reason was fairly obvious: the Jupiter program was too big, there were too many interwoven interests and decision-makers, the science and technology was obtuse and impossible to understand, and the real mystery of Jupiter was shrouded in secrecy and officially unavailable.

How could they narrow things down? How could they follow-through, and bring home the golden-fleece, with some kind of conclusive answers as to why so many different fragmented aspects of the 100-year old program with Jupiter and her moons, were going FUBAR (please see US military dictionaries, for 'Fouled Up Beyond All Recognition'). Deveroux had a passionate mind and imagination, part of why he was good at his job. But it was hard to grasp, so he took time out, and enlarged his scope of field, to see what the time ahead might worry them about next. And it wasn't music created by scientists.

His quarters aboard the 'Down' were similar to those of any of the crew. It was not a passenger ship. Two men shared a mid-sized room, entered by a sealed doorway. Inside were beds or bunks, very pleasant but also easy-to-use or clean, most any laundry or toiletries were disposable instantly on most of the ships and bases. The ship had 60-percent artificial gravity, so they could move about fairly normally. Deveroux's room-mate was a quiet, elderly man, a notable poet and philosopher, also on his first trip to the stars. The rooms had food, drink, communications, computers, minor view-stations, and other features. Not very fancy, but adequate.

"What to do, what to do?" Deveroux was thinking. *"Busy, busy, busy."* Maybe it was instinct, but one clue was from talking with his pal at the Seattle University, Podliakov. Something was known, or suspected, about Jupiter, from the earliest years of the program, and then hushed up, its advocates silenced, the research suddenly cancelled.

What was it? How important, what was the level of its influence or effect on the rest of the program? Had they learned somehow that Jupiter could be expected to blow up at some future date they could accurately predict? Or did they find some secret alien artifacts? Was there evidence of a massacre at Jupiter, like some early-era expedition that had somehow gone horribly off-course, or some odd genocide, with thousands of dead corpses frozen in lakes of sub-zero hydrogen-gas, hidden in Jupiter's legendary 'red eye storm'? Was there some magic substance they had found deep inside Jupiter, that would wipe out all mankind back on Earth, if revealed?

Whatever it was, Deveroux began to feel that this connection might explain a lot, and could help explain the other problem-areas. But he had no idea what the real truth here was. Only a few remaining space-program workers of advanced years would even know what he was talking about, and even fewer would have the original research data-available, if they dared to share it, probably still under penalty. So, it was frustrating.

Radio-waves fly faster than solid-material ships full of men and goods, so Deveroux was able to communicate with Angel-Face, aboard the 'Aerotica', a million miles ahead. He could also connect to case-manager, Vanessa, or 'V' with the PPP back home, (on a schedule).

He didn't want to spend a lot of time chatting about their PPP efforts on any radio-link. They were still supposedly undercover, the calls and links were routinely intercepted and reviewed by various powers. Somewhat later, however, 'V' got him a draft-list of a few names, as text-files (redacted to secure link radio-wave data). He didn't even notice when it appeared at first on his private computer-system, it was somewhat unofficial. So he could now read her message in private:

"Dev: some useful details for you, from research. These are five individuals, currently assigned to Jupiter Programs at various locations, on-site now, as you and AM arrive. Info below. These men were involved personally in early program research-exploration at Jupiter, as long as 50 or 60 years ago. They may be useful in your research. Here are the names:

-Gerry Reedly, age 86-years. Spectrum-analysis specialist, current location @ Program Alpha-Base/Amalthea.

-Montrose De Montrose, age 80-years. Planet mass-density evaluations and analysis, current location @ Program Alpha-Base/Amalthea.

-Nonnly Guitierrez, age 73-years. Early Jupiter mapping. Current location, Program Ganymede Base.

-Mort Philby, age 72-years. Early Jupiter mapping. Current location, Program Ganymede base.

-George Thomson, age 72-years. Early Jupiter mapping, specialist in materials and gasses. Current location, Program Europa base.

-Thanks, Dev'. This same list to Agent Mendoza aboard the 'Aerotica'. PPP-info file private/concealed transfer. Security coded, Verified. Best in your work. V-"

Deveroux was not surprised, they had talked about this before. It also jived with his chat with Prof. Podliakov, someone he trusted. *Excellent work, Vanessa,* he mused. This meant he would be busy finding these people, while at the asteroid base, and then hopefully moving by smaller ship to the large planet moon-orbiting stations. It would also possibly be a way he could start to understand the deeper mysteries of Jupiter, the strange things they had reported, including Brandeis' report, and ideas related to ancient extra-terrestrial activity at Jupiter.

Aliens. Alien minds. Bizarre ideas about the human local planetary right to leave Earth and explore local objects. Jupiter is big, lots of stuff. But not that big. The Milky Way Galaxy, and whatever other sentient inhabitants it may contain on other worlds, might not have had an interest in Jupiter, maybe long eons ago, other than its proximity to an inhabitable world, Earth, with its rivers and streams and oceans and fish and horses. *I guess with the alien mind, you just never know,* Deveroux mused peacefully. *Don't try this at home.*

He would spent more time with the man called Beautiful-Truth, and in his persona as 'Peter Finches', he was ready to offer some off-hand opinions about health-matters in the Jupiter Program.

"Astro-Health Group has no strong view or vested-interest in the outcome of the debate about genetic modification of astronauts and space-workers to improve their service in space," he was saying. The two men were hanging out at a 'bar' or 'club'-area, not much of a club, but they served liquor and other drinks and it had a larger-than-usual view port.

All they could see, gazing outward, was a portion of the ship, sort of a large tiered platform-wing structure leading down and down to some other structures on the external hull. But it was lighted for whatever reason, creating a view. Otherwise, little but the black-blue indigo of the abyss, and a few stars. The 'Down' had passed the mid-point on the way to the Amalthea. Anywhere on the ship, there was usually a hum or gentle rumbling, it was all so very large. The bar, known as the Skull-Cap Night-Cap Club, was pleasant and dimply lit. A few other staff and crew were also there.

"Well, yes, that is your area, not mine, I have no strong view either, it seems controversial," Beautiful-Truth said. "The human body maybe is not for a lot of genetic changes, yes?"

"The bastards would create slaves and grotesque mutant-men who don't even eat or breathe real air, just so they can work efficiently in space," Deveroux suggested coldly.

They paused, a moment as their thoughts united, perhaps unhappily at the idea.

"You know, I can apply the same technique we use to create music-compositions from entire planets and moons, to a single individual man or woman, from their body itself. It's different, but similar, obviously much smaller scale. You do a life-scan, and connect the vibrations to music sources for composing tunes and songs."

"Really? Amazing. No, I never thought of that. Well, Astro-Health Group is sending me up to review the programs currently in place, for compatibility with any future decisions on astronaut or space-worker genetics, DNA alterations. So I am just getting information they want. I'm reluctant about the idea, myself."

They sipped their drinks. Deveroux had a brandy and coffee. Beautiful-Truth had a Bloody Mary. They also had tempura-battered fish, rare enough there in deep-space, but as usual the food for space-travelers was always the best they could supply, another perk to chase away the boredom or fatigue.

"Too much information," Beautiful-Truth said. "These programs make no sense, to me, personally. The historic human form has been sufficient for workers in space for hundreds of years. What would they do? Create bodies that don't need air?"

"Uh, maybe not that far advanced with it all," Deveroux ('Finches') said. "The DNA can be modified early in life for added strength and blood simplicity for oxygen absorption. Other features, like lung capacity, fluid-retention, sleep-cycle, eye-sight light and dark needs. We are not really busy creating monsters, sir. But some think it so."

Another drink down, they were feeling happy and easy, casual. "Jupiter is a hellish planet, Mister Finches. My group of artists can create wonderful music from the vibrations. I'm sure Astro-Health is not creating monsters, for any work in space. Please, no."

"No, no," Deveroux emphasized. But he know, the Astro-Health Group role was not going to make things any easier. They had failed to change their cover-story soon enough to the food group.

CHAPTER 24: Mort Philby Is Dead

"But why? Who would try to kill him?"

"People who don't want your father talking to investigators about his early Jupiter years, with the program."

"But why? Who are these people who don't want others to know about pa's earlier research?"

"A large consortium of Earth-bound industrial and power-structure interests, who've been planning a war based on these findings, and other information, for about 23 years or more. They've invested a lot of time, money and effort, and they want their fucking war. So your father is just in the way."

"A war at Jupiter? A space-war? Wait a minute, who versus who? We're all one team. And they want my daddy dead because he can still remember the original source-material research?"

"Yes."

"I don't like that!!"

(Anonymous Tweedle-Dee, to unidentified Tweedle-Dum, logged as transmitted inter-program communications, Jupiter Alpha-Base/Amalthea, that same period).

Without a lot of fanfare, Deveroux began to realize that the work they had undertaken for the PPP at Jupiter, was perhaps endangering the lives of some or all of the five men Vanessa had indicated in her previous message. Morton Philby was suddenly found unconscious on the floor of his private quarters, at the Program Ganymede base. Daniel heard the story from Commander Martin Brandeis himself, the pilot in charge of the 'Down'. Philby was 73 years young; after he was found by a co-worker, he survived what was thought to be a geriatric heart-condition. But Brandies felt there was more to it. If anyone knew what the rumors throughout the Jupiter program ranks might mean, it would be somebody like Brandeis.

"This is from a friend in urgent-medical care at the Ganymede base, okay?" the Commander told Deveroux. Brandeis had a pleasant anteroom near enough to the ship's flight-deck helm to invite Deveroux to chat, but not on deck in front of the other crew. "Philby is one of the Old Men, the program has all kinds of old men, okay? For some work, geriatrics is not a problem, and these guys understand how stuff works out here, better than anyone alive. So they keep them on, well into their 80's, if they can, and mostly everyone takes very good care of them. So whoever poisoned him, he wasn't from up here."

"Poison?"

"They feel it was fast-acting poison through an oxygen-ventilator. Philby has a pre-cancer, at his age, on his lung, so his room is specially out-fitted for higher-than-usual oxygen levels. It's a medical, they can easily beat the cancer, it's not that. Ganymede is one of the most stable harvesting stations. It was on Europa that the 'Ferrous-2' went through that whole glorious crash. Philby monitors deep-planet activity readings from Jupiter, the main planet, from this station, for the Alpha-Base data-researchers. He's a cool old cat. Or, he was."

Deveroux listened to more details. He felt as if he had personally been accused of cold-blooded murder, though of course it wasn't. A relayed robotic assassin could fairly easily have been set-up to deposit some deadly vapors in Philby's life-sustain, even remote-controlled from elsewhere. Ganymede is a long ways from home. Philby's daughter worked at Amalthea in Human Resources, so she watched after him. Mort Philby had become rather frail, his skin was brittle, his bones wet and mossy.

The 'Down' was functioning on automatic-pilot, for perhaps another 20 million miles. A navigation was needed to target properly towards Alpha-Base. Even on auto-pilot, staff and crew monitored all systems constantly. A gentle buzzing, low-hum, nine months. Brandeis' office was stark and somehow bleak, he didn't go in for a lot of fancy decorations. His Command-Control Alternative Link was always nearby, a remote hook-up to all of the ship's major alert functions on many systems he could review at a glance. A hardback print copy of the 1970's novel, 'Watership Down' could be seen in a case of glass and wood.

"Try to visualize what I mean, Finches," Brandeis was saying. "The Ganymede base is a lot like this ship. It's as big, or bigger, overall. The loader-station orbits Ganymede, the third-largest object in our solar-system. Ships come up, ships go down, and grab or suck or pull or yank or break off stuff we want from Ganymede. It's stored and prepped, then loaded onto ships like this one. Then we haul it back to Earth. Mort Philby is just a dust speck in a wind-tunnel of eternity, he mostly works monitoring various aspects of changes going on with Jupiter itself, surface level and deeper, then that info is used by Ganymede station and back at Alpha Base. And of course its pretty dull."

"That's often the case with you folks out here in the deep space, isn't it?" Deveroux said.

Brandeis laughed. "Yeah, well. I kinda' see what you mean, so fine, no big deal, but, yeah, right. So Ganymede is run by a guy named Charles Benway, strictly a mining operation, shipping, loading. He's a good man, king of his planet, the Little Prince. But the old guy, Philby, is of no real use to him. Nothing much ever changes at Jupiter itself, but if it did, like a huge magnetic storm, they all need to know. So there he sits at his station, a few years anyway, he chats with his daughter by radio-link after-hours, sweet old guy. About three days ago, he didn't show up for work, or log on, since he works from his private suite. Some co-workers checked on him, he was unconscious and hardly breathing."

"Doesn't mean it was a hit, or a murder, Commander."

"Okay, believe what you need to. You're the curious one. The life-sustain systems to any rooms at these bases are completely secure to tampering. But because of his age, he has a supplemental adjustment, increasing oxygen levels in his private quarters. A hand-scanner from one of the security guys showed traces of unusual elements, the medical units recognized it. Someone or some thing simply knew enough about his private quarters, to taint his air supply. And THAT's not easy, they would have had to be very determined about doing that. Other things as well."

"And somebody in urgent-care medical on Ganymede told you all this, and you believed it?" Deveroux asked him.

"Yeah, a guy named Ubinicus Veritasmas, a physician or doctor, at Ganymede base. Very reliable, according to me."

The large whale-hauler, with her mostly empty belly, fell forward towards the asteroid belt through endless distances of total nothingness. She was only somewhat larger than a dust-speck, relative to single human being. In some strange way, the great distances seemed shorter and shorter, in their thoughts, like the back of something called the Universe, and they themselves, ever to seek it's front.

Morton Philby was face-down on a hospital bed, at the Ganymede base, with no real idea that 'Peter Finches' or anyone else of any consequence to himself personally, had any real interest in a heart attack, at his age, that he might suffer. And Philby had even less idea that an assassin would be hired to accomplish what Nature always seemed to nag him about. It just wasn't in his emotional vocabulary. Why? Science is science, its either true or not, they either understand it or they don't.

The small emergency room triage center, or medical berth, was large enough for a dozen or so patients at the same time, and had a modest staff. It also doubled as long-term care, if needed, since shipping anyone home from Ganymede was a long-form prospect, it took time, if they were seriously ill it was only worse. The sick hours were dragging by in a different space. Philby was almost unconscious, a nurse applying electrodes to his back and legs, to scan his respiratory.

In his thoughts, Mort tried to recall what had happened to him, again and again, like a mirror in his mind. He was working on the Jupiter scans, 1500 hours, a few hours into his shift. The scans were set to a certain electromagnetic signal range, a radio-spectrum analysis, to keep track of changing strata-rings on Jupiter, that would indicate giant storms and heaving waves of energy. And less impressive changes too. The screen was dull, nothing much going on, like a piece of slate or dark gray stone to his weary eyes. Jupiter is so large at that close range, that his measurements were like those of a gnat trying to analyze the Pacific ocean for signs of wetness or saltiness, with either form acceptable as long as the gnat didn't fall asleep, (which was preferred).

The images seemed to creep into a deep, limbic part of his awareness, where he could no longer block out the raw unreality of reality's unrealness. Like sheets of glass cutting across his psychic awareness, pain, a painful tightness in his breast or chest, and then he woke up in the hospital. But he never really regained full awareness until much later. In a certain sense, he took it all in, the doctors and nurses, their conversations and movements and medical practices and applications. In another way he was hardly conscious.

"He seems to have heavy palpitations, Doctor Veritasmas," said a nurse, or doctor's assistant. "The heart muscle is heaving and beating too fast." Veritismas was reviewing some charts and analysis of Philby's blood. A cross-segment analysis of the hemoglobin showed high levels of homocystine, caused by an unidentified and elusive chemical agent, that the doctor was certain was a naturally unavailable substance. In other words, 'not natural causes'. Mort wasn't dead, but any report the Doctor might need to file would have to indicate, 'not by natural causes'."

Mort tried to lift his head, he could only moan, drooling saliva onto the blue linen hospital bed pillow. By now he was sedated, and he a leading science expert with a finer intelligence than most. His career in the space program seemed like a rush of memories, first this part, then that part, first these people, then those people, first this view, then that view, wives, sons and daughters, education, births and deaths, loves and romance, sex, food, walking and talking, amusements and culture, Earth, his home, mountains and trees, cities, and people.

If his work was any value, he was content with that. It didn't matter. He had mostly forgotten it all. Nothing new to learn about, and no way to acquire new knowledge, including walking and talking, exercise, food and drink, people. New memories, new experiences. People. People. *What people? People? Where?*

"My god, doctor, the man is over 75 years old," said one of the other nurses. "I can't believe they have him working in deep space. He's in good shape. But the heart eventually gives out, it's true. I guess that's what happened. Poor old guy."

They worked on him some more, with compassion and all their skill. "That's NOT what really happened, nurse," the Doctor intimated.

"What then?

They now could start to finish up.

"Mort Philby," Doctor Veritasmas said. "An astro-science analyst from way back, from earlier in the program, back when they still didn't even know what they didn't even know, about Jupiter itself. I've met him a few times, he has some great cock-and-bull stories about the early program. Mostly no one believes him, but they're great stories. Look, let's run a second test on his heart-beat, with full spectrum-view. Use the Astro-Health Group machines, those are the best. It's possible we can restore his normal cardiac quickly if we move now, I need specific information to make some choices that can save his life. No cock-and-bull, nurse, okay?"

"Yes sir. But, doctor, why did this happen? Who would try to kill him?"

"People who don't want him talking to investigators about his early Jupiter years, with the program."

"But why? Why would these people not want others to know about the earlier research?"

"Big business back home, money interests, planning a war, or some armed conflict. A lot of time, money and effort, a lot of wealth ordering new weapons and machines and ships. So he was just in the way."

"A war? I never heard that one. We are all one. I don't think I like all that crap!!"

Philby suddenly heaved a bit, and there was a horrid sound under his blanket. The doctor and the nurse responded. There was an awful smell wafting into the room.

"Oh my god, he's shitting bricks. Oh no!"

"Get him cleaned up, nurse. Thank you," said Veritasmas. So she did.

CHAPTER 25: Doctor Montrose is Also Dead

Charles Benway, the Ganymede docking-station commander, was among those in the program who was very familiar with the disposition of H20 in the program hierarchy. The vast supplies of pure H20 at some of the Jupiter moons was indeed impressive. The moons Io, and Europa, for instance, were partly made up of high-purity deep-frozen H20. In large quantities, various levels of purity but mostly pristine. These objects were almost as large as the Earth itself, with all of its environmental problems.

The Jupiter Program by that time, was old enough that its texture and tone, for those traveling among them, was somehow dark and somewhat sleazy and a little creepy, if not very creepy. The technology, the machines and computers, the bright displays of flawless data and critical information, the communications and look-sharp chain-of-command, quasi-military, the mighty ships and engines, the great distances, the sense of power and mastery; it was deceptive, at first seeming a wonderland of perfection and total safety.

Perhaps it was the human element, perhaps the agedness of the systems. A big part of the reason the Program was 'so aged' was that it had taken Earth-science space-theory and astronautics, a very long time to set it all up, placing the bases, building the ships, training, learning as they went. But they were also confident they knew what they were doing. And they were still making improvements, and even building new systems and bases or platforms (such as work to repair the Europa-base following the wreck of the 'Ferrous-2').

Taken as a whole, the Jupiter Program, and associated programs (Mars, Earth's moon, the asteroid-belt, and early attempt to work on Venus), were like a many-tiered staircase or ladder, a string-of-pearls reaching outward, point-by-point, connecting Earth to her immediate neighbors. But to live and work there, inside the bases and ships, it may have seemed more like the dark, damp, filthy dirt holes and tunnels and rooted hillside-grasses and small berry-bushes, populated by rabbits, in the English countryside, in Richard Adam's tale, "Watership Down', the namesake of Brandeis' ship.

After a very long time, the program had more problems and difficulties than common-sense and safety-standards would allow. This created an opportunity for Deveroux and Angel-Face to enjoy their first voyage into deep space, and all that meant to each of them.

"It doesn't mean anything," Mendoza would say, reminding himself he was not the center of attention, as he sometimes felt would save him.

But Charles Benway also knew what only a few of the workers cared to discuss: something was wrong with the water-storage system at the Alpha-base. Alpha-base used much of the water they collected for use back home. Hydrogen-extraction techniques made the H20 valuable in a thousand ways, other than simple wetness. The Amalthea Alpha-base was storing vast amounts of water, in huge holding tanks, deep inside the rock asteroid itself, as they had for many decades of work in space.

The learning-curve was terrible slow, as far as Deveroux was concerned. He just didn't care for it all. He felt claustrophobic, he felt out-of-place, or as if he 'had nothing to do'. He found the rooms and toilets and eateries and view-ports interesting and even exciting at first, but later rather dull and even dismally redundant. And he soon learned that among the crew and staff, things were mostly light-and-easy, conversational, or transparent, but that some topics and ideas or inquiries, even innocently, could produce a strong negative response, like a door slammed in his face.

One did not ask about leadership's personal habits or personal lives, or ambitions. One did not ask about deaths, accidents, technical irregularities, or highly unlikely dangers (such as accidentally going off-course, or a hit by a meteor or comet).

One did not delve into topics about aliens from other worlds, alien minds, or religion and spirituality, death, the soul. So, like a good investigator, it was these areas he intended to concentrate on, as far as his job with the PPP was concerned. *"When the door slams shut, it slams shut for a reason,"* Deveroux told himself. And he was right, but those truths were still hidden.

The report about Mort Philby, the Ganymede-based spectrum-analysis scientist monitoring Jupiter's 'insides' and whirling winds of gaseous layers, who had suddenly become ill under suspicious circumstances, was a buzz among the Program regulars. As large as the Jupiter program was, this included some 3,000 or 4,000 men and women. One reason was that serious illness, infectious diseases, death, and certainly any 'suspicious circumstances', were rare indeed.

The space-work and ships and astronautics were all oriented towards complete safety at all times, for obvious reasons. Another reason was that Philby really was loved and cared for, people liked him, he was older, and sort of a resource and guidance, for a lot of them. The early space-program people seemed to have an attitude about it all, that appealed on many levels, when the newer workers were dreaming their dreams of serving mankind in space, and all the glory and wonder of space-travel and space-labors. A guy like Philby somehow made it worthwhile, the charm and adventurous explorer style of the science and technology, from the previous era, quite different from what the program had become; dark, dank, corrupt, full of ambitious lieutenants, heaving with enormous wealth and power, and yet 'so full-of-holes'.

"But what do they think is really going on, Charlie?" one of his seconds asked him during a lull. *"With the water tanks at the main base?"*

Benway shrugged. "We don't know," he said. "Levels have dropped dramatically, I mean, for a flawless system, in the past few months, it's strange. Maybe half a year. And we still haven't gotten to the bottom of it. On the surface of things, it seems like just some kind of speeded-up attrition. But we can't account for it all that way. There is normal attrition or minor loss. But these amounts are way, way up."

"And that's all?"

So, word of what had happened to Philby spread like wildfire, ship to ship, base to base, radio-wise, telemetry-wise, video-wise, like a chatter and a hush. Of course the incident was logged and recorded, and reviewed. Another failure. What had really happened? Sadly, many felt they already knew.

The journey to Amalthea-base for Mendoza aboard the 'Aerotica', was now much nearer their destination, only a few weeks away. The 'Down' was still months away, bigger and slower, though still of course very fast, both ships traversing many hundreds of millions of miles in the darkness. Doctor Wu, the Jupiter Program manager for space-based operations, was also aboard the 'Aerotica'. Mendoza found other opportunities to chat with him.

"The Astro-Health Group has no vested interest in genetic modification of your soldiers," he told Wu, there on the food-court deck of the 'Aerotica'.

This was much like a cafeteria, a safe-place for the passengers to enjoy one another's company, there with some of the crew, and other passengers. A ship like this, or the others, would maintain a sleep-cycle, day-for-night, with ship's lighting, and morning-cues, wake-up calls. Some of the passengers on this ship had chosen to be 'put to sleep', with long-term techniques that basically placed the body in a semi-comatose suspended state, with medical review and medical care. It made the passage seem much shorter for them.

For others, they could remain awake the whole time, and work on their computers or communicate far beyond the ship's confinements with associates even as far away as Earth or the Alpha-base and elsewhere in the system.

"We have no soldiers, they are only astronauts," Wu responded.

"I'm sorry, that's what I meant," Al said.

"No matter, Mister Gonzales," Wu said. "You should be informed, that's all. We have security, of course. But no soldiers."

They made up their plates of food, croissants with sugary syrup, oatmeal with butter and milk, orange juice, and a power-bar nutritional supplement that was effective for the rigors of space-travel. The low-gravity made the muscles weak, there was a certain amount of physical atrophy, the same type that Astro-Health Group was assigned to help with, over many years. Mendoza was only posing as an agent for this company, but he knew enough to maintain the illusion, though it was a thin disguise.

Doctor Menuda Wu was perhaps himself an example of how long-term space-work effected people. He was sort of a creepy guy, his skin and muscles had an odd look, his belly was somehow overly bloated, not fat, but extended, his arms were too smooth, hairless, he seemed sleazy somehow, and it was known to others that his lifestyle was oriented towards pleasures and excessive indulgence. He did no hard labor, he made no space-walks in the suits, and he could not pilot a ship.

It happened to Mendoza's thoughts about then, that he was somehow in the odd position of promoting a point-of-view he found detestable (the DNA modification of human beings for work in space), in conversation with the man in charge of the Jupiter Program. Also ironic was that Wu was somehow an unhealthy person, by appearances, given perhaps that he may have benefited personally from genetic-level physical assistance for his own long-term space work.

"Yes, well, at any rate, it is not something my company is certain about at all," Mendoza said. "We're looking into it, it is only a proposal. If it was done, your space-workers would be stronger, have a higher comfort level, and also a higher safety-level. You would see less fatigue, less worker complaints, fewer errors."

"All very desirable, I'm sure," said Doctor Wu. "As I said, when we arrive at Alpha-base, you'll be free to review our current health programs, study and gather information. I would request that you don't advertise this idea with staff and crew, at least not publicly. You're right, it's controversial, and I don't always want my people to be upset with anything that isn't confirmed."

"Certainly, of course," Mendoza assured him. They were seating themselves, starting to enjoy their food. One of the ship's crew was romancing a beautiful younger woman, apparently among the student passengers, with his bravado tales of being an astronaut and space-man. The girl ate it up, wide-eyed and delighted.

"Is that all?"

Benway paused and sighed. "Well, there are some kind of unwelcome life-forms involved. Some kind of slime-thing, bacteria or fungus, like fast-growing branches. They figure it arrived somehow in the base food-supply."

"Remind me to just have the fruit juice, then," the other one said. They both laughed.

As the 'Aerotica' moved closer to the Amalthea base, it irked Angelo Mendoza that it was all too obvious that he was pumping Dr. Wu for information, or his position on things.

"The program update news arriving on the computers say there was an unusual death at the Ganymede base," Mendoza commented. "Sad, I guess an older worker. The reports said he died after a week or so in the infirmary."

"I know about it," Wu said coldly, his mouth full of oatmeal, then wiping his chin with a cloth.

"What happened?"

"Not sure, he was a spectrum-analysis planet monitor, keeping track of changes on Jupiter itself. More than 70 years old, one of the very early program technicians. His heart failed, but there was some talk that it was not by natural causes. As if anything much was very natural out here at all anyway."

They munched their goodies for a bit in silence. "Sad," Mendoza said again. "What is your sense of it, as Program Manager, if I may ask?"

"I shouldn't really talk or make a judgment. The Program has all kinds of troubles right now, including inner disputes and feuds between various vested interests, the labor unions, the religionists, and outside powers. So, its not certain, but this old fellow may have made enemies of his own, that's all."

"But that's murder, isn't it, Doctor Wu?"

"Perhaps, perhaps not. It's not your concern, sir. I shouldn't speak of it now, but you asked, so I answered you. It's an administrative matter now. Let me enjoy my morning, if you please."

And so it was, and so they went.

Far ahead, the famous Solar-system asteroid belt yet ringed the star (Sol) between Mars and Jupiter as it had for ages and ages. Many millions of stone fragments, large and small, moved in orbit much as a planet would, spread out over a circle larger than the imagination could contain.

The Amalthea-base was built into the stone of a small, irregular planetoid. Amalethea was larger than the moons of Mars, not as large as Earth's moon, and smaller than most of the moons of Jupiter. This 'planetoid' was chosen because its orbital path was stable and predictable, in relation to Jupiter. So it could be used for all kinds of space-travel related purposes, as useful to work at Jupiter.

Ships could stop there first, re-fuel or repair, crews and staff could make berth and rest, exchange workers for shifts, or scheduled trips home. Communications here were more reliable, and also information sources and data-base, much of it from Earth's long-distance telescopes and radar, etc. As the planets moved in their orbits, sometimes the distance to Jupiter was longer, or shorter, from Alpha-base. But the system worked because the regular pattern of orbits was found to be favorable most of the time.

Bmmmmmm, bmmmmmm, bzzzzzrrrrrr, bmmmmm, brrrrr.

Days later, deep within the Alpha-base itself, one of the small robotic server-droids that populated the hallways, work-rooms and technical decks, wheeled along in its meaningless electronic song. The thing was the size of a small trash-can, it had automated guidance and directions. Anyone could tell it what to do at any time, as a universal resource for base-residents. They were programmable, and usually a specific task would go un-noticed by the general population. The thing wheeled along, humming and buzzing, blinking lights.

It was late in the 24-hour cycle, at the base, towards the third hour. The droid was rolling along a hallway in one of the resident living-areas, where the regulars had rooms. *Bmmmmmm, bmmmmmm, bzzzzzrrrrrr, bmmmmm, brrrrr.*

Montrose de Montrose, the 80 year-old planet-density researcher/specialist, was just waking up for his day. Montrose was also one of the 'Old Men', from the early program. The droid had been sent by an unknown person to kill him, much like Philby. Wrinkled as an old tree, not really cranky or mean-spirited, Montrose was unlike Philby to the extent that he spent all of his time at the Amalthea-base (Philby had been at Ganymede).

Older, also a 'national treasure' (for lack of a better term), Montrose was sidelined in his work and years to some esoteric research about the deep inner mass-density mysteries of Jupiter. It was a slow boat, anything they learned was boring unless a person had an interest in the data, and the analysis was on-going at various levels of methodology and accuracy. Really his work was superfluous and redundant, but called for a certain kind of knowledge and science-skill. He was good at other things, too, and enjoyed playing dominoes with the other men his age in his spare time.

His quarters were neat and tidy, but the bedding was un-made, he had been working with some papers that needed removal, and the bath-toilet facilities had not been cleaned. So he was expecting one of the server droids to show up eventually and help him with these tasks. The entry-door buzzed, a light blinked, so he touched a button and the doorway opened. He saw only the little droid-robot, then turned away as the machine rolled inside the room.

"Greetings Mister Montrose," came a voice from the thing. *Bmmmmmm, bmmmmmm, bzzzzzrrrrrr, bmmmmm, brrrrr.*

Within about five minutes, Montrose was dead, his body collapsed on the floor. The droid automatically wheeled back out of the room with the paper-trash and disposables. The droid had also made up the bed, and cleaned the toilet. Programmed well indeed, the thing slipped away down the long halls and work-areas at the base, the doorway to Montrose's room sliding shut behind it. Montrose's body was not discovered until a few hours later.

Bmmmmmm, bmmmmmm, bzzzzzrrrrrr, bmmmmm, brrrrr. No one noticed the droid at all, as it was identical to 200 others, all over the base.

CHAPTER 26: The Quantuum Mind

"By the way, I exist," the Quantuum-Mind, somewhere in a dream, somewhere by a river, somewhere by a tree.

Thadeus El Supremo Vente Viro Amore claims he is god, and in his dismissal, for lack of breath, or banishment, for dis-enjoyment, being also admired and loved, Second Commander Pearl had invited Dan Deveroux to meet with the Quantuum Mind, via an extended channeling session. *Quality questions queried quaintly.* Pearl didn't hate the intrusive qualities of religionists, such as in her role as Priestess of the spaceship Down's on-board Fellowship, (or, one of the leaders). She didn't hate god, as many seemed to. And she didn't hate poor Thadeus, though she knew it was not true.

Hatred ruined everything, and discomfort was a close second, they were related to love, especially in those meditations for which the Quantuum Mind Fellowship was so famous. As space-exploration expanded, in 200 or 300 years, despite the abusive tendencies of the traditional truths of the past, the 'space-religions' developed prayers and meditations that were considered especially insightful or inspired due to the nature of work there.

The El Supremo, he went peacefully off to sleep, it is his good-bye, for their style of deep-space worship and adoration (church and meetings). In other words, he didn't exist. But he was loved, revered and respected, *"go figure".* With space-travel, people like Menima Pearl, found a spacious place, within, for oracular-mystical experiences, that was attributed by many fans and adherents, to the qualities of the deep Indigo Abyss itself, or Mother Night, known as 'space' or Universe.

The promise of space travel included, for them, the opportunity of mind-expanding sharing, or communion. These returned to Earth as writings, poems, prayers, music, statements of truth, books, guidance, that were thought to eventually produce a new religion-system, suitable more specifically to communities in space, perhaps in more hundreds of years. Galactic imagery was a favorite among the children, (if any), and the Mothers were assured to normal forms of behavioral guidance.

"Thank you for inviting me," Deveroux said, entering with Pearl and three friends, for a gathering in private chambers aboard the 'Down' (still in-transit). Meeting with the Quantuum Mind was a cautious thing, for him, there 50 million miles from Earth. That's why they had guides and gurus, experienced enough to initiate the un-initiated with joy and welcome.

Essentially, from their point-of-view, Deveroux (and others), were helped to 'meet with a higher version of themselves', through a series of meditations and enlightenments, developed for the space-travelers. The same traditions were common back home, not really different at all, the 'higher self', or 'inner child', or 'divine wisdom', that men had sought, either within themselves or even very, very away from their bodies, for all of the endless human voyage.

"The experience would be mild," Pearl said softly but firmly. The others in the room included a man, and two women. There was very little to indicate any priestly orders or uniform art-work, the style was very free and easy, with some banners, bells, chimes, small water-fountains. Pearl's strange-looking idol, her 'minook' (similar perhaps in modern vernacular to the by then ancient Magic 8-Ball toy, more sophisticated), was also in attendance, as dead and unconscious or unaware as could be, but very jealous of Thadeus El Supremo Vente Viro Amore, a mere machine.

Deveroux understood they were to do some prayers and introductions, as a meeting he might enjoy, and he had some literature and reading on the group. It didn't bother him, he felt he could always walk away, or find himself as he always had since childhood, in any event of extremes, which they had heard of. The zealotry, or fear. They were creating the opposite: bliss, peace, joy, understanding, self-awareness.

Hatred ruins everything, Pearl repeated to herself idly. *Why would anyone ever create fear and cruelty? Just part of the whole, part of the whole, an accident, the Accidental Jurist, the judge, the executioner's song, the sick-twisted reformer, himself un-reformed.*

So, they sat in a circle, on pillows, removing their shoes. Bells and candles, and other pleasant features. The three membership guides hummed together, a melody or tune, familiar to them, somewhat odd, but somewhat regular, with a beat, and a vibration. Deveroux had dome some meditation, some techniques, part of his background with the PPP, as a global agency, was to be familiar enough with mind-science, that culture and groups or teachers, would not throw him off with some seductive ideas or controlling ways.

He was needed elsewhere, doing his 'truth work'. And, true enough, the Quantuum-Mind Fellowship might easily have seen him as an enemy. They were not popular at all with the Jupiter Program management. Deveroux was slowly coaxed into a mild trance-state, music, humming, esoteric affirmations, summoning. The Quantuum-Mind opened gently to him, he was welcoming himself, to himself, at the point of death, a sand-mite in the endless nothing. No up, no down, no air, no warmth, no food or drink. Yet all those things in perpetual provision.

Deveroux's higher-self was somewhat suspicious (as usual). *"Dan, listen, these witches and warlocks here may not feel your investigation will fail to burn their boats with Jupiter program administration, know what I mean?"*

"That's not why I've come, Higher Me," Deveroux said. "I'm on the other terms, you know, eternal peace, love, forgiveness, flowers and dandelions, that stuff. Mystical. I'm just being social. They're not witches and warlocks, and I'm no spy or Hollywood secret-agent, come to bust up the party, either."

"If I knew you were coming I'd have baked a cake," his Higher Self replied. *"I just meant they might have something else in mind for you personally, later, or as their guidance works its wonders. Like, they might put you off-track. I know, its an ugly thought and*

all that, but its only because I care. I'd drink your piss in a coal-mining accident, Dan, 100 feet underground. You know its true."

Deveroux's higher-self, his link to the Universal Mind, was very practical and determined. His was a strong mind, but also somewhat overly personal, they were very tight, but not romantic, frequently a missing element.

"Let's just hum a song and trip out on the galaxy pictures," he said. "I need to rest, I need inner refreshment, and I need to be unafraid of the Abyss. That's all your territory anyway, Quantuum Dan. The research will be fine. Maybe they'll throw me out a view-port, then we'll be together at last."

"There was another killing, Daniel, this one at the base. No one told you, it was another old man, with the early program, Montrose. You'll hear about it soon enough. So it's not like some of this information is not a kindness or useful towards business, as un-happy as exposing wrong and murders may be. And that's all your territory, man of truth."

The Quantuum Mind seemed to bounce a bit like a jello-plate that had been tickled by a fork, well on its way to something less physical.

"Peace, forgiveness, love, kindness, flowers, music," Dev's heart projected, both ways, the dark and the light, and stillness finally came. The Montrose death was rippling into whatever awareness was around, the doctors and security and staffers and schedules, management, the pilots, the moon-bases. Two down, and they hadn't even arrived at the main base. How it was the Higher Self knew about that? It seemed not to matter.

It was a secret (and no secrets were possible), that the so-called Old Men, might hold the keys to understanding previous secrets, deep within Jupiter, that were linked or connected to current Program failures. Or giant robot bug-brain monolithic archetypes of some magical alien architecture, swimming happily along in some stone-cold guarantee in oceans of helium and hydrogen gases beneath the colored waves of Jupiter's winds, pressurized to levels of mass-density almost equal to the Sun's. Pretty screwed up in any case, Deveroux knew.

"That woman is going to transcendental your ass right into a tidy little nightmare, Dev'," said his Higher Self, watching, watching, watching. *"She's extremely interested in sex with you, for some reason."*

"Really?"

"Well, that's the view from here. In the words of Solomon, 'Do not go into her, my son'."

"Meaningless, meaningless, meaningless. Way too butch, Higher Me. Way too military. I don't like her hair much, either. Probably a sexy animal in the sack."

After about an hour, the session ended, with the Fellowship Mind-Guides (including Pearl) tapping Dan's shoulder's. "Peter? Peter? Wake up. It's okay, you're back."

The Quantuum 'Q'-mind was again quiet, which was the goal: inner peace. Dan was surprised to find that he had indeed entered a state of bliss, under their guidance. He had no fear, it was just his job. The intelligence of the Higher Self was guardian at every level anyway, so it seemed counter-intuitive to set forth a conflict with the religionists. Of course he had sexual desire for Second Commander Menima Pearl, naturally, and he knew it, so did she. But they were not an easy mating at all. So there was tension.

Hatred ruins everything.

They spent more time together, there aboard the 'Down', the huge whale-hauler, speeding towards the asteroid-belt, that afternoon, according to the light-cycle cues. Pearl showed Deveroux some of the features of her idol, the all-knowing toy 'minook'. As antique and obsolete as it was, it was very unique as an art-work, like some sort of gaudy Tiffany lamp, ornate and opaque, hard to see through, because it was empty, yet always offering answers, puzzling ideas, insights, direction and guidance, koans, observations, notions, music, language.

Minook: a robotic spiritual tool, built for its beauty, not for its use. Perverse. An idol.

They talked more about the Quantuum-Mind Fellowship, her role as a Priestess, and their membership. It was a pretty simple deal. They had wanted to provide meaningful spiritual services for the space-workers. But, eventually, the high-end absolute science, the esoterica of the age, the nature of the Abyss itself, and the politics and social-needs of the space-workers, and other factors, found the groups to be perhaps rigid or odd or obsessive or not natural.

Like losing something they thought they could provide by parsing out its previous gifts, the way people felt about Earth and its waters and roads and cities and people, hills and trees, mountains, laughing children, dogs and cats. Worth the effort? Or just prolonging some agony of loss that could only be replaced by itself?

"Well, it's been fun, Commander," Deveroux said. "I was always interested in spirituality, but I sort of felt I didn't really need to be too dogmatic or into organized activities. But I appreciate your service here on this ship. I can see how it fits."

"I also steer the 'Down' from the flight-deck, three days a week," Pearl boasted. "Not Commander. Second Commander, Mister Finches. I am multi-skilled."

He took her hand and grasped her fingers affectionately just a moment. Then they separated to their different tasks and goals for the day, there in the high-tech tin can, with a few hundred others. The long hallways and doors, the foyer-areas and commons, with lights and vid-screen displays, opening again, each walking off, smiles and good-byes. The 'Down' buzzed and hummed.

Space-travel, what a blast, Deveroux told himself. *Maybe I was wrong.*

It was a while later, from his private quarters, he was able to set up a private radio-comm link. He first needed to hook-up with the PPP in New York/Long Island Authority. That process alone took nearly three hours, the link had to be routed, coded properly, then established both ways.

One of the staff at New York for PPP, would set up a second radio-comm link to the Amalthea-base, but not to the top office (Doctor Wu, or the Program Administration). Those would have been normal channels, but Dev' didn't want to tip his hand. For the moment, given the deaths of both Philby and Montrose, he wanted to talk with Alpha-Base Security, and move across the list of names provided by his case-manager Vanessa ('V'), with the idea of protection or security for the three remaining 'Old Men' on the list, and any other 'Old Men' they didn't know about. They now seemed in jeopardy of some kind.

So it took a while. It wasn't something he could handle on a so-called laptop, but he worked with the communications system on the 'Down', and various relays. He was authorized to do so as 'Peter Finches' with Astro-Health Group on a Private Communication. So, he slept a while, waking after another hour to an alert-signal, and then the link was ready.

"Go ahead, Finches. Amalthea-Base Security Office. This is Lo-Ann Ordz, Watch Commander. Please confirm?"

Deveroux quickly wiped the sleep from his eyes. The Quantuum-Mind bliss-state trance had been exceptionally relaxing. He slept like a baby while waiting for his radio-link to set up, many millions of miles between points. The PPP back home had done their job, Amalthea-Base Security was on-line. Total delay: six hours.

"Yes, hello? Alpha-Base Security please. This is Peter Finches with Astro-Health Group, aboard the 'Down'. The link here is by schedule, I believe. Who am I speaking with?"

"I'm the Watch-Commander, Mister Finches. How can I help you?"

So Deveroux lied to him, explaining that he had heard about the deaths of Mort Philby and Montrose de Montrose, and that Astro-Health Group had intended to contact both those men, before he left Earth, as part of his work. The science was about the long-term health-effects of Jupiter's massive gravity-well, on workers at orbital base-platforms on the moons, very near the main planet. So, with these uncommon deaths, he wanted base security to have the other names on his list, and to perhaps provide extra security protection for those men, given the other two suspicious deaths. Philby and Montrose had died weeks apart, the 'Down' was no more than 60 days out from the Alpha-base. It was a white lie, a slow dance.

The Watch Commander (Ordz) considered the claims, and agreed to formally receive Deveroux's 'list'. There were only three names remaining on it, now, among the living: Gerry Reedly, Nonnly Guitierrez, and George Thomson. Officer Ordz also recommended some other names, and they spent time by radio talking it out. There were several early program older workers who might have similar characteristics to the dead, and might better be watched, until the 'Down' arrived. Not that Deveroux would protect them at all. He just wanted to talk with them.

"This is very rare for us, you understand, Mister Finches?" Officer Ordz was saying, the private-secure radio-link between them buzzing and alive. "Mostly my job is boring and dull, nothing ever happens out here, and if it does, its big trouble almost no matter how minor. That's how the Alpha-Base is. So, we're looking at these deaths for now as natural causes only. Both those men were very old."

"Well, you know your job, of course," Deveroux said. "It just came to mind that both men were on my list of interviews for Astro-Health Group, and now of course that won't happen. Paranoia, I guess, but I just thought your office would want the other names on my list, and maybe back me up so I have someone to talk to when the 'Down' finally arrives."

There was a lapse between the voices, one on the distant space-ship, one on an asteroid base, far, far ahead. "Confirmed, Finches, I have your info, it will move up the ladder for review and implementation. I can't really work directly with you after that. Security here has to do its own thing."

"Sure," Deveroux answered. "I'm familiar with the Space Authority." He was there in his quarters. The University teacher, the poet, that he shared the room with was reading a book quietly, listening to him absently with one ear.

"Thank you, Officer Ordz. I will contact you when the 'Down' arrives. Best luck with your work."

Then they ended the link. Dev's room-mate shifted his shoulders a bit. He was a writer-philosopher of some kind of fame back home, sent out to gaze at the stars and dream up answers for death, hell and the grave. Or the economies of Earth, the price of tea in China, and the chronology of chickens and eggs. He was an older man, they shared the room for the voyage, Deveroux didn't even know his name.

"You could die out here, Daniel Deveroux, do you think it's so?" the man said, relaxing from reading his book, from seated nearby in a chair.

"Even a second or third time, my friend, true," Daniel answered dryly, wondering what was ahead. "Even a second or third time. Excuse me."

CHAPTER 27: Operation Odyssey

"If we can really understand the problem, the answer comes out of it. Because the answer is not separated from the problem."
Jidda Krishnamurti, 1895-1986, Indian religious philosopher.

Doctor Wu, the Jupiter Program Manager for Deep-Space Operations, sometimes felt himself a 'wonderful counselor', with a superior wisdom all his own, and a way about things that others admired. Like anyone, he was liked by some, disliked by others, but above all he loved himself, and was a true lover of self, and a sleaze-bag with a gazillion-dollar super-tech powerhouse operation at his command, or at least, so he supposed. Reality, in terms of who this man really was, his relationships or style, perhaps different. The Jupiter Resources Program was 'out of control', but this was normal by that time, era 2,413-14AD/CE (Common Era). It was all so complicated, after all.

The 'Aerotica' would make space-dock at the Alpha-Base within only another few days. Menuda Wu was returning to his work-place, along with the other passengers. He had been Earth-side for some 'business' he didn't want to discuss, with anyone, for it being problematic and highly classified, and also perhaps never going to happen anyway. Something was brewing, in the 'red eye storm' that was both ageless and possibly even terrifying, but why worry? His own nest was feathered well, his safety was assured, he had only to bide his time and ride it out, keep with his people and power-base, and things would go whichever way they went.

So, for the big-picture view he felt entitled to, the reports of the un-natural deaths (Philby and Montrose), were merely a bother and an administrative trouble, temporary and meaningless. It would blow over, they'd cover it up as before. The clock would run out on any outrage or criminal charges or investigation, and he could get back to work. Life goes on.

From his private quarters on-board the needle-craft high-speed flight to Amalthea, Wu could work on various tasks he needed done. It was a good time to fiddle with some sensitive data, he judged. His portable Command-Desk Communications servo was at-hand. At the moment, he wanted to put to bed some of the logged-records on his visit to Earth. The personal quarters on a ship like this were adequate but not extravagant (unlike his rooms at the Alpha-Base, which could be luxurious). He had the Command Device opened and was working his tasks like this:

"In-House Only (classified Wu by penalty): Jup. Resource Command data-log 679G-2b: Re: M. Wu to confirm, OPERATION ODYSSEY/Earth-Planetary Galaxy Posture (will/will not) prep. to JP installation/orbital path, flight-command/control for newly developed OO ships/fleet, target-dates for tech-work/labor as listed:---"

This particular agreement went on for several pages in detail, listing targeted building-construction/installation dates and schedules for a new (and highly secret) effort known as 'Operation Odyssey'. Earth Galactic Posture (EGP) was one of that era's many space-based industrial-groups/agencies. This one was essentially organized so that any local planetary exploration would have a cogent or understandable approach to far-distant Milky Way Galaxy events (such as nova-star explosions, black hole evaporations, galactic-core dark-matter gravity wave ripples, and, yes, 'other inhabited worlds' much like Earth, though these were un-confirmed).

The agreement here was secret, and part of why Doctor Wu had traveled to Earth. EGP wanted permission and cooperation to start installations at JP bases and orbital stations, for high-tech command control flight-monitoring devices and machines, for a fleet of newly developed ships. Only Doctor Wu, and EGP insiders and a few others, knew that these ships were specially created fighter-ships, with highly-intense and new-generation technology weapons. Additionally, the other very secret detail, was that these machines would be operated and piloted by disposable human clones.

JULIAN PHILLIPS & TOM LUONG

Wu only had to set forth the administrative tasks needed so the work could go ahead. It would take many months, and was only a first-step toward the overall Operation Odyssey conspiracy. Conspiracy? His language usage may have been vague, but it simplified Wu's understanding. *Something was brewing, in the 'red eye storm' that was both ageless and terrifying.*

So, he had to go through their installation goals, and then figure out his end of the management, connecting to all the locations, crews, orbital stations, people and gear-needed, making sure it was available, and the labor-crew transports and schedules, and not incidentally keeping the true goals secret, at the same time. Another similar task on his Command Device, as he worked, ran like this:

"In-House Only (classified Wu by penalty): Jup. Resource Command data-log 671003-35x: Re: M. Wu: OO meeting-conference, (Milan, Europa-Hispanola/Basque)-in attendance/present Earth-President/Supreme Planet Commander Thadeus Vente El Viro Amore: M. Wu, OO planning team; witnesses. OPERATION ODYSSEY(OO) Year 14 Galactic Posture report and update, dated—"

This secret file was also lengthy. It was essentially the secretarial 'minutes' of an EGP conference (in the city of Milan, formerly in Spain), where the big-shots could lay out their plans and goals without invasive transparency or outsider witnesses, media, other government, etc. This went on and on, it was highly classified. In the world at that period, or future era, secrets meant very little, mostly because the high-science programs and new-technology developments were so completely misunderstood by the Average Joe, so the Average Joe didn't really give a damn, or even keep track. This definitely made things easier. On the other hand, without review, these 'demi-gods of progress' often lost their way.

Wu perused the files liesurely; the meeting was not boring. After almost 15 years of work, EGP was 'ready' for what they anticipated to be some sort of inter-stellar, Galactic-level event that may or may not involve affairs at Jupiter. No one really understood what it was all about, or what this so-called event was supposed to be. Earth's 'galactic posture' program was supposed to prepare them for whatever else was going in the Milky Way Galaxy. But, they simply didn't know what was going on elsewhere in the Milky Way Galaxy (which was also a mystery).

So, for whatever reason, they fully intended to be ready, and this involved the new fleet of ships, and Amalthea-Base, and the planet Jupiter, and the other Solar system planets and inhabitants. Doctor Wu needed to review the conference-meeting minutes in detail so his operation would know ahead of time, what was needed, over the next months and years, for Jupiter Program cooperation and the satisfaction of long-term goals for the EGP groups.

So, for example, during this meeting, the conferees were troubled about whether or not the Jupiter Resource Program would be able to provide enough of the 'Helium-3' elements and raw-materials from her moons, that the new fleet of ships used as a fuel. How much Helium-3 was available? What type was it? How much conversion-purification was needed for the raw-materials to be ready as fuel for the new ships? Could the raw-substances be prepared as fuel in deep-space using deep-space industries, or would it need to go all the way back to Earth, before it could be used?

So, Doctor Wu had to be able to answer these questions, and then scope out the Alpha-Base asteroid JP resources, including man-power and brain-power. *Don't plan ahead, my father said,* he thought to himself. Just do it.

That work alone, just answering those questions in truthful detail, might take him weeks. So, there in-transit aboard the 'Aerotica', with a few day's off from home (at the Alpha Base), he spent hours messing with these and other secret files. Wu was a competent administrator, he enjoyed his work. But it was boring and very technical.

Hours passed, and a secure-line radio-link communication from Alpha-base was routed to Wu's Command Link servo. It was one of the base security commanders, a woman named Cyrolia Linsom Ee. Basically, a cop, with the Space Authority.

"Yes, sir, Doctor. We've followed up on every available witness or medical source on the deaths of the two men, Philby and Montrose. It only took a few days, there really were no witnesses. Philby died of heart-failure, apparently something in his ventilation. Montrose was found dead in his room, medical said it was from shock. Security feels both deaths were possibly murders or assassinations, sir."

Wu waved his hand idly. He was exhausted, and wanted to enjoy his evening-cue hours. "Yes, yes, I know," he said. "They were both older science-tech guys, supposedly targeted for some special knowledge from the early years. Yes, so sad."

"Yes, sir," replied Cyrolia, who was security lead on the case for Alpha-base. They communicated by radio-link, yet millions of miles apart. The 'Aerotica' sped through the Abyss, a silent steel falcon of astonishing power and velocity. Their communication was secure, un-filtered by other space-program links or monitors. "We have very few of these sorts of deaths," Cyrolia continued. "The mortality-rate out here is next-to-zero, as you know, a safety matter. We prefer they go home to die, naturally."

"Of course, Officer Ee," Wu said. "We also prefer they are not murdered, right? This is a mess, you understand? As base-manager, I can't tolerate even the fractional rumor of this kind of affair. Our unions and space-workers and malcontents or angry officials are already showing up at my office with weapons. It's your job to control this crap. What do you propose?"

"All our security is standard, Doctor," Cyrolia said. "Standard protection and guards are already in place."

There was a long pause. The radio-link buzzed lightly, it was a solid connection using a radar-technology micro-wave frequency. "All right, Officer Linsom. Listen. Here's what I want. Months ago, there was an incident, such as I just described, I was threatened with harm at the base by one of the union men, who felt the 'Ferrous-2' crash and other incidents were some kind of personal affront. An angry person. Do you recall what I mean?"

"Yes, sir," officer Linsom answered. "Security tracked it down to a small group of space-walker miners, the external suit-men who go down to the planets. They worked off Europa, where the 'Ferrous-2' wreck happened. About three or four men, they were using drugs, too, I believe, and very excited and angry, they wanted revenge."

"Correct, Linsom," Wu said. "There is another player here, however. The touring musician, do you know who I mean?"

"Yes, yes, I do. You mean the man they call Beautiful Truth Amakmid-Cornerstone? With the Large Globe Object Music Society. They make electronic music from scans of the planet."

"No, no, not him," Wu said. "The other man. Janus Marciel Penieur. He's an entertainer, a guitar-player, he does a show, travels the system, you know him?"

Another pause. "Oh yes, I believe so. I saw his act, he's good, very skilled and funny. Just a touring guitar act, with his clone."

There was a pause on the invisible radio-link.

"Arrest him, and charge him with these killings. Do it right away." Doctor Wu gave the order.

"Sir?"

"Arrest the guitar player, Marciel Penieur, and hold him under charges of arranging the deaths of the two men, Philby and Montrose. That's an order, Security Officer Linsom."

"But, Doctor Wu, if I may," Cyrolia said over the radio-link from Alpha-base. "He had no connection whatsoever to these deaths, he didn't know them or have any grudge or motivation. None of our investigation indicated his involvement whatsoever."

"I realize that, Officer," Wu replied. "Please do as I have directed, report back to me by regular transmission log when you have him in custody. I want him arrested and in confinement on these same charges. You can consider that a command directive from myself. I will detail any explanations later."

A long pause. "Yes, sir, Mister Wu. Our department will comply as you have said. Linsom out, end transmission, Alpha-base security, Space Authority at 2591."

The radio-link dropped out to dead-air. A light buzzing, a 'status-available' beep-tone, then nothing. Doctor Wu shut down the devices on his side of the call. Wu wanted the sudden deaths and rumors of attacks, and the idea of a plot against these older workers and science-researchers, to be over with quickly. The deaths were a nuisance, a bother, and would only mean more trouble and unrest.   The guitar-player, Penieur, had been arrested before. In fact, he was regularly charged with just about anything that normal Alpha-base security could not easily deal with or handle by actually charging or convicting the real criminals or violators. Wu smiled. It was a little-known part of the entertainer's show. Fall guy for greater powers, as-available. Because Penieur was a 'nobody', they could arrest him, charge him, hold him as a prisoner, interrogate him, and then when things quieted down, release him back into the Jupiter Program population, and declare the case solved, without informing base-regulars, who would forget it all by the time they were done.

The guitar-player didn't mind, and was paid well, for helping out. Otherwise, long-term Jupiter-program workers, highly-trained and valued, would be wasting their time, since crime was so rare among them, more seen as an entitlement, and contributing to overall program corruption. In other words, Wu routinely cut them slack on such things, letting it go, even the use of threats and weapons, as a kind of reward, to quell further upsets, given the program workers were among the astronautic elite-class and very needed at their regular jobs. A guitar-player is one thing; a deep-space moon-mining collector-ship pilot or crew-man was another. And any real murder investigation would proceed nehind-the-scenes anyway.

"The man's situation is expendable," Wu thought to himself. "I'm far too important in my other work to fuck around with this junk. Let him sit in jail a while, then we'll just hide him away on Europa or Io, no one will care. Then the deaths or killings will blow over. Maybe they'll enjoy his guitar playing on one of the smaller moon operations, gets dull over there."

Then he took his leave to enjoy the evening.

CHAPTER 28: Always Night

"But Commander Wu, I'm not at all sure what you're asking me is even possible," said Amakmid Beautiful Truth, much later.

"I am a Commander, yes, in rank, here at the base," Wu said. But the proper formal address or salute is not 'Commander'. But, anything is possible, my friend, it's a simple request."

"Well, simple in theory. Or, as a friend of mine used to say, 'it's easy means, it's not always easy', where new proposals are concerned. But he was a restaurant chef, so..."

"Think of it as a specially-prepared meal for me and my guests, then, if you please. You have plenty of time. We can talk about it more. I just wondered, you know, I am a healthy man, despite my long years in space," Wu answered him. "The technology you represent is intriguing."

Beautiful Truth paused. His thoughts were elsewhere. "I imagine I'll be looking at the project however you wish me to, sir," he said.

"Then we'll get along just fine. You're dismissed, thank you."

Doctor Wu's radio-link to Alpha-base Security Officer Cyrolia Linsom Ee, was not intercepted or 'tapped'. However, Mendoza, also aboard the 'Aerotica', had a source at Amalthea, who later filled him in, about Wu's idea to ignore the slayings and arrest an innocent man.

The source here was one of Vanessa Signo's insiders, working backwards through the PPP, back home, unbeknownst to others. *Truly a corrupt and rotten thing to do,* Mendoza observed. He had never heard of the guitar-player, Janus Penieur, but it made sense. The deaths of Philby and Montrose also more-or-less fit the scenario. It seemed that the early years of the Jupiter program and the research about the planet, back then, included some information, that 'wanted badly to be concealed'.

"I wonder what the hell it is?" Mendoza self-reflected in private.

Al's own research and ideas about what the PPP may eventually find out about the Jupiter program, were changing as they went along. Almost a year of prep-time before launching into space, and now some five months on a voyage to the asteroids. He and Deveroux (who was aboard the slower transport-hauler the 'Down'), could communicate as well, as the ships moved through the depths. They'd heard details from Commander Brandeis' report, and those matters were quite true, (the union complaints, the technology failures and leadership neglect, the space-religions and their intrusive mysticism and superstitions seeping into the fabric of life for workers in space, over many years, the wreck of the 'Ferrous-2' and the blame no one would claim, and other complaints), all more or less confirmed and documented.

Brandeis did a good job. Now, it was starting to look like those problem-areas were to be over-shadowed by a deeper and darker mystery concerning the large main planet Jupiter itself, and something known or speculated about, deep inside the huge gas giant.

Jupiter had a dark, evil quality, it was so humongous as to startle even experienced space-workers, gazing overhead from a loading platform or small ship, filling the entire view, overwhelming. More than 60 small and large moons or objects were zig-zagging around Jupiter, some went one way, some went another, some faster, some slower. And the giant herself, turning on its planetary axis in only ten short hours, was extremely fast, given its size, 300 times as large as home. Mendoza wasn't looking forward to a close-encounter with Jupiter, from some small ship or observation platform.

So, perhaps they would find out, or maybe not, or maybe it was all some kind of delirious scam, some plot or cruelly motivated action, even that of other-worldly 'aliens'. God help them all where that kind of business was concerned. For Mendoza, it only meant he wanted to go slowly and cautiously, about it all. He wanted to get back into the loving arms of his woman, Marta in Belize, and spend time with his kids again, and

THE JUPITER PLAN: ANARCHY AT AMALTHEA

feel the solid Earth under his feet, and breathe the fresh mountain air by the oceans of the Gulf. The Jupiter Program was not his life's work, it didn't matter quite that much to him personally.

Falling head-first into a mystery so dark and deep, it held no real appeal. If they were killing people because of something they supposedly knew or might reveal to others, that was not really anything new. *As old as the hills.*

Some time later, he and Deveroux found they could also talk, connected by a similar radio-link. The two ships were very distant, but technology and science had advanced far enough, by that era, that with a few key-strokes and set-up time, they could chat pretty much as much as they wanted.

"Al, good to hear from you. How's your journeying? Are we having fun yet?" Deveroux was saying.

"I got the luxury cruise, you got the fish-boat," Mendoza said. "It's fine here, they have it all worked out very well. Boring, takes too long, you know. I guess I was thinking it was an over-nighter for some reason, like a regular jet-air transport, subconsciously. Boy was I wrong."

"The Land of Always-Night, Al," Deveroux said. "Such is space. Dark out there."

So, they chatted as friends will do. They both knew about the slayings of Philby and Montrose. They hadn't a clue about the Earth Galaxy Posture groups and meetings and agreements that Wu was working with. Sad-but-true, the PPP was not in the business of following people around and spying on them. Wu was a respected citizen. The EGP meeting was also respected and basically approved. Like a gigantic puzzle, they might have made sense of things, if they had known. But they didn't.

At this point, they both knew their chatting was also probably monitored. Their cover ID's as Astro-Health Group science-guys, meant very little. It helped, but they had to assume there were really no secrets they could depend on keeping.

"Philby and the other man, Montrose, what do you think?" said Al, inquiring of his partner.

"Al, please," Deveroux said, the radio-link buzzing, a few unseen ears or recording devices also listening-in. "I already knew this sort of thing was going on. After speaking with Professor Podliakov, in Seattle, he's a friend. Richard Podliakov, his father, Eldon, is still with the Jupiter Program after many years."

"The Seattle professor is the son of a staffer in the Jupiter Program now?"

"Yes. It was his opinion, it's the same thing. Same damn thing. The early Jupiter Program required deep-interior planet scans and inquiry of a scientific nature, many years ago. The science wanted to know what the hell was down there."

"Sensible, I guess," Mendoza replied.

"Podliakov told me what they thought they found was hidden away, a long time ago, I mean, a really long time ago. Thousands of years, or longer. Eons. He said it wasn't something he personally understood, but it was significant. He said Jupiter is a shit-hole for hells with no end, and that there was something inside the damn thing, such as alien artifacts, or giant machinery, and that it made no sense to we mere mortals."

"What? Oh, come on. This is not known."

"They could not go ahead for a thousand years, like some sort of pre-ordained ticking celestial time-clock that would eventually go off. But he also admitted he didn't understand it and never would. And this is a Ph.D. University Astronomer. Richard Podliakov, we worked together for a few years at least. In Seattle."

"Before your death. Fine," Al said. "So what's fucking up the Jupiter Program, is Jupiter itself, is that it? Like they simply didn't understand the planet would not accommodate the bases and work they wanted to do. Incompatible. Maybe too much gravity. Maybe too hostile or some strange attraction, like a pull or oddity of nature, that now has become a problem. What about that?"

The radio-link buzzed and hummed. "Al, what I felt was strongly delivered to me from Podliakov, when I spoke with him, was that it just doesn't matter, from a science point-of-view, or in terms of actually solving problems," Deveroux said. "In other words, it can't be understood, so why bother trying to blame the damn planet itself? If we go home, and report to 'V' that the under-laying cause of the Jupiter Program's difficulties, is that the planet Jupiter itself has some previously unknown negative overall effect, what good would it do?"

"Well, if it's true, they might..."

"But would that even be something we could prove? They can't shut everything down. And they won't. It may be true enough, in some sense, and maybe the remaining older-generation early era program researchers could help explain it all. But so what? It comes back full circle, and the program screws itself. Go figure."

"Sounds familiar," Mendoza replied. "A snake with its tail in its mouth."

They spoke more, looking forward to finally meeting up again at the Alpha-Base at Amalthea. Deveroux and Al had worked together quite a few years. But this was a strange case, even for the PPP. They knew they would need to depend on each other as things went ahead. A positive attitude was essential, besides friendship and professional approach.

Deveroux had his own curiosity to apprehend, there on the 'Down', as a guest, but not really very welcome. The radio-link dropped away, and then went dead. They would rendezvous later at the base. It entered his mind as the call ended, that if anyone was monitoring the communication, it may not have seemed related much to the Astro-Health Group.

The space-ship 'Down' could not but fall ahead into the emptiness, her moving target smaller than the head of a pin, in the endless nothing, and yet precisely on-course, by the strong hand's guidance of Commander Brandeis. Deveroux also was able to spend time with Brandeis, who had much more to say about things.

"I should not talk about my Second Commander Menima Pearl, on-the-record anyway," said Brandeis. They were talking some time later, at one of the ancillary control rooms, for a function Brandeis needed to perform as a daily matter in operating the Down's navigations. "We've worked together for three or four years. She's a decent person, very efficient and well-trained. But I've felt her involvement in religion has had unintended consequences."

"How so?"

The Commander was analyzing the daily navigations through a machine linked to their telemetry and long-distance trajectory. "I fear she may be mad, Mister Finches," he said blandly. "She feels the weight of her potent beliefs as if channeling. Is that what they call it? She has told me privately, the so-called Quantuum Mind of hers is connected to distant star-clusters. Distant cultures. They speak to her, and through her, and the news isn't good. But it's madness, its unreliable, mysticism. And she believes in it, I guess because she has to. So, it's her business, her personal hell, I suppose. But it works through the ranks. There's a connection there, but its hers, and she doesn't tell what she knows."

"Do you feel she may be trouble for you? Insanity in a commanding officer of a space-craft like this?"

"No, she hides it well, I guess, as she should. She's much too smart. But from what I've heard of her, the Quantuum-Mind can go to the black, full of hatred, cruel, sinister, lusting for something only it could comprehend, deep within Jupiter. And that effects her, of course. But the program doesn't need a squadron of people who feel the same way. It's like some dark cabal, or conspiracy. Like a special knowledge or unique, special secret insight they feel they alone hold, and then circulate throughout the other workers out here, and officers too. A thing like that can spread like wild-fire."

"And now the deaths, the murders," Deveroux added. "I see your point maybe, not certain about that type of thing. Needs a doctor, I guess."

"There's been criminality in the program before, Mister Finches. The program is old, at this point. Years before, an individual was found to be embezzling very large amounts of wealth over a few years, from the system. He really raked it in, huge amounts of wealth. There was another case where a commanding officer was killed by a crew-member over the love of a woman, also a long time ago. And then we get the drug-use, sad but true, and the rowdy ones. Just the way it is."

He paused, satisfied his gyroscopic navigations were precise enough for the ship's regular travel. "So tell me, then, Mister Finches," he said. "Why does Astro-Health Group want to know about my Second Commander's mental health anyway?"

"You know why, Brandeis," Deveroux said. "It's a health matter."

"Oh, that!" Brandeis replied, laughing. "The whole program is fucked up, Daniel Deveroux. You and your partner and the Planetary Program Proficiency won't be saving us. But I won't blow your cover. Good day, sir. I have my duties, you know."

Deveroux was also dreaming, he felt. Here he was, dead himself or deceased, according to official records in Seattle, anyway. This didn't improve his social life. He had a dread, it followed him around or became part of his awareness. Pearl's seances, channeling the Super-Mind, didn't really comfort him. It was eerie, and the strength or focus he knew he needed for the job he'd been sent to do, came and went. Maybe he was just getting old. *Was the journey really getting longer rather than shorter? It seemed they would never arrive.*

He slept that night (by the sleep-cycle cues), restlessly, after some other chores. Mendoza also clued him in about the guitar-player, Penieur. He found it laughable. The poor guy was totally innocent!!

"Wisdom is whatever works, when the chips are down," Deveroux told himself. "But that's not the way it works, for a guy like that. He's a pawn, a fall guy, framed."

He longed for it all to be over, as far as their work for the PPP on the Jupiter program. He now was feeling much like Mendoza, that there really was no answer, they would never find a solution or single-bullet remedy, or sinister conspiracy behind it all, that could be dealt with effectively. They'd stumbled into a mess, and could only muddle through until they found their way back home, if they ever did.

"The food's good, anyway," he observed. Then he fell out into what sleep he could find.

CHAPTER 29: Player Guitar

"Don't shoot me, I'm just the piano-player,"
--Elton John, 20th Century musician

You, you, you, who? Who was it? Why? How could they do that? Where are they now? Why? You, you, you. Why?

Like a really mixed up dream, with choosers and losers, and more cruel bargains than a body can tolerate, a tortuous route, there ahead, hanging above Earth in a mind, but not in a truth, the Jupiter Program, these vapors of communal human heart floated, bloated, and gloated, as they always had, as story, events, people, places.

Cyrolia Linsome, the Alpha-Base security woman, had disobeyed the Base Commander Wu. The musician, Peniuer, was not arrested right away. Instead, Linsome directed staff, and also participated herself, to run data-discovery on the one they called 'Peniuer, Janus Marciel'.

"He's innocent of any charges against him," Cyrolia spoke softly to her co-commander, there at the far, far distant Amalthea-Base at that time. Her associate was a hefty young buck of an astronaut named Le Van Ho, an Asian man about age 42 years. Van Ho was also with the Alpha-base security, as what they called a 'responder' (an authority operative with weapons and passage-keys).

These two, one might have said to a friend, 400 years from this writing, are Galaxy Babies, Galaxy Men and Women. They naturally see things differently than we do now. An innocence went with them, the same essential human rights values that motivated Deveroux and even Mendoza.

Why? You, you, you. Where are they now? Who was it? How could they do that? How can we un-do that? Is it possible? Why? Who? You? You?

What they found was not very interesting. Penieur was a popular musician back on Earth, with a significantly large audience. He became famous by following himself around with a video camera, for a few years, as he traveled the world playing his guitar and doing shows. He actually did this by having himself cloned, which was very controversial even then (4,415-18AD/CE). But, fame was not a concept in most cultures that very much resembled itself, by that time in history, in the West.

One might have thought that Penieur was a ghost, yet one known, or familiar. So his data-file and background were easy to learn about. A resident of the Thailand peninsula most of his life, he was nevertheless racially described as a Caucasian with Latin blood-qualilties, from hundreds of years of the Peniuer families. The only reason he was touring the Jupiter Program platforms, for a year or so at a time, was because he had a sponsor in the program.

It all worked well, his act was very well-received and creative. He had an education at University, a background in aero-nautics and languages. He had taken up travel with his guitar and gear, as a form of personal spiritual quest-journey. The videos went wild as a viral-global infectious video joy, that many found delightful. Yet, he was unknown to far more than ever knew him, or his music.

Van Ho was stunned by Commander Wu's order. "He must be pretending we should arrest and jail the musician, for charges other than the deaths of Philby and Montrose," he said.

"No," Cyrolia said. "Those are the charges Wu wanted."

"Why?"

"Well, when Wu was young, maybe someone bounced him on his fucking head, Number Two," She added. "I don't care. I have a job to do."

Linsome could be that way. Security at Alpha-Base was mostly lax, and meant very little, everyone knew the costs of violence in that sensitive and safety-conscious environment (such as a gun-shot that caused a leak or venting of air-supply). But at times, things got intense, and many other very clever weapons were easily available. Like field mice lingering years in dark wet tunnels on a hillside, the entire Jupiter Program suffered from an odd claustrophobia, with many parts, on other hillsides, and other mice, and other rabbits, so-to-speak. And as observed in many social-studies, this created an anxiety. And plenty of Helium-3 for back home, as well. Galaxy Babies. Galaxy Rats. Galaxy homes.

Van Ho processed the notion with little outward emotion. But inwardly, he saw his work and job compromised. It meant they, too, were now complicit, and that their work would be outside the comforts of the normal program-born justice system. And even for their group within the Alpha-Base command structure, it simply made things more difficult, and more dangerous. As police officers they were not supposed to care, and be courageous, instead, and use good judgment enforcing the rules. Van Ho could sense that it was a set-up, a scam, and this was foriegn to him, he found it hard to grasp.

"It makes no sense, Commander Linsome," he said. "It makes no sense."

Penieur could be found about then, relaxing before a performance. He was a thin, wiry-looking man, mid-tall or taller, with a specific and intentional wilderness appeal, like some backwoods boy. There was something about it, it became sort of a focus, with his act, and other so-called entertainments, among the Jupiter Program workers and staff. The obvious Idea was the reminder of Earth's basic and essential nature.

The guy could play, his guitar plucking was both classic and very unique and new. His act was within the Country-Western universe, but he worked it through as standard songs of Earth's best guitar sounds and styles. By the time his shows ended, Marciel was obviously 'just another musician', in the thoughts of audiences, mostly space-workers. A tough crowd. But he left an pleasant impression, and this moved him forward in the popular imagination, 700-millions miles from Earth.

What is a brothel or a bar like in the years ahead, almost half a millenium? Dark? With music? Fearful of angry people? Drugs or alcohol? Food? Women with sexual standards that include taking money for sex? Sleeping rooms in the back? Low-lite or opaque viewports to gaze upon the abyss, and Jupiter, like a wall of colors, stunning at times to view and hypnotic as well. A certain warmth? A weapons policy? Bouncers bigger than you? For Peniuer, who had played shows in places like the Indian sub-continent, Japan, North Europe, Russia, and Brazil, it was old hat. He played the moon-bases (Earth's moon) for years. Like a musical Buddha-Matreiya, joy and sorrow followed him among the stars and planets, a wanderer.

Linsome was a very practical person. So she sent some men down to where Marciel could be found, to arrest him and take him into custody. It all happened like this:

Penieur was at the bar, having his typical drink, a hot coffee with whisky or brandy. He was chatting with a gorgeous young bar-maid, very busty and attractive of course. They were talking about how the shows back home could be enjoyed at the Alpha-base in real-time with a little preparation, via the Elsewhere Eye HoloCast. But it was Marciel's opinion that there was no substitute for live-performance shows.

"Have you ever heard of a man named Bob Hope, a comedian in the late 20th Century?" Penieur was saying.

"Well, yes, I think so," the bar-gal said. "Why?"

"He was very popular and did thousands of shows with music, comedy and dancing girls, for the military troops and soldiers during at least two major Western wars," Penieur said. "You know, to cheer them up, they were getting shot at, you know."

"Oh," she said. "Like a charity. Yeah, I heard about that."

"Yes, but it was the live-performance proximity of favorite entertainers and celebrities, and the girls and so on, that really made his shows work so well with the boys," the musician said. "That's why."

Just then, three armed security guards quickly entered the bar. Most of the regular space-workers in a social environment, would dress as they pleased, or to attract the opposite sex, or for whatever reason they might choose to look the way they looked, all sorts of outfits were popular. Some of the women, for instance, would go topless, and bolder men wore only a cod-piece and odd armaments or togas.

But Space Authority security guards were identified with uniforms, pale-colored pull-over coveralls with insignia and bracelet-designed communications, and small arms, or weapons. The insignia looked like an angry green turtle, a lightning bolt, and a discarded oxygen helmet, and were found on one shoulder, the breast of the suit, and in larger detail on back with lettering: BASE SECURITY. The weapons used were a simple shot-gun style heavy-powered 'bag-shot' compressed air rifles, sufficient to put down almost any man, but not lethal. And also Taser-type non-lethal stun guns, they called a TASP (Temporary Alphawave Stun Pistol).

The security team looked around coyly for a moment, then spied Marciel at the bar and quickly approached him. He seemed dully unaware at first, then tense, then angry.

"Marciel Penieur? Base security sir, can I please see your pass?"

He paused in discussion abot 20th Century comedian Bob Hope with the sexy bar-gal. "Pardon me? My pass? Why? My ID is in my music case, and also at my residence at the Ganymede facility."

"Get the music case, then, please sir. This is a security matter. Are you the musician Penieur?"

"No, I am not," he lied. "You have the wrong man. Why? Did he do something? People say we look alike."

The security lead smiled. He already knew he had his man. "You are under arrest for falsifying information to a Space-Authority Agent, sir. Hold out your hands."

"I will not, you Nazi," Peniuer replied darkly.

"You do not have a choice, sir," the lead man said. Then he waved his hand toward the other two men. "Take him."

There was a scuffle. The two security men grabbed the musician roughly. He was not a big man, not hefty-size or bulky, not exceptionally strong or a fighter. They had no trouble settling him down. Penieur cried out, over and over: *"Bleachers! Bleachers! Bleachers!"*

Then TASP was applied to great effect, and Marciel slumped over unconscious. He was not meant to be harmed by them. They took his hands, cuffed him and marked each hand on the back of the palms with a bright blue-purple indica ink from a plastic sponge.

The bar-gal seemed to enjoy it all, as did other patrons of that particular bar-venue, known as the 'The Color Wall Bar', for the sake of that particular view of Jupiter itself through an oddly entrancing 'real window'.

"No guitar show tonight," someone commented, and got a laugh. The room was a buzz, then the security men hustled Peniuer out into the transit hallway, and then gone.

"Name that tune," someone else said.

"What did he do?" the bar gal asked. But no one could answer. Then they just shrugged it off, wondering, as gossip and chit-chat scuttlebutt peaked. Peniuer had friends, of course, and the entertainment manager for the system watched over him and took care of his needs, including transport and papers. But he traveled alone, it was a solo act (except for his clone). No thirteen disciples or entourage. He also often videoed his work and shows or people he met, which were sent back home to fans and other outlets.

Later, the musician was revived by medical staff and nurses in a holding cell, a 'jail'. Not a medical facility, but similar enough, for a deep-space platform like the Alpha-Base at Amalthea. It was very spare and boring, with only a bed-space bunk, a near-enough toilet-shower utility, a small storage locker, and an entrance sealed with thick doors and electronic locks, and some sheets and pillows and magnetic attachments for movement.

There was no way out unless someone came and got him and could unlock the doors, which required authority at least equal to that which had placed him there. At the moment, because of the effect of the so-called TASP (Temporary Alpha-Wave Pistol), two medical workers (nurses) were making sure the jolt didn't threaten to kill him.

They gave him juice, ice-packs and strong odors from ammonia, a very common element among the moons of Jupiter, with a powerful-overwhelming scent. The TASP generally dropped the human nervous system into a sort of trance-stupor, good for an hour or so on almost anyone. But it was seldom fatal.

"Marciel? Marciel, wake up," the male physician attending was saying loudly. They shook him a bit, from where he had been laid prone. He started to come around. As they knew, Jupiter-system management could place a man into unconscious semi-comatose 'deep sleep' for as long as a year at a time, without cause or explanation. The TASP was different, only a non-lethal crowd control item, suitable for space-work. Anyone under its influence, upon waking, had a very acute headache for a few hours.

"Hey, wait, wait, what the heck is this?" Marciel was saying. Then he rested and took a few controlled deep breaths, four in, two out, and then repeated for a minute or so. Oxygen flooded his senses, his mind woke up, then he looked around as his body responded. The two medical staff held him up, he sat up on the edge of the bed-bunk.

"This is holding cell 1442 Alpha-Base Quansut Quad, Mister Penieur," the man said. "I am a medical worker. You are being held on charges. That's all I can tell you right now."

"Gawd, my head!! What the hell was that??!!"

"Drink some juice, sir," the other medical staff said. Peniuer sipped the drink, it was very helpful. Then he grabbed his forehead with one hand and fell back on the bunk. The lower-gravity (throughout the base at 65-percent) made this seem like it all was happening in a kid's bounce-house.

"Hey, I am in pain here, okay? Seriously."

The two medical staff spent more time assuring him and checking his vital signs, he would live.

"I want base entertainment, understand? Base amusements manager, Brodesky Gentlemans-Nest, on upper deck offices. I insist, they can't do this. Contact her immediately, please. Will you do that? Promise me. Promise me now!"

The medical workes chuckled a bit. "An ombudsman will be assigned to you and you will be informed, Marciel. I saw your show at the Color Bar a few weeks ago. You are very talented."

"Fuck you, you lousy screw. Get my agent! Brodesky Gentlemans-Nest. I want an attorney!!"

More laughter, then they left the holding cell, the electronic doors buzzed and blipped, and then closed with a thick-strong 'click', unmistakably making Marciel Janus Peniuer a prisoner at the Alpha-Base 'jail'-system. He moaned a bit, then tried to sleep.

CHAPTER 30: Welcome To Amalthea

"A mathematical equation is different than a formula, pre-programmed for some sort of use, or application by an operator. I can tell the difference by looking at the equation or math-sequence, sometimes. It's the same with different languages," Le Van Ho, Second Commander, Amalthea-Base Space Authority Security Team.

Maybe it was just something that couldn't be expressed. It seemed like there are souls in the Universe, to whom captivity, is somehow an anticipated state. Say for instance, your mother's-mother's-mother's-mother's-mother and on back into the deep lagoons of a person's geneology, and even into the pre-historic ooze, food and nutrition and life-stream manifestation, age after age, a mollusk. An ocean snail or under-sea waterlife. Food-chain, 'you are what you eat', translation: your mother has claustrophobia because she was originally formed in the planetary history, ages and ages past, as a delightfully colorful and healthy undersea mollusk, called a Tiger Slug (now extinct).

Those kind of people, those kind of souls. The difficulty lay in the time required to explain, (to your mom), that this type of sea creature has no hard-shell, no tightly locked calciated shell halves for Venus to balance; it was a shell-less, spongiform type. For the sake of the image, though comforted in reality, it was a jail-cell in her thoughts, that only the truth would relieve. Bio-Oceanic Phylum, but not a language, usually.

Security Man Le Van Ho's mother was back on Earth, in Singapore. The Space-Authority was probably the most powerful agency of all the deep-space programs at that time, in terms of essential muscle, police power. The reason was, there had to be standards equal to all levels of the various explorations, for safety and orderly long-view goals, to process into general success. Still not military, anathema to the heart-and-soul of the space-astronauts and scientists, but the final word always fell to the

Space-Authority concerning any matter of more serious weight or consequences.

Things like launches, satellite placements, building programs, routes and schedules, staffing and office-holders, labor, and security or conflicts, crashes, deaths, accidents, and more. For Le Van Ho, his work was very like that of a high-tech cop, also moving throughout the 'rabbit-tunnels' all over the J-Program platforms and stations. His mother was very proud, she really had no idea. But it was a great honor.

"*Ban-len, li-chi chen. Bebe kum-lu man, ling chow-ching, hib-jing-jung, kola te mo tung. Ling tai-chow, bano-hap en-Byan. Kin yu kow, jung hin, jung maior. La henya muti hey. Komela, den-dung, treline, do vid nam, kaomo kuna. Cuma-kum hati hiya ere, ailon becklame. Ban yu, hanya annoya, da hangun, van kish-niche,*" his mother told him before he left Earth, on his most recent tour-of-duty.

It was a language only they shared or understood, only this man, a soldier on a space-station, an Asian man, and his mother, a fabrics laborer in Singapore. Spoken between them only, in this way, it resembled no other language, from the past, than perhaps Korean or Japanese, but it was neither. He sent her money regularly, like any good boy would.

Van Ho, desired wistfully to talk with his mother about the fact that the musician, Peniuer, supposedly had a 'clone'. If Penieur had a clone, how could they arrest this man, without complications based on the liberties of an identical twin? Much less complications resulting from his innocence!! Shouldn't they arrest them both?

And as an intelligent person, Van Ho had no doubt personally, that the traveling musician was innocent of the deaths, the two J-Program Old Men. In his thoughts, his mother only said to him: "*Yen, for basai, bayoon he fawtwaho. Chun yamaho yan, venico, sing-yoi, bin chow-chow, no vent yo no, chin chow-chow. Do haut yen soon, yo chitzu, yo du sai, me you bofosan, the Lord is On Your Side.*"

The 'Areotica' had managed a flawless docking procedure about a day before. It may be hard to visualize, the deep indigo Abyss, not black at all, speckled with stars, *"nice outfit, diamonds too!"* From the position of a small ship approaching the Amalthea, the prior vista now changed. This was also rare, the distances being so vast, with literally nothing to see. No scenic route. The pebbles, stones, rocks, boulders, gigantic mountainous planetoids, and huge asteroids such as the one where Alpha-Base had been built, were spread like the very best butter the Milky Way could produce from her silky-white tresses, there at the edge or corner of the Galaxy.

So it appeared like plates or ledges and stones, floating, like styrofoam peanuts in a Halloween apple-bobbing tub. Yet vast, huge, bordering the infinite, evidence in anyone's sight of view or knowledge, of some kind of event, more ageless and timeless, and less known or understood, than any other, at least at first view. The Alice-In-Wonderland Effect eventually wore off. The 'Aerotica' drifted down as easily as any hawk or large bird, with pilot Rolf Deneuri handling the transit deftly.

The Alpha-Base was on the 'top' of the very large Amalthea planetoid, which was known unofficially as the Rosebud Stone. It was spectacular even without the base. Rosebud Stone was an unusual shape, smoother, and reddish or deep brown in coloring, as if the material of the stone itself were of another type altogether than most of the billions and billions of others. It was determined to be 'jasper', normally greenish, but discolored by the vacuum and cold of deep space, and this a reddish hue, when it could be seen, for there was little ambient light at all, from the Sun.

Yet a strange glow filled any object in space, when viewed externally. A ship, a moon, a planet, so it was deceptive, the long slender rays of Earth's Sun hardly even breaching the dark, dark blue beyond, to touch the million reflective surfaces with its life-sustaining warmth. And there was no life here, none, until Man walked upon this stone, perhaps 50 miles in width and length, rounded but with higher parts, and a long flat lower part, such that builders and planners could fabricate and design the Alpha-Base, as if tucked into the higher stone parts, with a plateau useful for many space-travel purposes, many miles long, before the face of its facade upon the emptiness, like a strong, hard statement: *'We Are Here'*, that none could mistake, nor deny.

The Alpha-Base itself was a series of towers, towers-within-towers, and rows of long-low circular and cavernous main-to-middle areas it would be hard to call 'buildings', but there were several, very large. From these were extended several connecting tunnels and roadway-trams, moving outward to other types of structures, holding tanks, docking machinery, launch-pads and hangars. There were very large and very bright 'kleig lights' that ringed the entire area of the base, organized by colors of the rainbow for incoming ships and vessels to easily navigate, and also as beacons that could be applied for radar, microwave, main radio, and other types of transmissions.

Antennas, small ships, loading areas, trams, machinery, and somewhere atop it all, Commander Wu, despised by many, feared by none: *welcome home.* A Security man like Van Ho meant nothing to him, and Peniuer, the musician they had arrested on false charges, meant even less. Lovers, parties, attacks on his critics and enemies, drugs and encounters, and sleep, were his food of choice. And he chose the same diet for himself again and again, and now merely the best and superior edibles and meals, or big dinners. The man also consumed people, among those who knew him, but as one would overwhelm a new friend, and he only had 'new friends'. But this was also a mystery.

The deep-built H20 reservoirs down below the Alpha-Base itself, comprised a system of 30 very large holding tanks, directly under the base, built into the deep stone of the so-called Rosebud asteroid. Each was about the size of a common Earth-side football stadium, they were domed on top with a mechanical tressle-and-buttress method, that allowed for workers to move about and rig things like pumps and loading connections, or analysis, and so on. The same wing-set system was in place at many of the orbiting moon-bases, they worked like longish tubular people-movers, that men could ride inside, and maneuver all over the huge tanks.

"Reservoir 22 roll-over team, please advise," on the inter-comm they all used.

"22 Rollover here, ready. Go ahead," was the reply.

The team of three men had set out in one of the wing-set riders, rolling almost as if in flight, along the mechanical bridgework tressles, in the dome encasing Reservoir 22 from above. Below, a deep pool of pure water, shimmering like silver, lighted only as essential, with rays of artificial light from the wing-set rider, glowing downward. It went down hundreds of feet. Their only job at the moment, to perform a spectrogram. The wing-set hummed, gliding ghostly, three men inside.

Then there was a sound, like steel-on-steel, the wingset gyro-stabilizers shuddered.

Rolf, the pilot in charge, could dock the 'Aerotica' at one of the Rosebud Stone's longer platform spaces, designed for that type of ship. The larger ships and whale-hauler transports, could only dock at some distance and shuttle down. Amalthea was not a loading point for materials and substances from the planets or moons (helium, hydrogen, ice or H20, many others). But Jupiter was often 'in view', though sometimes not, depending upon the relative orbits. From Alpha-Base, Jupiter, even as large as it was, seemed about the size of Earth's moon, viewed from a Seattle beach in the summertime, with beer and chips.

It was of course beautiful, and her many moons, worn like a dancing necklace she could never release, 'in slow-mo', they would say, laughing. But they knew, her wisdom was not mocked, Jupiter was a Fat Woman who might somehow accidentally kill them at any instant. But her helium and hydrogen were among the purest and most efficient and clean sources of high-level chemical energy-fuel, that had ever been discovered by Man, with all his cheerful learning, or not-so-cheerful, 'out of gas'. Alpha-Base also benefited and was sustained by products and raw-materials from the moons of Jupiter. The base was the preferred customer, first in line for the water, and especially the hydrogen.

Mendoza was thrilled to make port at long last, at the Alpha-Base. Angel-Face had never traveled much at all into space. At this era in Earth-history, it was still exceedingly rare, and a privilege. Of course, the docking and navigations of the fast-ship or 'needle shot' he had flown within as a passenger, were obscured to him during the actual process, now several long and tedious months in-transit, little to recall of the journey but a vibrational-hum and buzz the ship made wherever one went.

There were bumps, jolts, rocking, a sense of motion that swooned into a nauseating lack of center or confusing miasma-vortex of non-Euclidian logical spacial-relatedness. Then things settled down. In a few hours, Al was ready, and was directed to an air-lock, to move inside the stunning and Oz-like Alpha-Base itself.

By this time, Mendoza had taken up the habit of making personal photographs of everything, which was approved. Few had visited The Castle, but it was acceptable to regard the place with a certain reverence, born of the airless and hostile Abyss, so unlike Earth and home, as to seem ever-dangerous to any of them, at any moment. Even a moment outside the protective life-sustaining environments of the various stations, the transports and main facilities, or ships or walker-suits, and any living thing would suffer instant death, by freezing, and suffocation. And some of the bio-science arts leaders had even by then proposed that certain life-forms could indeed survive, but not Man. *No big loss,* Al thought speculatively. *Nice rock.*

Angel-Face wanted to rest, but it didn't seem very likely. He was not on a vacation, none of them were. Within a few hours, he was booked and logged as a visitor at Alpha-Base Admissions: Alberto Mortissimo (Gonzales), a Ph.D. Genetic Research Specialist and Health-Care Applications Proficiency Examiner for the Astro-Health Corporation Group. The experience was such as entering a huge transit facility, leading inwards to a vast complex of service-oriented space-based marvels of technology and knowledge. Now he was truly one of the rabbits the tunnels of

'Watership Down', so far from home, and he knew it, and it encroached upon his senses and spirit.

Mendoza was more at home shucking pig-meal corn and salt-blocks among the chickens and street children of Marta's farm-land he had bought for her in Belize, on the Gulf of Mexico. A plastic wrist-band, and a neck-worn plastic digital pass-card on a nylon cord, were granted to 'Al' as a new long-term visitor, at Alpha-Base. No beautiful women, or Oz munchkins, were forth-coming or dispatched to show him around, however. There was an Introductory Protocol for Visitors, but it was assumed that Al's Earthside pre-travel training and preparation, had explained the basics, otherwise he could hurt himself, or be a harm to others accidentally, by ignorance.

In general no one ever arrived at the Alpha-Base without sufficient preparation and screening of some kind. No 'tourists' in funny shorts, no 'out-of-towners' with ice-cream cones and gawking children on a leash, snapping photos, no school-bus mentality with college-kids sneaking a buzz from a pill or a toke in the toilets in secret, on breaks, no tour-guides moving crowds through with gaudy and inaccurate descriptions of various sights, though a loud-speaker hand-mic, *"This way please, and on your right, you can see the third largest moon of Jupiter, on this view-screen, one-at-a-time, please."* The place was all-business.

The other Planetary Program Proficiency researcher on this particular assignment, his partner Dan Deveroux, was yet millions of miles away, on his way there, aboard the 'Down'. They had been able to remain connected enough through communications by radio and micro-signal (wave), such that Devroux was already appraised of Al's arrival at the asteroid-base.

Mendoza had no real idea what was going on, concerning the musician in custody, by that day-and-date, charged and incarcerated regarding the so-called killings. Even if it had been officially explained to him, he wouldn't have understood much about it, because it made no sense. Peniuer's arrest was a fraud, he truly was connected in no way at all to the deaths of Philby and Montrose, whatsoever.

Menodza also did not realize that one of the passengers also disembarking from aboard the 'Aerotica', at that time, who he thought all along was a student of some kind, was in fact an Unauthorized Cloned Person, or Unofficial DNA-Replicant, created from the genetic cellular supply provided many years ago, from the very physical-organic body of Marciel Janus Penieur, the musician, now there in a jail on the space-station, and feeling somewhat sad or blue about the whole thing. For just a moment, the two men passed in one of the hallways during the Admissions Process.

"Excuse me, is this the way?" Menodza asked him.

"No," the Peniuer-Clone said. "There is no way, that way. This way." He pointed.

"Oh, thanks, appreciated," said Mendoza. Then they each went about their business, as everyone did there, even the robot-droids and semi-aware animal-plant life-form creatures found here and there, working as organic cleanliness and sanitation absorption units and amusing 'pets'. Clean-up crew. Mutants. Crap-suckers with a sense of humor. Hortas. Men here sometimes felt the same way.

"Rollover? Rollover, are you there?"

The men inside the wingset rider were frightened, as the disturbance knocked loose one of the thing's rail-locks, and it began to slow, and then droop over, and then began to disconnect from their only way back to safety.

"Rollover 22, please respond?" from the comm-link.

CHAPTER 31: Twin Buddhas

"No, I never did. I never would, it's just not the real me. I'm not that kind of person, sorry, you must have the wrong man,"-- Adolph Hitler, World War 2 German dictator, 1930-40's, to a young woman at a brothel in Amsterdam.

The guitar-player, with his many travels, had found it almost Universally true, that an entertainer, or one hired for amusement's sake (including but not limited to prostitutes), was very seldom a truly important person. In Tel Aviv, Marciel could earn a few hundred units a week, in Western Units, but he had to be careful how he represented his own opinion; in Russia, the DJ-Clubs and Dance Bars, were a realm in which being lost forever might not have seemed half-bad, but the pay was poor; in Mumbai, Indian-Subcontinent, without putting a snake up his nose, the novelty wore off quickly. A 'cloned' musician doing old-time US-Western favorites and rock songs, with video of his travels, and the mystery behind his 'replicant' of himself, an identical twin. Yeah, they did duets, too, two-part harmony.

Seldom a very important person. By year 2,415, fame was somewhat another form, than in original years when Mother Media was young (Jung). A celebrity entertainer moved a certain amount of good-will and general happiness, a premium ingredient in the space program from the start. It would have been clear to an outsider, that celebrity talent were groomed and prepped, for their social roles, as one would handle a prize livestock animal, in this future world, yet unwritten.

So, someone was very clever about Penieur's career, and had him cloned. Literally, cloned, in terms of the use of his original DNA, but not a 'natal-clone', the only authentic sort. Otherwise, Peniuer-One would be 50-years old, and Penieur-Two would be only a baby, at the same time. Not working. The other method used Marciel's DNA, mated to one of the plant-animal clean-up life-forms, known as 'Mafu-Forms', such that an apparent new life, identical in appearance and style to 50-year old Janus Peniuer, was born.

"And I love him, and he loves me," Janus would often tell curious fans.

Not Mini-Me, but second self. In many ways, for Marciel, he was committed to the idea that his twin did not exist, in such capacity as its termination or death, would relieve many pressures and painful circumstances, and not just for himself, but for the life-form he was using. Clones were very stylish, and quality clones were expensive. But a very wealthy or powerful individual who could have the technique accomplished properly, could elude many Earth and Space-Authority agents and laws, police-forces, and enemies. This contributed to Marciel's legend (in his own mind).

But it backfired, as Doctor Wu had wanted all along.

Al 'Angel-Face' Mendoza didn't know Penieur, or his clone. The identical twin was also a stranger to him, though the creature had apparently been aboard the 'Aerotica' on Al's flight, anonymously. The creature? The 'IT' (identical twin), the clone, the plant-animal man, truly human in appearance and every manner, but much more a spare body, and much-despised, in general. The technique was basically illegal, Penieur had sources.

It was thought to be cruel and unethical, and was similar to the genetic-modification program Astro-Health Group was considering for the J-Program, as their PPP front, or cover-story. But Mendoza was well-aware of the case with Philby and Montrose.

So it was only a matter of time before they might encounter each other, excepting that Marciel was still held immobile in a cell, the jail. Al didn't know this, either.

The musician slept. Then he woke up again, measuring his headache. He had heard of the TASP. It was a mean jolt, but non-lethal. Then he fell out to sleep again, then woke, the light patterns in his cell had changed. He took a shower-bath, and used the toilets, then he tried to sleep again. Later, a food tray was delivered through the main locked doorway.

"I want an attorney," Marciel told the guard with the food-tray, without any hesitation. "What are the charges?"

"You are charged with the murders of two men, about a month ago," the orderly said. He was a hefty, even chubby, medical-staff and jailer. "Hey, I saw your guitar show. You're good!"

"Probably somebody else, not me," he said.

"What, you got a twin? Ha-ha! Yeah, I watched the '*Twin-Buddha Song-Walk*', on media. Is it really you? That was great!"

"Get me a representative, or your Higher-Up's will change your name in about a week, in your sleep, without telling you. Understand? *Cid Bixi Mimim*. I want to talk with your base amusements-manager, Brodesky Gentlemans-Nest, immediately. I'm scheduled for several shows soon. And I demand an attorney or ombudsman, immediately. Thank you, you can go."

"Sure. Don't you even want to know who you killed?"

"I never killed nobody, young man. I never killed a man, a woman, or child, or any human being, ever, in all my born days, and that's a fact," Marciel said. "Piss off. These charges are false as frog-hair, junior. *Cid Bixi Mimim*. Do as I said."

The guard left, miffed and upset, the automatic door locked again.

Mendoza found his way eventually to another local partition-block bar, a hang-out called the Elephant Hut. Life at Alpha-Base, was intentionally oriented towards the comfort and relief of the astronauts. From all the loading platform bases and stations, about 12 in all (at Ganymede, Io, Europa, Copernica, and other large moons, bigger than planet Earth), trained and skilled, approved-certified labor, was almost always treated as an elite and privileged class among them. So, the Alpha-Base had bars and clubs and pleasure dens for all sorts of dubiously hedonist activities, all approved and also clean.

But 'boys will be boys', and it was sometimes a wild scene. The Elephant Hut was of a complexion that Mendoza found familiar, after he had settled in. It was also near his assigned quarters. He ordered a fairly strong alcohol-opiate mixed drink, and some food. The journey was quite long, he was exhausted, space-travel was different and had many psychological and physiological effects.

The fact that Marciel Janus Penieur-2 (the clone) had traveled aboard the 'Aerotica', along with the others (such as Alpha-Base Commander Wu), was not something Menodza had much concern about, during the voyage. Perhaps why he paid no attention to him. The real Marciel had not been charged, he was a very minor celebrity-entertainer, in Mendoza's world, and he wasn't even much of a fan (he had heard of him, though, in some off-hand way, as a passing faddish music act).

So, neither Deveroux nor Mendoza would have thought to seek out the musician, out of thousands of people, regarding their investigation for the PPP. Additionally, the clone, while in every way resembling, acting, speaking, thinking, dressing, eating and drinking, as any other human being would, seemed only to Mendoza as if another passenger, aboard the in-transit 'Aerotica'. They had little contact with each other, Al may have spoken with him at meals or while doing other things on the six-month voyage to the asteroids. But there was no interaction related to who this person was, why he was headed to the Alpha-Base, who he worked for, etc.

For many people working in deep-space in those days, their affairs were often private or even extremely covert, or handled for huge sums of money and industry back home, so it was common for such people to keep things to themselves. But with the apparent murders of Philby and Montrose, Penieur was at least now of some interest, at this point in their PPP investigation.

"Cowboy guitar player, big deal," Mendoza was thinking. *"Okay, he's got a clone. He didn't kill those men, and neither did his double."*

Mendoza paused over his drink, the alcohol-opiate mix was very relaxing and soothing, but like any drink, dulled his senses. The Elephant Hut was longer than it was wide, with seats and a long bar, video screens, and port-views that were telescope-targeted to reveal the endless spinning nightmare of Jupiter itself, about the size from there at that season of Earth's moon, never visible to the naked human eye in all eternity, but quite beautiful in its way.

The bar-gal was a sensationally busty woman, topless except for nipple-stars, and a long purple-reddish hair-piece that fell delightfully over her upper torso, and a colored skirt beneath, with an ornamental belt, necklaces, rings. Her named was Cindy. A job on a space-station was a real accomplishment, for her, and she knew how to handle all sorts. Mendoza seemed to her to be just a businessman. She smiled, he smiled back.

"Hi. I'm new," Mendoza said. "Only three days here. What's the real-deal on how to handle myself in this place? Clue me in, would you? Oh, I'm Al. What's your name?"

"Cindy," she said. *"Cid Bixi Mimim."*

"Yes, *Cid Bixi Mimim.*"

"Well, Al," Cindy started to talk, her voice low, moving near him from behind the bar, cleaning some glasses. "It's very simple. Alpha-Base is home, for maybe 1,000 people, day-in and day-out. Your standard space-station or space-travel protocol applies at all times, every where you go, and whatever you find yourself needing to do. Protocol saves lives, keeps everyone safe. Other than that, please enjoy. You get your business done and brag about it when you get home. We like to enjoy ourselves while we can out here."

Mendoza seemed shy a bit, for whatever reason. Maybe it was the partial nudity (fairly common). "I've never even been to the moon, only went up a couple of times in Earth orbit," he said.

Cindy laughed a bit and winked at him slyly. "You'll get used to it," she said, then went back to her work.

Standard protocol. His mind returned to the month or so of training and preparation the PPP had arranged for he and Deveroux, prior to their departure. There was a lot to know: safety-zones, emergency compliance, suspended animation states and required agreements, external space-suit learning curve, food and drink, how to negotiate a partly null-gravity environment, or a totally null-gravity environment (in some cases, such as docking procedures or passenger transfers), life-sustain topics like what to do if the air-supply shuts down suddenly in a small space of some kind, health-issues like psychosis, fears, stress, exhaustion, illness of almost any type, and also details about their specific trip, the locations and positions of the planetary objects, some basic science about Jupiter and the ships, and the schedule and arrangements for their return voyage back to Earth.

For Mendoza, at that point in his life, he wondered in a sad way if he would ever return to Earth at all, or perhaps die out in space, for whatever reason. Visions and memories of his kids, and Marta, back on the salty beaches of Belize, sustained him. And thoughts of his other wife, too, a wonderful woman. But he felt depressed. The investigation was not going well, they really had very little to go on, and it seemed as if the entire assignment would eventually reach only a hard, cold, dead-end, with little or nothing of value accomplished.

Why? Because Jupiter was a dark mystery, and would likely remain so.

Just then another man sat down at the bar. It was Marciel Janus Penieur. Or was it? The man was mid-height, not bulky or heavy, more like a thinly-built old tree, small, his skin tanned or naturally dark, bony, his features somewhat hawkish, his dress somewhat rustic or Old Earth styles.

"Can I have the Blue Moon?" he asked the bar-gal, Cindy. "Do you have that one?"

Cindy glanced his way. There were other men and women enjoying a respite at the Elephant Hut at that hour. It seemed an easy time, all sorts of business and activities going on, each to his task. The ventilation air-supply was cool and fresh. Al wondered, had Marciel been released from the holding cell?

"Blue Moon with Orange, or without Orange?" Cindy asked him.

"Without Orange," he replied. "I like it without Orange, please."

"Just a minute."

Penieur seated himself close enough to Mendoza that to avoid a greeting may have seemed hostile. "Hello, how you doing?" Mendoza said. "I'm Al Mortissimo, with Astro-Health. Glad to know you, always enjoy new friends in a strange place, how about you?"

"Howdy," Marciel said. "Just getting a drink."

"The Blue Moon is a hard wheat-ale, yes?"

"Yeah, it's like a thick wheat-and-barley or hops brew, lots of bubbles. If they use orange in it, it sort of spoils the original flavor, if you ask me, but they have two kinds. They're both good. It's also an old song, from the 20th Century."

Mendoza observed dispassionately that the man had not provided his name. "How do you like the view?" he asked Marciel (or the 'other' Marciel). He indicated the view-screen with the real-time image of Jupiter.

"It's a fucked-up planet, my friend," the man said. Cindy now brought his drink. Mendoza lifted his glass.

"Here's to Planet Jupiter," he said.

"Cheers," Marciel-2 answered back, and they both downed their drinks a bit, resting from chit-chat. Mendoza relaxed. Whoever the man was, it meant very little to him just then. Deveroux would join him in about two weeks, at Alpha-Base. Their basic approach would be to track down and interview various individuals, those who hadn't been killed, anyway. They would also be getting into some of the classified data-base sources, if possible, on related topics. They would also be reviewing whatever they could find out about basic J-Program management and decision-making.

The trouble was, all of the same information was available elsewhere, there was nothing new about any of it. So how could they serve their function with PPP, and source-out what was supposedly wrong with the whole J-Program empire? If they already knew? Had the PPP been duped, and the two partners been sent on a wild-goose chase, to uncover nothing, or only what they were supposed to find out, for high-powers and masters with agendas of their own?

Perhaps this stranger knows, Mendoza thought to himself in silence, sipping his own drink a bit more, enjoying the various view-screens about the bar. He glanced excitedly at Cindy's boobs, concealing his lust as well as he could. The woman may as well have been a super-model. They all were, for her job. He thought to ask the stranger what the women were like at the base, as if he possibly knew. As if something like that ever changed much.

Marciel-2, the plant-animal creation, or clone, hunched over his brew, silent to himself as well. If he understood or knew, that his other half was being held in a confinement-cell, on bogus murder charges, elsewhere at the base, he didn't show it.

"Wow, look at that," he said momentarily, lifting his hand to point at the view-screen with the images of Jupiter. Mendoza shared the view, and they could see for just an instant, on one side of Jupiter in a sort of halo-effect where the edge of the planet was framed against the deep indigo, there was a very small flare of white light, as if of some small or even tiny gaseous explosion had happened by nature's design, some random burning or flare-up from within the planet.

Relative to Jupiter itself, the sudden flare appeared very small. In reality, it may have been the size of Earth's moon. They could only see if for a short moment, on the telescopic-video view, and then it faded quickly, and was gone.

"Hmmm, some kind of flare or flash, I guess," Mendoza said. "Cool, I wonder what it is?"

"Mmmm, probably meaningless," the clone-man said.

"Probably, yeah," Al returned, and they went back to their drinks and the humiliating solace of Cindy's tits, bouncing a bit as she worked, just out of view.

CHAPTER 32: So Drink Up

As Marciel-2 was saying, chatting with Angelo: "No one knows, no one is supposed to know, that's why it's called a secret. The science-guys have speculated about it for even more than a hundred years, since the earliest years of the program. Is it hot? Is it cold? What's down there? Earth, and our little corner of the Galaxy, we're pretty small potatoes on the Galactic-scale of stuff happening now, and way-nothing for the Universe at large, you agree? The Milky Way Galaxy is mundane, really a shabby bit of work, just a basic spiral-form, a few billion stars, not even very exciting at all."

"Except for Cindy," Al joked.

Marciel-2 laughed, Cindy, the buxom server at the Elephant Bar at base, just ignored them. "Thrilling enough to perpetuate the species, I guess," he said.

"Fart-head," Cindy called back at them. "Perpetuate the species all by yourself for all I care." They laughed more in a good-natured way, enjoying their drinks.

"You look at Jupiter, and you think you see," Marciel continued. "You don't see it at all. This is the second largest object in our Solar system, second only to the Sun itself, and our Sun is a rather small star, not very impressive. But it's like Goldilocks and the Three-Bears, just right for life on Earth to appear and flourish, right?"

"Sure, sure, I always liked it," Mendoza replied.

"Nice and comfy-cozy," Marciel went on. By now both of them were on their second and third beverages. "So, the main idea is that we're not alone in the Universe, by whatever name, or god or deity, or race of angels, or alien planet-makers, or accidental design that includes this sort of thing, or even your dreams and notions that you even exist at all. Sort of like, a gentle awakening, for Earth's population, which may take thousands of years to work through, or even not at all. There were similar ideas about Earth's moon, when we first went there, and Venus and the other planets. Like, a message-in-a-bottle, from a million years ago. So, that's what they're hiding, that's the big secret, that's the big deal about Jupiter, if you ask me. What do you think? True? False? True enough?"

Mendoza hummed and closed his eyes, for a moment. "Well, I can see that philosophy, I guess. But, just because we found a big, giant, ugly planet, that no one could ever live on, near enough to Earth that 5,000 years of science and education could finally arrive us way out here to harvest loads of free hydrogen,I mean, just because it's there, it's no reason to conclude anything extra-ordinary about what we already assumed to be true about Earth and ourselves, right? What if it was just a stick or a brick, or a soda bottle, or a big giant flower, or a big giant snake or a giant monkey? See? I mean, in and of itself, Jupiter tells us nothing about Earth or Earth-origins. Its just a big ball of gas. So what?"

They paused, their philosophies like tripping balls. "That wouldn't be true of historic Earth religions and myths," Penieur-2 said. "All of the planets had associated gods and deities, and heroes and legends, long before anyone ever even dreamed of clearly looking at them truthfully, or accurate scientific examination, because it was impossible. The thing is, we didn't come to Jupiter to find out whether or not it was really and truly a Greek legend, or really and truly a Roman sub-deity, like Zeus, or Jove, or Jehovah. We wouldn't research that because it was all dismissed long, long ago, as empty legends and meaningless superstition, with all due respect. So we wouldn't look there, right? But those parts of the human mind, the collective human spirit and larger human venture, its all still there and very active, underneath, beneath the surface. For my money, anyway."

Like your bad-boy boner for this sexy bar-maid, in your pants, Jack, Mendoza was thinking. *Hidden underneath it all.* He enjoyed talking with Marciel, or so he thought. The man was florid and well-informed, and very enthusiastic and intelligent. But Al's mind was elsewhere.

He found it hard to concentrate on the clone's explanations, as wild and outlandish as they were. So he just listened a while, wondering what he'd be doing for the next few months, for the PPP, there in the Abyss. Cindy would be off her work shift soon anyway, and another woman just like her would take her place. And other men just like the two of them would court her attention and pleasures, as it had always been. It meant nothing, yet it meant so much.

"Big deal, dude," Mendoza replied to what the clone was saying, lifting another quaff of beer. "It's a legendary object in space, true, with a very long history of human thought and emotions and various religious temples and oblations and worship. Big deal. What does all that have to do with anything? You don't think it's going to explode, do you? So who cares? All we want is the hydrogen, and other useful materials."

Marciel-2 just sat quietly, then drained his Blue Moon ale. "Hey, bar-gal?" he said. "Cindy, dear? How about another?"

Cindy moved away from helping another customer further down the bar, towards them. "Another Blue Moon, no Orange? Your official limit is five, based on your biological profile from when you came in, mister. And you're on three already. Sure, okay, give me a minute." Then she went to get him his fourth drink. By now Mendoza had spent more than an hour just hanging out there.

"I bet you think I know something," Marciel-2 said. "You probably think I have some secret answer, or I know about the real conspiracy here, or maybe I have some special connection to some huge thing going on right under our noses."

"No," Mendoza said. "I hardly know you. But, man-to-man, I'd assume you have a curiosity, like anyone would. It is fascinating."

"Throw me a bone," the clone said. "Here's one for you, think about this. The deep interior of Jupiter is extremely high-density, very high mass-weight, hot, from the pressure, but not fissionable, like the Sun, not a nuclear furnace, just extremely dense gases compressed into a thick stew of pressure-cooker intensity. Boring? No one could ever go there, no living thing could survive, no man could voyage there in a submarine diving-bell, no ship could penetrate it. Even our X-rays and micro-waves and various scanners have trouble with it."

"Common knowledge, dude," Mendoza said.

"Yeah, and of course the thing is also very old, as old as the Universe itself, as old as the Big Bang. But Earth, with its life-forms, its trees and birds and horses and cows, would be younger than Jupiter, right?"

"Why would you say that?"

"Because of Earth's life-form manifestations," Marciel-2 replied.

"But why wouldn't we be equally as old, or even older? Especially human beings."

"Because life-forms on Earth clearly had some kind of a beginning," the clone said. "We can date this, roughly, but even the religions place the planetary start-up, Garden of Eden or early formations, life itself, as having some sort of origin in time, then some early period, dinosaurs, like that. In other words, because life on Earth, cows and dogs and horses and trees and fish, is a rare things, certainly for Jupiter, we can only assume it all somehow started, and will all somehow one day end. Evolution from fish and lizards, the details don't matter."

Mendoza waited for more from the man. Cindy brought him his drink, and Al also ordered another of his own, the alcohol-opiate drink he was enjoying, called a 'Punchy Drunky', for whatever reason, a local favorite. "Why am I even listening to you?" Menodza said, turning a sharp gaze at the duplicated man, so full of ideas.

"Because I'm amusing," the clone replied. "So, wait, there's more to this scenario here, okay? What I mean is..."

"What's your name, anyway?"

Then Mendoza was at least somewhat stunned, to witness as MJP, (Marciel Janus Penieur), the original, yet identical in appearance to Marciel-2, the clone, (who he had been speaking with), now walked up slowly towards them, from the bar-room entrance. They were truly identical twins, except for their clothing. The musician had somehow by that hour been released from the jail-cell where he was being held. He seemed upset, tired.

"My name is Marciel Janus Penieur," he said, now at the bar as well and grabbing a seat, but it was not the clone who answered him. "I play guitar."

"Our name," the clone offered as well. "We play guitar."

Even the bar-gal, Cindy, found this un-settling, and others gawked a bit. Penieur was familiar with the effect he had on people when appearing in public with his clone. It was, in fact, the basis of his entire fame and fortune as a musician, and of course it was a scam or joke, in a way.

But looking at the two of them, there at the bar, it was certainly the case, Gemini was with them, they were identical twins, down to their hair-color, eye-color, the facial features, with the same type of general physique, the same musculature and even similar veins in the forearms. The original was dressed somewhat differently, but in the same style (a sort of rustic early-period Americana, to go with the Country-Music). He reached out and hugged his clone affectionately, slapping him on the back like a brother, laughing a bit, they had been apart a long time.

"You're a fucking clone!" Mendoza said, also laughing, now that the actual circumstance was clear.

"Well, with any luck, if she goes for it," the clone said. "No two things are identical."

Cindy scowled. "Forget it, bozo," she said. "And him too."

"Two for one!" Marciel complained, but she just turned away. He'd heard the jokes before.

"Marcy! What the hell happened? Base logs couldn't locate you when the 'Aerotica' dropped us off!" the clone-man now was saying, also agitated, legitimately concerned.

"I was detained," his father said. "Don't ask."

"I was in-transit anyway, after Las Vegas, for half a year, on-schedule. Fun time. How did they treat you? What was the problem? I'm only three days here, and for two days it's like you don't exist on official base-records at all. I thought you were floating home like a popsicle or something. You were supposed to be doing shows at the moon platforms by now, I thought."

"I don't want to talk about it," came his answer. "I was framed for a crime, at the convenience of greater powers. What are you drinking?"

Things went on this way a bit more at the bar there at the Amalthea-base serving the greater Jupiter materials-harvesting program. Original Penieur had been released, as expected, having nothing whatever to do with any murder charges. Space-Authority Second Commander Le Van Ho, and his superior, Cyrolia Linsome Ee, could now proceed with solving the deaths, off-the-record, which management preferred.

It was Wu's way of preventing a lot of public scandal and diversion. And also a suitably time-consuming sub-plot, explaining the musician's imaginary motives for the deaths, as the primary suspect in the case, cooked up in exquisite detail, satisfactory at least to the deep-space justice system, for the time-being. The musician of course was not happy about it at all, but he could let it slide. He knew full well he'd be cleared of all charges in a few weeks or months. Unless he was needed to accuse of some other horrid crime, he supposed grimly. And no one needed to know he'd been paid a large sum in secret for the favor to the administration.

The Space-Authority also habitually changed his name, or that of others, on official records to make the process of convicting a criminal who didn't necessarily exist, of a very real crime, supposedly, for the good of all concerned, *'in the interests of justice'*. This also made things hard on the clone-Penieur, in terms of re-connecting with his original DNA (his dad, Marciel-1), when he arrived at the base along with Mendoza. *Cid Bixi Mimim.*

A day or so later, Mendoza could again reach his partner, Deveroux, by radio-phone, through the base communications center. Deveroux was still traveling with the staff and crew of the 'Down', the huge whale-hauler transport, far, far off in deep-space.

"It was just an accident," Mendoza told him. "I was at a bar. I must have been talking with him for hours, we drank a lot, too, that night, a day or so ago. Daniel, the guy is a clone! It wasn't even the same person!"

The radio-link made some strange noises and metallic buzzing sounds, linking the two men. "A clone? You were at a bar talking with a man you thought was the one they said killed the two Old Men research-science guys from the early years? And he was a clone of the real man who they said did it? Are you sure this makes any sense, Angel-Face? Seriously."

"I know, I know," Al replied. The two partners needed to stay in-touch about the PPP work anyway. "But then the real guy showed up, and there were two of them!! The way it worked out, neither one of them killed anyone. It was a frame-up."

Another pause. "Certainly works out well for the reputation of the Space-Authority," Deveroux said. "Okay, well, so what? You know, I've run into clones here and there along the way, they are nice people, like identical twins, that's all. They don't live long. The only thing is, we still want to know about the deaths of these men, the science-guys, I forget their names right now."

"Philby and Montrose," Mendoza added.

"I agree your guitar player didn't kill anyone," Deveroux said.

"Highly unlikely. But whoever did, and their actual motives, this relates to our work here, you think so? Understand me, the killings were probably not accidents, not just like that, not these two particular men, not at this time, and not in the way they died, and not considering the overall space-program appreciation for law-and-order, or safety standards anyway."

"Then I'll be getting to know the guitar-player, I suppose," Angel-Face said.

The radio-link strained over the great distances. "That would be good, until I arrive at Alpha-base," Deveroux said. "Just make sure it's the real guy, Al. Find a way to tell them apart or something, and find out what the guy knows, concerning the notorious and sleazy Doctor Wu, and why the program puts up with such stunning inefficiency and injustice. Shit, the duplicate was back on Earth was in Las Vegas or something, at the time, is that right? He probably knows nothing. Maybe the real Janus Penieur can help us track it back to its source, and we can get a lot more done, later."

"Too bad he's not a girl," Mendoza joked.

"Two girls, Al," Deveroux said. "Gemini." Then they talked more and the radio-chat line was then truncated by a signal. The moons of Jupiter danced between them all, there in their high-tech rabbit-tunnels, floating around like warm, fuzzy and furry experimental lab-gerbils, waiting for their next meal of nutritious green pellets.

CHAPTER 33: Musical Mafu

"Happy birthday, Mister Finches. Or is it, the 'other' Mister Finches?" Le Van Ho, Second Commander, Base-Security. 2,414-AD/CE, to Daniel Deveroux, upon his arrival at the Rosebud Stone/J-Program Alpha-Base.

The only guy among them who was already dead was Deveroux. At least, officially, that he knew of. It's a simple matter to set both of the Mario Brothers free, and the princess of the 20th century video-game, (if she happens to have frozen to death, a common fate there in deep-space). Truth can help there, for sure. And for Daniel, it would have been far less liberating in the pursuit of his own best interests, for the PPP to suppose an aggressive inquiry, regarding the Jupiter-program. *It's not a game, don't try this at home,* Deveroux thought silently to himself.

The 'Down' had at last navigated to the vicinity of the Rosebud Stone, but because of its size, was placed into a sympathetic orbit-pathway: i.e., moving into orbit around a large global object like Ganymede, the ship would match the general and far huger orbit of the main-trajectory center-of-gravity (Sol). The circumference of that circle might easily be calculated at a huge distance, but their navigation-computers and people, could fairly easily estimate how this was done. The Captain of the 'Down' (Commander Brandeis), worked with his staff and the Alpha-Base navigators as well.

After a few hours, a shuttle took Deveroux and others over to the base. He connected with his partner by arrangement. This all took a few days. *It's over,* Deveroux knew wordlessly. *We've arrived both in fine shape. All that remains is to solve the shattered mirror of complexities we've been sent to research, and then a similar trip back home to Earth.*

Maybe in another couple of years, Terran authorities would adopt to recuse the matter, and bring Deveroux back to life, legally, anyway. When Deveroux finally caught up with Mendoza,

Angel-Face had signed them both up for guitar-classes, taught by Peniuer. The classes would be taught twice a week, in a multi-purpose room, at one of the lower-deck quarters at Alpha-Base.

"Guitar classes? Really?"

"Yes, really. For you, we were considering a harp."

"Very funny."

This was part of what Marciel had been brought to the base to do. The program would never really waste his talents by only doing amusements and shows at the rest-areas (bars). So, between shows, he taught music classes, a much more efficient use of his time, and also very popular with the Alpha-Base staff. It was Mendoza's idea that this would provide he and Dev' with a cover while doing other secret chores for the PPP (as Alpha Health Group experts on a completely unrelated assignment).

They also felt that given the musician's tendency to get arrested, that Janus Penieur might hold some key or other that would move things forward. The Jupiter case had been almost two years in-the-making, the researchers were both sick of it, and now very far from home indeed.

A guitar class? How like Angel-Face, part of the reason the two men were excellent partners. He had a talent was for this sort of conflict management. A guitar class was innocent, harmless, broadly beneficial, pleasant enough, and included the legendary success of the 'hide-in-plain-sight' strategy, a very old gum-shoe deal anyway. But a guy like Al took to it very easily, second-nature to him. Deveroux had his own tricks-of-the-trade as well.

To sweeten the pot, Marcial Penieur, the guitar-teacher, was now at least on the stage for players in the bigger drama of what the PPP was seeking. Not so true of the Astro-Health Group, but on the other hand, another layer of falsehood that certainly may help or heal appearances (making all they would be doing much easier for them, anyway), was the musician's rare and dis-ordinary use of a true cloned self, albeit a 'non-natal' clone. This because the Astro-Health Group cover was specifically oriented towards DNA modification proposals for the men at work in space.

Peter Finches and Alberto Gonzales Mortissimo had found a home-away-from-home, and at least some lower-level sympathy for their real purposes. After all, no one ever really approved of any killings, and Philby and Montrose were both much-beloved. *We are the good guys, after all,* Deveroux's inner-narrator jabber-jockeyed his thoughts back-to-center.

Deveroux's friend, Professor Richard Podliakov at the University of Seattle, had a connection there as well. His father, Eldon Podliakov, was among the long-term, very early Jupiter Program science leaders, who was a resident of the Alpha-Base then. All of the rest from Vanessa's list, who still lived, (Reedly, age 86-years, spectrum-analyst; Nonnly Guitierrez, Jupiter mapping, age 73-years; Thomson, age 72, also mapping), were also rounded up for the same guitar class.

"Pretty smart," Mendoza said. "The disguise may just work."

"If they agree," Deveroux added. "Not everyone is eager ot learn to play guitar."

"Oh, of course they'll agree."

The Elder Podliakov was 90 solar-orbits in agedness. In those future days, a healthy man might live to a lifespan extended even close to 150 years. Even at its very earliest years, by 2,415-16, the program was only perhaps 120 years in-operation. But it was a new low for these priceless men-of-science to be considered assassinated, as they now suspected. Mort Philby and Montrose de Montrose would never know, and they cared even less now then they ever did during a long service to the space-program at many levels.

As a dead person himself, Deveroux felt a certain kindship with both of them. But as one still walking, the association was more with the robots at the base, or the Mafu-Forms. Not even fully alive. He wondered how the Marciel Peniuer clone could ever teach a guitar class, being essentially non-human. Or was he?

There were a few others in the class. Mendoza had worked with the two Penieur's to gather the various information needed to contact each of these men.

The others were people that any of those on Vanessa's list felt may also know what Deveroux and Mendoza were trying to figure out: the hidden-mystery deep within Jupiter itself, that had been hinted at by early research, and then suddenly hidden away, locked up, as somehow 'too hot to handle', many years ago.

"I have no musical talent," Deveroux told his pal. "I can play a harmonica, that's about it."

"All the better to learn to play a guitar, then," Mendoza said dryly. "Come on, old friend. This is an easy way in the door here, the point-of-entry. These are the people we wanted to find. I realize there is other work to do. But there is not even any ship leaving for the moons any time soon, I checked. You have to schedule that officially by permission. It's not a taxi-cab."

"The method goes to looking where we last understood we intended to find out what we already know," Deveroux said. "Again."

"Then why look there? Isn't that redundant?"

"I don't know."

"See? Start with ignorance."

"All right, Al, all right, fine," Deveroux said. "But let's not lose sight of our goals. We can talk to these guys all day if we want, in other settings."

"Not necessarily. This is more efficient and more fun, loosen up, you'll get gravity-sickness," Mendoza said.

"Ahhhhh!" Deveroux threw his hands up, angrily. They'd had enough, and there was no more to be had, until they turned the right stone over, in the right way. So, they'd pretend to play guitar (or take lessons), between bits of effort to convince anyone else at the base that they were actually working for the Astro-Health Group. For things to work, Penieur, the musician, had to be in on things. But Deveroux figured he wouldn't care much for it being a falsehood anyway, given the false charges against him. That is, 'sympatico'.

Penieur's guitar-class was scheduled ahead a week. They were to be held in a large multi-purpose room. At the first class, there were ten people, seven men and three women. Peniuer had a seat at the front of the room, with his guitar and an era-modern floor amp. The set-up produced a very clear-sweet guitar sound, acoustic-electric, old-fashioned but flawless in play-action and sound-tone, not really loud at all. He also had his song books, and so on. Each of the students had a similar set-up to work with, or brought their own. Some would use head-phones, some would be learning with no electronic at all, 'un-plugged' (a favorite).

Classes were two hours per session. Penieur went over all the basic guitar-playing principles: chords, strumming patterns and beat, or rhythm, lead-lines of flat-picking styles for melody, guitar-tuning and guitar-care, classic songs, vocals and performance, and so on. Some of the class already knew some basics. Deveroux was among those who knew very little, a slow-starter or late-bloomer for rock-n-rollers. But he gave it his best.

"Let's work through with an E-minor chord, then we're setting up a beat with that, a basic four-four rock beat. Then you'll alternate back-and-forth to the A-minor. But to simplify, for beginners, you can play the A-minor like this," Marciel was saying, as he demonstrated.

A small on-site video projected his left hand position on the fretted guitar neck, at the A-minor, so the entire group could see his fingers, on a video screen. At the same time, a 'tab-map' showed the tablature-diagram for the precise fingering. Marciel basically showed them how to play the A-minor exactly as they played the E-minor, requiring only two fingers, so it was very simple. Some of the old men's hands were weak, or their fingers arthritic, the knuckles bulging out painfully. The class tried to imitate his fingering as he said.

"See? It's just like the E-minor," he said. "So it's not a true A-minor. It has a somewhat different ring or tonal-quality, referred to as a 'diminished' chord. But it sounds great as a counter-balance to the E-minor, and it's a quick-start on what I want you to learn."

All nine students attempted the pattern at the same time, but completely liberated of any actual togetherness; in other words, a total mess. A cacophony of music and guitar strumming or plucked strings poured out into the large class-area. Chaos, ruined tones and obliterated actual music, with no regard of tempo or suitable mix, yet somehow a delight to Penieur. E-minor, then A-minor (diminished), over and over, nine guitar-players, each above the age of 70 years-old.

"Look at him smiling," Mendoza chatted with Deveroux. "He loves it. He knows its a hopeless mess and nothing could please him more."

"His guitar class looks very similar to some Alcoholics Anonymous meetings I've been to," Deveroux joked. But, after a while, Marciel's system started to make sense. One at a time, or two at a time, students would move closer to his spot, and play the same chords, only now he could add an electronic beat, and a simple recorded bass-line. So even at their very first guitar lesson, the Old Men of the Jupiter Program's early-era research teams, could pull off some reasonable blues-rock guitar bits, and it worked well. Marciel would drop in a fancy lead-line of flat-picking, it sounded great.

"All right!! Dig it!!" one of the guitar students let go a cheer. It was Nonnly Guitierrez. The old guy was having the time of his life, laughing. He had never done any serious blues-jam rock-n-roll before, or not much, even at his age. And the beat goes on.

"One and two and three and four, count the measures," Penieur was saying. "Boom-boom-boom and boom. That's it."

For the base management, and Doctor Wu's office, the guitar class was the smallest of potatoes, really just a blip on their general scheduling. The older workers were not by any means retired, it wasn't an Old Folks Home. They all still did very meaningful work, and kept regular work hours with regular assignments. But mostly they monitored various changing aspects of Jupiter's multi-level gaseous nature, for fluctuations that told other departments important details (like the movement of huge oceans of hydrogen at the planet surface).

The cover was working. Now they could speak freely with the Old Men, without attracting a lot of intrusion.

Peter Finches and Mortissimo Gonzales had business for Astro-Health that was approved and legitimate. Ships to and from the moons of Jupiter took time to arrive from points-afar. No one was really concerned for how they spent their time, and a guitar class was well within the sort of thing that kept up their morale at the Alpha-base. At closer review, Menuda Wu, the Alpha-Base Director and Manager for all deep-space operations for Jupiter, may have discerned from the names of those enjoying the guitar class, that there was more going on here. Since his return to the base, however, Wu had been busy with parties and his nudist lifestyle activities with friends. He had also been inspired recently to try out the famous 'Circulator' sex-toy for two, just as a lark, to see what it was like. A different drummer.

Between hours, during the second guitar-class session, Deveroux was able to speak with Eldon Podliakov. This man was the father of the friend at Seattle University. Eldon had been credited with basic research and ideas on some aspects of the artificial gravity used at all the deep-space bases and on the ships, a very demanding and complex technical discipline with astonishing new varieties of mechanics and energy used to manipulate or appear to manipulate gravity-like conditions. And then he also did years of mapping work for Jupiter overall, which he compared to mapping a kid's swimming pool from a moving high altitude jet-airplane, to determine the height and speed of each tiny ripple in the two-foot deep plastic pool, and the chemical composition, heat, and purity of the water.

"What is Operation Odyssey, Professor?" Deveroux later asked him privately. Eldon Podliakov frowned, squinting.

They enjoyed hot-chocolate between guitar sets. "How do I know you're the same Daniel Deveroux my son studied with at University of Seattle?" Podliakov asked him squarely, before answering. He was tough for an old guy, he looked like a small tree full of knees and elbows and a bark, too. "I will only discuss that topic, or any research of mine at Jupiter prior to year-2,360, with that man. My son's friend, Daniel Deveroux. I do not know anyone named Peter Finches with Astro-health Group. Sorry."

"But Doctor Podliakov, that's me," Deveroux tried to explain. "I'm Deveroux, I can prove it. I know your son, I saw him before we left Earth, at the University, Seattle region."

"That's impossible," the much older man said. "You're dead."

"Exactly, and I can prove it, too," Deveroux said. "That's how you'll know. I have the official death-certificate and body-disposal records."

The conversation continued like this a bit more, for that session. Marciel was trying to organize the study of an old Johnny Cash tune, which was at this era about 450 years old.

CHAPTER 34: Secret Science

"Cold, isn't it?"--Daniel Deveroux, Planetary Program Proficiency researcher, at Alpha-Base asteroid, year 2,415, to an anonymous resident, referring to a frozen ice-cream dessert served at meals.

So what's the big deal? What's the deep, dark secret? What was the shocking truth about the Jupiter Program they were searching for? As PPP investigators Mendoza and Deveroux were learning, the sad facts-of-the-matter, and the depressing realization they were facing, was much as Daniel's pal at the University of Seattle (the younger Professor Richard Podliakov, Ph.D. astronomy) had tried to warn him about. Planet Jupiter was a mystery, especially its deep-interior, and would likely remain so.

Research and science, even over many years, could only reliably demonstrate some basic facts: a huge ball of tempestuous gasses, a super-dense interior of incredible pressures by virtue of the giant gravity-well created by so much 'matter' in one place, and more than 60 moons, large and small, the whole mess whirling around and around at incredible speeds, with even the moons going in opposite directions to one another. And now, at about year 2,415-AD/CE, for the perhaps obvious reason that somebody 'knew something they shouldn't know', some of the very early program-researchers and science-guys were suffering mysterious and unlikely deaths, probably murdered.

Who would they 'rub out' next? Were the two PPP investigators on the same 'hit-list'? And why was the hidden information significant enough to kill people? It seemed at least likely enough, that something else was going on, than a shallow surface-examination would reveal. And whatever it was, it was just the sort of thing the PPP wanted to know, with their only legitimate goal being to help or assist greater authorities concerned with failures within the vast J-Program.

A second and third guitar class came and went. Some of the students really had no interest in playing guitar, and were hard to seduce into the music lessons. With proper identification, Eldon Podliakov eventually began to trust Deveroux (who was not really dead, only 'officially deceased' by legal-status, for his own safety). Jupiter-Program management hadn't quite caught on, that Peniuer's guitar-classes was somewhat a conspiracy of their own, with more than seven of the surviving early Jupiter-Program science-researchers attending.

"I hope none of these old birds are going to end up dead or targeted by some robot killing-machine, because of us," Mendoza intimated to his partner.

"I hope not, too," Deveroux replied. "There are other information sources as well, on some of the orbiting moon-bases, millions of miles from here. No one person has all the answers."

A dread that had followed them for much of the journey now seemed palpable, like a foggy fear. "Please, just shut up and play your harmonica, would you?" Menodza said. "This case is killing me."

But they were learning much more than just guitar-chords. Later, Deveroux was able to draw out the Elder Podliakov again, on some of the topics of interest. The old guy was cautious.

"It was never confirmed, Mister Deveroux," he was saying. "I mean, Finches. In our wildest dreams, back in the day, what we felt we had discovered was bizarre enough to both hide it away, or deny, and also to doubt completely. So, I don't want to mislead you that I actually know anything. But I think I can see now that the classified information is ready to be exposed. Too many lives are at stake, too many people hurt or harmed."

The old man handled the wooden guitar as he would hold a child. He wasn't much for learning to play, rather clumsy with the basic strumming and chords. His old hands were thick and wrinkled, strong enough, but not agile or fast. So he just strummed the instrument back-and-forth dully, like a practice-motion only. They were seated in the class-session room, near each other on chairs. Music filled the room, a cacaphony, a litany of lost-in-space novice guitar players.

"All right, it was classified 60 years ago, I understand," Deveroux said in a hush. "Can you give me a hint? We're working for global authority program proficiency to correct some of this."

"Sure," Podliakov said. "Sure. We thought we had found very large objects, deep inside the planet, that resembled mechanical things. That were not natural-shapes. In other words, we felt we had correctly identified very large shapes inside Jupiter, under incredible mass-density pressure, that were shaped as if by intelligence, probably extra-terrestrial alien people, ages ago. Happy now?"

Dev' didn't answer. He blew a bit on his harmonica, trying to match the harmonies of Podliakov's repetitive strumming: a C-major chord. They paused in a meek melody, as Podliakov's claim sank in. Other students were also practicing some of the lessons provided by Marciel Penieur for that day, (or his clone, it no longer seemed to matter which one was running the class). *Must be nice to have an identical twin,* Deveroux found himself thinking. *Even better than concealing my identity among the dead. Much more proficient.*

He hit a sour note on the harmonica and stopped. A solid 30 years junior to Podliakov, what the old man had just told him settled into his thoughts. Angel-Face was also chatting up the natives at the same time, discretely and casual.

"Okay," Deveroux said weakly. "Very large, giant shapes or machinery of alien origin, deep inside planet Jupiter."

"Something like that, young man," the older man said. Their conversation was not much overheard, but others in the music-class had by then comprehended who Mendoza and Deveroux represented, unlike the Alpha-base command structure, at least officially. *A global authority program proficiency.*

Podliakov continued. "It was long ago. Early in the program. We needed as much information about Jupiter as possible. My area was gravity, or gravitational theory. As the materials-program we now use for fuel and substances gathering was being created, there was a window of maybe five or ten years where research about Jupiter itself was intense. And this idea came out, but again, I caution you, not exactly proven, not perfectly certain at all. And also, this is classified. I am under orders not to tell you this. That's why I wanted to make certain you were, well, dead. Ha-ha. You know. I need to know that the man I came to see, or, vice-versa, was you. I can be hurt, too. I'm not invulnerable."

"And I guess there is the idea that the other two, Philby and Montrose, were killed by somebody guarding this information? What do you think?"

"I don't know. Do you believe it?"

"No," Deveroux said. "Our music-teacher here was who they arrested, and then released."

"Not necessarily," Podliakov observed. "I think its the clone teaching class today. But, yeah, he tends to get arrested whenever base-command needs a scapegoat. Works out. I think they even pay him, when he's on tour. An informal committed under-acheiver. A fall guy."

He cast an eye towards where Marciel Janus was working one-on-one with the man called Nonnly Guitierrez, also a rather senior space-worker (scientist), one of the Old Men. True enough, it would be difficult to distinguish the clone from the natal-born earth-man. Janus was showing Guitierrez how to hold the guitar properly while making chords. Other students worked in pairs.

"But, sir, please," Deveroux said to his older friend. "I can see where the proof of something like this is startling or shocking or whatever. Alien artifacts. Earth has known a lot of ideas about off-world ideas and peoples and even civilizations, for many years. Sort of old news, there. Not proven, but good information and research, planet-worlds with livable environments, water, appropriate size or heat. But why would anyone among our group here, from within the program, take the deadly step of killing anyone to keep a new rumour a secret? Murder, I mean. It still doesn't add up, the motivation is insufficient, if you ask me."

"It never is, is it?" Podliakov said coldly. "When is the motivation ever enough? Or too little? It's not like us."

So they let it go. Podliakov couldn't actually reveal the underlying causes of program failures, any accusations of killing anyone, or the exact nature of any alien artifacts inside Jupiter. Deveroux knew better than to push it. In a foyer-room beyond the main room where they were doing the music, Angel-Face was talking with George Thomson, another early researcher, as they took a break for some snacks, away from the others. Thomson was 73 years-old and had been doing exhaustive maps of the ever-changing complexion of Jupiter 40 years previously. He was laughing and eating a scone-pastry, with hot coffee.

"Montrose the Monster!" he was saying, to Al. "They called him that, he had a temper sometimes, the guy would get angry about statistical data and what he felt were on-going math-assumptions no one else agreed with. He said they made the same mistake for years at a time, and he never let anyone working on the same research forget it! Poor sucker! Oh well, happier now then before, I guess, in a way. You know. Home forever. Poor guy. I liked him."

"Yeah, well, but why, George? Why? It's sad. The space-program is never like that. That's not us. You guys are the best, team-mates, brothers. Why? To stop a rumour?" Mendoza also had a coffee.

"Face the music, Mister Gonzales, or whatever your real name is," Thomson said. "Face the music."

"It's only called music when everyone is playing the same tune, sir," he replied. "Isn't that true?"

They paused. The guitars nearby in the next room were a mixed bag of songs they could sometimes recognize. An amateur mess of practice and repeated exercises, all at once. They could also hear laughter inside and talking, and then the electronic prompter machines that assisted the lessons, beeping or making signal tones.

"Okay, well, let me ask you, privately, Mr. Thomson," Al intimated. "What do you hear about the so-called Galactic Posture group, and Operation Odyssey? Have you heard about that? Do you know anyone involved?"

"Well, it's the same type of thing," Thomson said. "I don't really know Operation Odyssey, I guess that's new. But the Posture Group, same thing, it only refers to Earth's long-term plans if any extra-terrestrials from other worlds ever happen to show up, or do anything here at home base. Or even non-sentient galactic events, like exploding stars. But it's a do-nothing group because it's never happened, and probably never will."

Mendoza closed his eyes again for a moment, humming to himself, gathering his thoughts. "Right, I'm trying to process what you mean," he said.

"They plan for things like what Earth will do if there is a First-Contact type situation, or what Earth would do if there was some sort of off-world attack, or what Earth would do if there was a big Galactic-core event, or a cause-and-effect relationship to a super-nova nearby. You know, like a billion-year plan. I guess if the Galactic Center started to go, or blew up. Or sometimes they go for topics like dark-matter and dimensional realms. To be honest, it's rather nutty and they seem to be a bunch of bored esoteric intellectuals dreaming of other worlds. So they map out Earth's so-called 'posture' for various scenarios. It's not a big secret. It just never happens."

"Is that good, or bad?"

"Good, I guess," Thomson replied. "Waste of time. I like reality just the way it is. Pretty stable overall. Except for my stomach trouble. Besides, why are you so concerned? What's your interest? Or just curious? " He laughed again and finished his pastry-snack.

Al lowered his voice. "I won't deceive you, sir," he said. "You know what the PPP is, and what we do. It's not a big secret, either, but we work under-the-radar."

"PPP? Yeah, somebody mentioned that," Thomson said. "I won't blow your cover. World authority program analysis. We're proficient enough, in my opinion, way the hell out here at the rim of the worlds, in the asteroid belt. Watch your back, though, young man. Management doesn't appreciate a lot of hostile investigation."

"We search where we last understood what we didn't know we already knew," Mendoza joked.

"Search well then, my friend," Thomson said. "Oh, and here's a rumor you might like: Operation Odyssey is supposedly on some kind of war footing. And we don't even have an enemy. How's that for proficient? See you around, I have to practice my fingering on that mutant's guitar tablature."

Then Thomson went back inside the main room. He sensed not to linger too long chatting with Mendoza, it might appear inappropriate. And of course it was.

Within another 30-minutes, Space Authority Second Commander Le Van Ho, along with two other armed security men, and his superior officer, Commander Cyrolia Linsome Ee, quickly entered the music-room, unannounced, and methodically shut down the entire group. Penieur was again out-raged.

"I have authorization for this music class here, officer," he complained very vocally. "Check your scheduling! We're approved!"

Those in attendance knew what to expect, and calmly gathered thier items. "I'm sorry, sir," Van Ho said. "We have orders to cease-and-desist of this venue use, at this time. You will please comply, and shut down your equipment. Your music students can go, but we need your records of those in attendance, please, sir."

"But why? It's only our fourth class. This is all under Rest-and-Recreation. It's an approved activity!"

"I don't know, Mister Penieur. You would please comply. I also need to know the activities of your Mafu-clone, please, what, where and when, if you could, sir. Thank you."

"Bull-shit, Commander! But yeah, I understand," the music teacher replied. By now everyone was breaking up from what they were doing, placing guitar and music items in the cases, moving out the doorway to their quarters. There was grumbling and moaning about the mix-up, but everyone cooperated. Deveroux and Mendoza could already see what was next.

The Security Commander, Linsom, a very beautiful and physically attractive woman, but totally devoted to her work, approached the two PPP partners. "Mister Finches, and Mister Mortissimo, with Astro-Health Group, please?" she said, as all the rest witnessed.

"Yes, I'm Finches," Deveroux answered. "What is this? We were just enjoying a break here to relax. We've traveled all the way from Earth only two weeks ago. We're both with Astro Health, this is my co-worker, Gonzales."

Cyrolia smiled, she seemed humiliated somehow, yet in control. "You need to come with me," she said. "Both of you. Base-Commander Menuda Wu has requested a meeting right away, with you both."

"What? Why?"

"We've already met, Miss," Angel Face said. "Doctor Wu and I spoke aboard the 'Aerotica', in-transit."

The Security chief paused. "Please get your things," she said. "I will accompany you myself. It's not far, maybe an hour from here by lift. Uhm, look, base-management knows who you are and who you work for, gentlemen, so please do not resist or disguise your research for Planet Proficiency any longer, it doesn't help things. You are global agency, well, so am I. Thank you. You have three minutes. Thank you."

She turned away and attended to the evacuation of the room, talking to Van Ho, her second. All Space Authority officers were known by their uniforms and gear, it was somewhat intimidating. Podliakov, Nonnly, Thomson, and the others old men, were already nearly out the door, glancing back. It was obviously a bust. Al and Daniel paused, they had a few things with them (the harmonica, Al's guitar in a case, a small computer-device, a long-distance remote activator, a small bag).

"Fuck," said Angel-Face. "We're screwed."

Then they followed the Security Command for Alpha-base Space Authority out of the music room, towards the lift-system that would move them into Menuda Wu's private domain, high above them all.

CHAPTER 35: Entering Wu's Lair

Cid Bixi Mimim, Deveroux self-reflected, as the Space-Authority officers, led by Security Commander Cyrolia Linsom Ee, the athletic beauty in her spandex-style work uniform with its clips and gears, buttons and weapons, (an impressive front for a woman charged with controlling others, often big, husky men).

"Cid Bixi Kiss My Ass," Mendoza intoned beneath his audible breathed words. They moved along the base corridors.

Little more than a cultural fad, during this period in the future, the term *'Cid Bixi Mimim'* was a popular greeting. It implied a sort of telepathic knowledge, that people for some reason felt functioned between strangers, to know one another's names, without speaking a word. A sort of 'Pax Vobiscim' inter-being salute to some unity between them all, yet informal, triggering a recognition. It began with some of the late-era Extra-Sensory Perception discoveries. *Spread the word,* Deveroux thought, in a dour mood.

And certainly anyone would excuse an impolite or hard moment, for the two PPP investigators. At this point, moving through the long barren hallways and lift-station elevators, towards the top-most 'head-quarters' where the base-manager, and manager for the entire Jupiter Program, had summoned them. Clearly, their 'cover' had been blown, the music class with the Old School Illuminati Science-Guys and their secrets, busted like a crap-game in a Chicago alleyway behind some gin-joint in 1923 (AD).

Linsom Ee, and two other security men, were on their way, with Mendoza and Deveroux free to attend, but not free to dismiss the meeting. It was a power-move of some kind for Menuda Wu. They more or less had expected at some point to be confronted about the research, it was clearly a threat to the power-structure. Probably nothing harsh, but their work was now rather like prisoners in a large penal-colony, trying to research the prison-system itself, and unwelcome to the task. But they were not prisoners, they were guests, even honored guests as representatives of the Astro-Health Group. The dishonor was less so by virtue of the fact that the PPP was a known-entity, not easily dismissed by such as even Wu. But it was a conflict, at this point, to be sure.

"This way, gentlemen," Security Commander Linsom Ee recommended. A doorway-gate opened automatically with a wave of her hand, implanted with a 'universal key' magnetic-emitter signal chip that could control any gate or doorway or locker or ship's helm-controls, locked or unlocked, anywhere within the entire program-system, including the moon-bases at Jupiter.

Angelo was also silent, perhaps considering what Doctor Wu may be busy about now, with them. Daniel was also doing his own psyche-up preparatory mental gymnastics: what Eldon Podliakov and others had been telling him was trying to settle and still within him.

Giant alien structures and forms deep inside Jupiter? How many? How big? Made of what? What was their purpose? Were these structures confirmed to exist? What other details did they know? Was there a way to determine the age of the objects? Why was the discovery hidden or concealed from public knowledge? How deep were the objects, inside the huge gas-giant planet? Were the objects like machinery, with some sort of operation or energy-programs? Or more like art-work, statues, obelisks?

They had already come some distance through the maze that was the inner-world of the Rosebud Stone Amalthea Alpha-base. Some of the corridors at upper levels, were passed through on small electric carts. Others had moving floors, like treadmills, they stood in one spot and moved forward by automated-motion.

Deveroux didn't like Doctor Wu, he was powerful and creepy. The elevator-lift moved upwards as the four of them found positions inside, it was large and spacious enough for even one of the people-carts. Attack, or escape from the Space-Authority officers was not even a notion. What would be, would be. Wu probably would not have them killed. Maybe locked up? They could only wait-and-see. What did he really want? The elevator lift hummed, there was a movement.

"Officer? Please, a word," said Deveroux. Cyrolia shifted her stance. "Just a question about our status, here, would you?"

"I am only delivering you to Doctor Wu's private chambers for an interview, Mister Finches," she said. "I have no other pertinent information. You are not being held in contempt or under arrest, you will not be harmed."

"What are our rights as far as any compliance?" Angelo now asked her.

"You are both under the authority of the base-command, as you know," she answered curtly. "Read the base-protocol manuals you received before you left Earth. We have complete authority out here, for obvious reasons. We could have you put to sleep and shipped back to Earth in suspended-animation, unconscious, but alive and unharmed. Human rights and freedom from fear is based on Earth-Citizenship Rights, obviously. Just act naturally. Be yourself."

She chuckled a light, giddy laugh. As if they would be tortured, or mind-probed. Program morale and camaraderie would never sustain such treatment, they were all on the same team. The elevator-lift rumbled a bit, then slowed to stop: Level 3, Level 4, Level 5, Level 6, through to Level 18. The wide automatic doorway opened. They exited together, none of the security men had weapons-drawn. There was another short hallway, and then a much larger gateway or entrance. This was Menuda Wu's lair, his hide-out or den. Few visited unless invited, and many were invited under the pretense of secrecy or personal favoritism.

"This way," Linsom Ee spoke. There was an external audio-intercomm. She pressed a button, like a door-bell. There was a beep-tone alert, then a female voice from within.

"Please identify," the voice said.

"Security here, this is Commander Linsom. Request from base-command, Doctor Wu has invited two men here, Peter Finches and Alberto Gonzales, with Astro Health Group. I have two other security with me. Request to enter, please".

Deveroux and Mendoza stood patiently, maybe a bit upset by the whole thing. The doorway was more ornate, more decorated then many elsewhere at the base, but clearly locked and thickly protected. Darth Vader's fotress? The gates of the Emerald City of Oz? The doorway to the sub-terrainian mines of Moria?

Another alert-beep tone. *Booooop!*

"Entry granted, Security," the female audio-voice said. "Please enjoy. A concierge will escort you inside, you can wait in the guest-foyer. Thank you."

The double-doors now automatically opened like the facade of some theater. They moved inside.

What they saw now was perhaps more obviously some kind of very extensive party or orgy, than the cold-dreary technical control-room they might have expected. There was a large main area, with marble decks, polished stations of real wood, pools and fountains, decorations and real plant-life. Tall view-ports with video-views to external points such as nearby asteroids, planet Jupiter, several distance-telescopic views of Ganymede, Europa, etc. Some of the large screens showed the ships, the 'Down' and the 'Aerotica', and at least one other Deveroux didn't recognize. The 'cell' smelled of spices and fragrance. There was music, from somewhere, with a sort of jazz-style beat, *boom-boom-boom,* steady and deep, rocking-roller. The security escorts moved to the foyer area where they could wait, taking it all in.

Then they noticed what was going on, inside, it could not be ignored. The larger hall was occupied by perhaps ten or 15 male and female guests. It was without a doubt some kind of orgy going on. Most if not all were nude, showing their athletic bodies pridefully with robes and towels, walking around with drinks, together or apart as couples here and there. Massage tables were in-use even as they entered, with people getting muscle work, sighing with pleasure. Some of the water-pools were steaming with hot vapor, and couples or small groups in each. There was food and drink, such as found at the bars for the regulars. Some were in at least a pre-sexual coitus, or love-making. Deveroux smiled, and counted at least five of the Circulator Group Sex-

Enhancement devices in-use or on-site. Others seemed involved in other activities. Mendoza had to laugh.

"All right! Party!" he said.

"Maybe not, Al, but I understand your motivation here to be sure. I should have brought a date. Be aware, there may be strong drugs here."

Then the concierge or escort-guide found them. "They're all yours," Cyrolia announced. "Cid Bixi Mimim."

"Cid Bixi Mimim, officer," their escort responded. Then the security team dismissed, turning away and heading back out the main entrance. "This way, Finches. If you wish to disrobe, there is a locker-room on this side."

They started to walk through the scene, other doorways were on either side. Some of the guests paused in their passions long enough to smile or wave as if to welcome them to a pleasure-paradise that had grown perhaps stale.

"Is nudity required? We didn't expect this kind of, uh, party, you know."

"Whatever makes you comfortable," the escort said. He was a strong-looking male, in a white robe-cloth, with a sort of wreath in his hair, and slippers, somewhat effeminate, but sexually attractive. He also had some kind of hand-held electronic note-pad, a guest-list application. Sensuality was obviously the standard, like a nudist camp. Deveroux and Mendoza knew the territory like any healthy men would. But this was business. Or was it? Doctor Wu was full of tricks, it seemed. What did he want?

They chose to buy a little time and disrobe in the locker-area, or prep-room. It was opened to the main room and not locked, with common hot showers, wooden benches, dressing areas, mirrors, toilets. They removed their normal attire, placing the basic jump-suits aside, then finding clean terry-cloth robes for each of them, slippers. They could also fresh-up with toilet items and water. Others came and went.

"Welcome to paradise," Mendoza said. "See? I told you. You died and went to heaven. Ha-ha. Good for you."

"I can get laid back home, Al," Deveroux said. "Aren't you a married man? Please, yeah. Please enjoy. Where the hell is he, anyway?"

"Jerking off in private watching alien porn-o," Al joked.

Then they returned together to the main-hallway party area. Wu was nowhere to be seen. The concierge was waiting.

"Professor Wu has asked you to join him in one of the pools," she said. "This way, if you please."

They finally found him, a ways off in a large hot-pool. Wu was already in the hot water, his robe and towels to one side. Two gorgeous women were with him, nude, laughing. One of the Circulator Sex Toy devices was at the side of the pool, the plush wrap-material idle and inactive, in small puddles of water.

"Please enjoy," the male guide said. Then he turned and walked away.

"All right, all right, welcome, Deveroux and Angelo Mendoza of Planetary Proficiency. Yes, yes, come join us. This is Celia and Teresa."

The two women said hello timidly. Deveroux and Mendoza slipped easily into the tepid water on stone steps. It was all very pleasant, of course.

"So you tried it out?" Mendoza said enthusiastically. "The machine orgasm device? Remember me, from Deneuri's ship on the voyage over? We talked about it."

"Ecstasy, brutal, animal pleasure, unbearable pleasure," Wu said. The two girls with him laughed. "Right, girls?"

"Works for me," Teresa said. She was dark-skinned, small boobs, thin red-brown lips. "Hey good-looking."

"Hi," Mendoza said, getting into the mood a bit. Deveroux also found it humorous. He and Vanessa had also tried the thing out too, back in San Francisco, before the adventure had even begun. Some things never change.

"Perhaps you're wondering why I've asked you here to my party," Doctor Wu said. He was not impressive in-the-raw, but it was clear he controlled much more than the conversation. The man was sly. It seemed like he could get whatever he wanted, whenever he wanted it, however he wanted it, here in his kingdom. A dictator with an appetite and lust, to be sure. Not the best for transparent inquiry into his failures.

Deveroux covered his privates with his hands until his humilty was concealed beneath the warm, dark water and steam.

"Seems okay," he said. "Nice. Thanks for the invite."

Doctor Wu had a drink of some kind by the side, and sipped through a straw. There was a fragrance around the pool, and flowers here and there. "It's a going-away party," he said. "For you."

"Oh? I'm leaving?"

"Yes, you are, Mister Deveroux, and soon. And your partner as well. So, for now, let's just chat and enjoy some pleasures together here in my little world, before you go. You like sex?"

The two women with Doctor Wu laughed, giggling. Teresa reach over to Wu as a lover would, and they embraced and kissed, her hands reaching beneath the water towards his center. Deveroux had to wonder if the base-manager was in complete control of his actions, the man was overly of a desire, loose, perhaps drugged, hardly the strong, dedicated leader at all.

His comment hit like a lead-weight, for Deveroux and Mendoza.

"You have two days, gentlemen," Wu said between stroking and fleshy touching. "I'm having you placed in a sleep-state, and you will be deported back home."

"But why? You can't do that! We're on official business here! Our visit was approved. Planet Proficiency has clearance at all levels of any global authority for our work! I formally resist this action, sir! The deep-sleep transport is only for emergencies, we could be harmed, it is not regular at all. You will hear about this from PPP, and the authority back home. Your programs here at Jupiter Resources are under investigation for numerous failures and corrupt practices." Deveroux was livid.

"Corrupt practices? Like this?" said Wu's assigned sex-play partner, Teresa. She then began a slobbery and sensual fellatio with Wu, as if to mock them.

The hall was filled with music and sighs. *Cid Bixi Mimim.*

CHAPTER 36: Big Secret Part Two, Three and Four

"Apogee, perogee, whatever will become of me?"
--Marciel Peniuer, travelling musician

True enough, Marcus (that is, Marciel, or 'Marcus' as his friends called him), had never been invited to such as Dr. Wu's party, or to his private rooms. But both he and his clone-form, had normal enough sexual interests. It was of great interest also to most anyone at the base, approved or disapproved of orgies, era 2,418 AD/CE, what went on. For Deveroux and Mendoza, it represented a voyage home, and a 'partial death'. Wu was swift and brutal: they were being deported, and he didn't need a reason. Once the sleazy Jupiter Program manager had a clear understanding of the direction Dev and his partner were heading with their research, he now seemed poised like a hawk on a branch, to end their work. This alone showed they were on the right track, but clearly, Wu's power-base interests were elsewhere. Very elsewhere. He had something to hide.

"Who do you love, right now?" Marciel asked Deveroux, later.

"Hmm? Oh, my jesus, I don't know. You? These bar-gals? I have a woman back home, you know. Myself, too, I guess, somehow."

"Tell me about her."

"Oh, she's confusing."

"True, true. You know, in the case of my double-identity, part of my own plan was to teach him to love. Do you think it's possible?"

"Of course, why not? That's easy, isn't it?"

"Depends how you take it, I guess."

Daniel was at the end of his hope for any reasonable conclusion to the investigation. This didn't mean he would pick up a weapon, but he had picked up a long-distance secure line to Earth and that very woman, Vanessa Signo.

Where do we go when we're lost, inside, outside, up or down, high or low? And where does she go? Are we not there now? Poor little 'hope', the last to die, for being so intent on what she always knew was true. See you again? Me? Soon? Woman, the wound that never heals. He wanted not to need her.

He communicated with her as he had earlier, the devices were only personal links to a much larger and more powerful radio-telescope radio system that connected Alpha-base and the ships to Earth. 'V' was 777 million miles away. And, yeah, she knew other men.

The line crackled. "Don't worry, Daniel," she said. "We'll get you out, somehow."

"His motion is illegal, Vanessa," Dev' said.

"A law unto himself. Translation: lawless," Vanessa said. "We knew this already. And don't be such a big chicken. The deep-sleep states are perfectly safe. He will probably only employ the method, if you're resisting. You sleep like a baby, for about nine months. I may be able to get you a wake-up call, earlier than arrival home. He sensed you as a threat. And, also, my dear, what was it you were onto that caused him to deport you? What did you learn?"

The line crackled. "Everyone hates Jupiter, and the feeling is mutual," he joked. "Jupiter and its children suck."

A long pause. "Surely you're not suggesting the planet Jupiter may have feelings and emotions of its own, like a person would?"

The line crackled. "Now that's stupid, isn't it? But I do."

The communications were secure enough, a dedicated line, based on the PPP authority. And Deveroux was right, his agency was very strong in terms of getting things done, including the defense and protection of any field-agents 'out there', such as himself and Angelo. The other cases they had considered for those years of their lives, had included looking into deep-ocean fish farms, and a war or two. Those cases had the advantage in his own thoughts, of performing their labors back home. There in the depth of outer space, things were not so easy. So they both knew,

PPP could throw her weight around with this one if she wanted to, but Wu would prevail.

"Get stupid," she said. "Sometimes our natural state of ignorance really is our bliss."

"Vanessa, if I don't get back alive..."

"No, no, you're already officially dead, Agent Deveroux. So, you never, ever, ever, will, really get home, at all, I mean, right?"

"I will sit at your feet, woman and adore you," he said. "The Big-Secret, Part One, looks like this: some 40 or even 60 years ago, these old science birds very seriously felt they had identified something of an origin other than nature's, deep inside Jupiter. Giant machines, perhaps best describes the data."

"Yeah, we knew that," Vanessa said by the private-communication radio, Alpha-base to Earth (Vanessa was in New York). "Part Two?"

"Big Secret, Part Two, Part Three, four and on. The Galactic Posture Committee has been preparing for a war or serious conflict, with the supposed far-distant creators of these machines, for some time," Deveroux told her as faithfully as he could. "They are creating a large fleet of war-ships, for battle and conflict, destruction."

"We knew that, too," Vanessa said.

"Great," Deveroux said. "The battle-ships will be piloted by genetically modified clones, or what they call mafu-form soldiers or pilots, mutants. The whole thing is choreographed to a timeline schedule, measured in eons. The Quantuum-Mind space-cult supposedly is their mind-link to distant alien beings, connected to the deep-interior machines of inner Jupiter. That's how they decided on the point-in-time when they expect it all to..."

"Of course!! At long last to reveal that Jupiter has a mind and feelings and a heart, and a life of her own, that the planet is alive and has a little house of her own, with a picket fence, and moon-kids, and puppy-dogs, and smiles with loving thoughts," Vanessa said. "And that planet Jupiter cooks great linguini-spaghetti, from a recipe she got from planet Saturn."

"Uh, no," he answered. "We don't exactly know that."

"Let me write this down," Vanessa said.

A long pause. "Well, just, for crying out loud, Vanessa. You knew, you said. We knew. The agency."

"Not me," came the link. "Someone with a bigger brain, I guess. Not me. The agency, maybe. Some other source."

"A sleeping giant they think, is waking up," he sent by his voice, that slenderest of ropes. "Sleeping inside a giant."

"That way lies madness," Vanessa said.

The radio-link buzzed across the infinite-spaces between. They talked more, mostly it seemed a jest to them, as individuals, to believe themselves a small part of such a drama. Work for the PPP could indeed be depressing. Here was this huge, expensive program, with many lives involved, and much benefit and blessing to all, salvation for much of Earth's energy needs. PPP could indeed assist, the agency's main job was non-biased truth, applied to various problems. How hard is that?

But the so-called human factor, the Doctor Wu factor, fear, lust, greed, selfish intentions and motives, meant that even basic facts and figures or simple science-data, were easily over-turned, stood on their heads in service of new masters. And some of the giants were cruel, it seemed so, one of them being known as 'love'.

Vanessa was not passionate about Deveroux. He was an attractive man, but modern life was so busy, and they enjoyed their loving as a casual affair, not serious, not intense, just pleasant and fun. But her mothering instincts tended to his sympathy, for all his suave-coolness, the man was just like everyone else, and pressures and pain, and especially loneliness, wore on her lover-friend like sand-paper on a heart of balsa wood. A war from within planet Jupiter? Well, it always something, isn't it, sweet-heart?

Where do we go, when the pain is like that? Where does she go? How much strength or faith is needed? Is there ever enough? He felt like a pawn in a game played by a kindly old god, moving him to here or there. And a pawn cannot see the bigger game, to be sure.

Deveroux and Mendoza were told more formally of their status, a day or so passing. They had only been guests at the Amalthea-base by that time, for about 80 days and nights, of which there were no naturally-occurring light-cycles, only the artificial life-sustain at the base or on the ships.

The messages were delivered by electronic private computer conduit: *BASE-ADVISE: circuit-schedule Rosebud Amalthea residents Peter Finches/Alberto M. Gonzales, (aka, D.Deveroux, Angelo Mendoza) Day-date A-42, official transmit: Office of Transport and Host-status/Security-Authority C.L. Ee. Guests scheduled for unconscious transport return to Earth, via high-speed transport-ship, 'Aerotica', (R. Deneuri); please report to up-load docking platform at the Level-Four waiting-prep arena by A-44, hour 1120 or earlier, with personal effects. Guests will advise/inform receiving parties, Astro Health Group, Program Planetary Proficiency agency, other. Msg.-delivered, approved-certified. CBM.*

Where do we go? Deveroux mused. We go home, of course. And it wasn't as if they weren't ready. Space-travel was tedious and full of discomfort, as he had thought. His body was growing weak with an odd kind of fatigue, by the time he arrived home, he might not even be able to walk normally, or Mendoza either. Only to learn or prove what other powerful people already knew, but wished to conceal. How often was that the case in their work?

There were other details. The five Old Men, the science-guys who legitimately cared, loyal Earth-men and program participants, were now a 'conspiracy', in the minds of some. Following the bust-up at the harmless-enough guitar-music class, it was business-as-usual for everyone involved. Gutierrez, Reedly, Thomson, Podliakov and another man, and a few others, were considered a 'black list', now. Deveroux also felt burdened that he had in part caused that. Not to mention the fates and families of poor Philby and Montrose.

"Who do you love, right now?" Marciel asked Deveroux, again. Angel-Face had joined them, along with Peniuer-2, the identical clone (rather like sitting and chatting with identical twins). They found some hours to rest, in Deveroux's private chambers/room. Marciel and his clone were not young men, they were both withered and knarly-looking, thin and boney. The clone was over to one side of the room, looking at a monitor and chuckling lightly to himself. Angel-Face had cleverly found a way to make a hot-soup for them.

"Look at this!" the clone was now saying. "It's the system news-alert. Something they're calling Operation Odyssey is ramping up for supplies and warehoused goods, starting today. They are shipping in a bunch of stuff, too, including people! I wonder what they're up to?"

"Real people? Or temporary people? Everything connected," Mendoza said. "Have some bread with this soup I made, try it."

"That's their war," Deveroux observed, now looking at the monitor with the clone. "Operation Odyssey."

"A war with no enemy," the original Marciel said. "Great idea, we can just all fight ourselves!"

"There's only us here, and I don't want to fight you," his duplicate said. "Hanging here like Christmas ornaments in a giant tree full of Chinese lanterns. Isn't it true an insidious enemy would pit us against each other, cause confusion, brother-against-brother, create fear and suspicion? How hard could it be?"

"I personally feel the space-religions are not helping," Deveroux said. "It's all well-and-good if they have esoteric beliefs. But the people I met, they truly felt that far-distant sources were able to move accurate and truthful information about the goals of the Universe Masters or whatever, regarding Jupiter and our little solar-system and Earth. Super-minds, super-beings. Something oppressive about it. Operation Odyssey is not happening all at once anyway. It's been in the works for many years. These choices were made long ago."

They agreed on a few things, amongst themselves. As it looked certain that the PPP was being ex-communicated (deported) by Doctor Wu, they wanted to avoid harm or accusation against the others, the Five Old Men. If they were also sent home, it may certainly be a good thing, they were old indeed.

Continued work in space, the stress was obvious. Even with a 150-year average lifespan now possible, at that era, the effort was perhaps pointless, and their endangerment worse for the PPP's diswelcome.

So they agreed to protect the older science-guys and their secrets, as best they could, and Vanessa's PPP authority-link agreed. They also agreed that although they were 'black-listed', they would remain loyal to the basic needs of those involved in the Jupiter space-program in general, rather than court disaster, as rebels or antagonists. The so-called 'conspiracy' was nothing of the sort, they only meant well and to do what was right in the context of their obvious roles. The conflict, if it ever even happened, could wait. Fear can wait. Death can wait.

They did their pow-wow, in private. Various sources of information trickled their way. *"I wonder what it's all about?"* Deveroux could only ask himself. It was only with a small awakening jolt, that he realized sadly that he could not answer that question, and probably never would.

CHAPTER 37: The Fearful Song They Sing

"I'm afraid to go sing my own songs," Penieur told his double-self.

"Why? You're okay, right?" said Marciel-2.

"I think they plan to blame me for the actual placement of the alien-artifacts deep with Jupiter, eons ago," he said sadly.

His clone paused, he was working on the tuner-pegs on his guitar.

"Well, was it you? I mean, I may be just a clone, but, did you do it? Did you somehow reincarnate from some ages-past dealio with a bunch of anti-human freaks? You should have told me!!"

The musician paused. "Well, I hope not, hell no. I never did nothing like that, that's a lot of work, expensive, too. Why the hell would I?"

The other one went back to work on the tuner peg, for it having moved off its center-point.

It had been many, many years, even hundreds of years, by then. To awaken to war is one thing. Earth's many children having dawned upon so many wars, that times of peace were forgotten, but not entirely. As early as the mid-1900's, the Age of Aquarius groups, and the various world governments, had felt fairly confident that it was at least plausible to consider the many UFO observations as significant enough, that plans to wage war against any off-world beings, were laid. If the global Earth powers-that-be understood an even highly unlikely superior-force threat of some advanced kind or other, then they would, and did, create the concept of defending the planet, in whatever way seemed best to them. Rockets, missiles, astronauts, telescopes, observatories. But, that was then.

It was understood just how unlikely such a conflict was. Given the immense distance between stars, so much further than the distance from Earth to Jupiter as to be unimaginable, why bother? Exploration of course was a motive, but Earth was of such little interest as far as Galactic affairs may or may not have been concerned, that as a destination it made no sense. Why? Galactic Unity? Inter-planetary overlord world-thieves and slave-keepers, or oppressive controlling powers, mad for some rare, dwindling resource, like water?

But the conspirators were telling what truth they felt they knew. Operation Odyssey was an extension of those old fearful ideas. Even now 400 years past the 2000-year benchmark, if the research was confirmed prior to any damages, it was still a terrifying idea.

The needless preparations looked like this:

For about 20 years, certain industrial societies and unions, along with the science-community and space-industry, had developed plans for a unique fighting force for work in the deep, as an anticipated response to whatever was truly supposed to happen eventually at Jupiter, 'from within'. This all took place, as far as any building and construction, under the authority of what remained of Earth's military, including all nations, and even Earth's so-called ultimate powers. Typically, the whole thing was concealed.

By predicting the future of Jupiter's interior contents, the leadership felt they 'knew something evil was going to happen'. Space-religions, such as that led by people like Menima Pearl, though doubted as merely superstitious, became powerful conduits or channels for 'distant thoughts'. And not true thoughts or happy thoughts. So, they had started building the new ships and new defense means, figuring to attend some bothersome conflict, near or around Jupiter's orbital path. And yet, as time passed, it seemed literally to be the case, there was no 'enemy' at all. And if there never was, so much the better, no one really wanted to obliterate all their other work anyway.

An entire high-tech Earth armada, with no enemy.

The layers of conceit and arrogant reasoning from fear were deeper still. There were assumptions about the origins of mankind itself, references to deity or god or gods, angelic types, creator beings, who for some reason from long ago, had an interest in planet Earth, and its people. Shocking as it may have seemed, some kind of reunion was suggested, with helps or assistance offered in peace, perhaps, for our struggling world, here at the corner of the Milky Way Galaxy, never even in her known history to have seriously assumed she was not but alone.

What defense could there possibly be, or what sort of successful organized war in space? Once again, Mankind would give it a try. Whoever they were, if they showed up, without a doubt the choice had been made to fight. They were strangers. They were much more advanced. They had superior technology and knowledge. They may be creepy or have sinister motives. All the old fears, insects, lizards, flying dragons, invisible vapors, and a few new fears as well.

Twenty years ahead-of-the-curve, a blink of the eye in celestial time, about 2,000 of these 'new-generation' war-ships were created, for work in space. These were more like the 'Aerotica', smaller and faster, than the gigantic whale-hauler, the 'Down', or the 'Ferrous-2'. The planners assumed other details and fearful scenarios, based on other details, based on other supposed information that was supposedly true, yet hidden, fearful, and dreaded. If the interior of Jupiter held other types of ships, space-vessels, perhaps robotic, were they still vulnerable to normal physical limitations? Could they still be 'shot down'?

This is what these 2,000-plus flyers intended for. Fighters. Given that even the 'Aerotica' was very large, perhaps three or four football fields in length and width, and it was true, they would not be much use, if in-flight qualities such as distance and speed, were not sufficient to their task. They wouldn't move to Jupiter directly from Earth, and instead were even then being stocked or warehoused at various locations in the Sol-System, where fuel, control-management, crisis or attack commanders, and other war-based technical facilities, were waiting. In a way, aspects of the Jupiter Program as a whole had been discretely adopted as space-military, without much fuss or resistance at all.

Maybe it was because fears of alien-attacks were utterly baseless and irrational.

Known as 'Solos', the space-fighter vehicles resembled warlike human art-forms. The designers seemed more impressed of the symbolic, as if to frighten or intimidate the intruders. So they were sleek and beautiful, with designs like birds-of-prey, and fancy art-work, which most space-ship pilots would laugh at. Artistic beauty served no purpose, no function, in most space-travel ships. Yet, beautiful, outwardly, stunning even, steel-and-polish, exotic metals, rare engine-types at the rear and underneath, bold designs and florid flaming wings with rapacious claws for melded embossed etched lines on their sides. Dragons, vipers, snakes, winged Valkeries.

It was the new old way of thinking, that the attack was also symbolic, or moved towards the mystical.

Inside the space-fighter Solos, however, a different story. A perversity was at work, also long-planned. The so-called pilots, these OO-Solo Fighters, with their various skills, were replaced by 'clones', or the mafu-types of life-resembling 'beings', much like Penieur-2 (Marciel's clone, the musician). Even more hideous, the mafu's were regarded by executive order as non-human, so the clones were thus expendable, with the toasted-glory of the kind-hearted leadership, offered like a meaningless virtue, that this would 'save lives'. Temporary men.

Genetic-experimentation, DNA manipulation creating generations of non-human 'clones', both natal-clones and the secondary type, were followed for science-protocol. Any ethical concerns were swept away. The ships would be controlled, operated and navigated, by these 'meaningless blobs of desire'. They could also think, and feel, and know, at least for a time. Disposable human beings, created for destruction after use. Cannon-fodder to the stars. Pathetic parodies of normal human beings, destined only for disposal. And those were the good guys, the Earth forces. The pilots were known as Solo Men, for lack of a better term, or some called them, 'mafu-men'.

Whatever genius was at work, it all continued like a long, sad march to infinity. It wasn't hard, any longer, by that time in history, to conceal even very extensive labor efforts. Industrial plants and well-paid workers were organized at various locations or cities and nations, back home (Earth). The ship-designers and technicians knew their tasks well, from much other previous work in space-travel design and systems. They knew the power-sources and thruster-engines (the Huschcroft Ion Accelerator was still favored). The high-speeds navigations had become mere programmable computer applications, the stress-level hull structure materials and metals, the communications and flight-means, were all well-known technology.

The Solo Men were trained and prepped apart from the construction of the ships, with more than 20 years to invent their bizarre roles or duties. A casual observer, reviewing a training camp for these soldiers and pilots, somewhere back home, or at the Moon base, called Planet View, may have assumed it was only another military-style business. The Solo Men needed only to pilot the ships as if they were robots themselves, only in this case organic, similar enough to flesh-and-blood humanity for highly skilled functions.

Thus, Operation Odyssey operated secretly, for many, many years, building towards the day-and-date provided by the mystics of the space-religions within the Jupiter Program, coupled with so-called science observations. Was any of it valid? Like bored and lonely kings and queens, often wealthy beyond the dreams of the common man, leadership at the level of Global Posture related to Galactic beliefs in other worlds, had made choices, and more choices, and other choices, commanding their powers blindly, only to hold onto what tiny grasp they had on the best interests of billions of people. Only to hold onto fear, the bottom-line of all their power.

What they were doing was evil, and they knew it. But they had a good excuse, after all.

Marciel-2 dug with pliers again into the hard wood of his guitar's tuning-head, the part above the strings at the top where a player moves his hands to form chords. The tuning peg could be removed with a screw, and then dropped out, leaving an empty hole, once the steel-bronze strings were removed first. But for this problem, the wood itself was blocked by oily grit and dirt, and corroded washer-rings. The collected dust gathered there on the guitar, during his travels to Las Vegas and other places for tourists, back home.

He dug into the wood with the pliers, to pull out what was stuck, if he could. It was an older guitar, acoustic-style, a classic Martin, very rare by year 2415 AD/CE.. But they were low-tech things of wooden frames and brass frets, shaped like women, and needed care from time-to-time.

"Okay, Marcus, okay," he was saying. To see the two of them together was like spending time with a fun-house mirror, which also walked and talked. "We call that guilt. Does it really matter if you were somehow involved in the beginning of the Universe or whatever, I mean, you personally? That's a good one, Marcus! Alexander Graham Bell sends his greetings."

"Well, of course not, Number Two," he said. "But if I happen to accuse myself, I might spend too much time trying to understand it, see? Not like they care. It's just handy, convenient."

"Not for us," his clone said. "Why don't I just play your guitar, instead of this one?"

"What's that? Why?"

"Oh, the tuning peg is screwed up," the clone answered. The two of them were closer than brothers, one might say, and genuinely friends. Marcus often felt sad about the clone, sort of like a strange good-bye. The ultimate ego-trip. But, at the same time, for a musician, a good career choice. "I pulled it out okay, but the little spiral-thing that rolls the tuner, is off-kilter. I can either put it back the way it is, or see if I can change this one so it works better, maybe with a file or small knife."

"That's a valuable acoustic guitar, my friend, please treat it well. No music stores up here to buy strings or tuner pegs, either. Two years for a shipment from Earth for something like that. And a Martin! My gawd, don't damage it!"

Marciel mused restlessly about his claim-to-fame, and why it was that good men like the two PPP agency representatives could be pushed around by half-assed dictators like Doctor Wu. And of course, he knew. Marciel was just a musician, a nobody. But everyone at the Alpha-base and beyond, were curious about current affairs, distinct from normal operations.

"I'm afraid to sing my songs, Marcus," he said sadly. "Isn't that the saddest thing? Fear of my own opinion, my own poetry, my own words."

The other one there kept with his effort on the guitar, an old hand-made variety, with inlay. Guitar players know, a Martin guitar is the best, for an acoustic. There was a subtle competition between the two. Like a star-born Gemini, neither very distinct from the other. Yet most people could tell them apart with a simple external view. "Yeah, sad, I'm crying," the clone said. "Boo-hoo-hoo, poor pitiful me."

"Do you hate yourself, sometimes, Marcus?"

The other paused. "No," he answered, smiling. "No. But I don't like to suffer."

CHAPTER 38: Beautiful Truth

"And now he's a sweet, saintly individual,"--Anonymous

If Jupiter wasn't a wisdom, of some kind, than what was it? If it was not there for a reason, a cause that brought it to manifest in the Sol-System, such as the heating and cooling gasses of the early inferno that somehow created the Universe, by appearances, then neither the Earth was a wisdom, nor its people, by virtue two cosmic objects so astonishingly different. Or was it? What does one tell us about the other?

Earth the greater (in her type of experience), Jupiter the 'unintended', yet side-by-side. The necessary, flagrant good humor of an inhabited, populated world, (home of the legendary Hebrew National Franks and Wieners brand of quality kosher foods). Take your pick and call it home.

But not maybe only a home for men, (translation): an alien-humanoid distant cousin or awareness like unto our own, had maybe already arrived, even very, very long ago. Zeus, Jove, and their son Agamemnon, Grecko-Roman legends and myths, Thor, Hercules, and Mars, Mercury, Neptune, with ancient roots in Italy, Greece, and Europe. The anthropomorphic Man-God, *(cruel, cruel, that one, like Thadeus La Viro Vente El Amore, El Supremo)*. And so to recognize, even in themselves, the megalo-maniac grandiose, ego-driven, power-tripping worship-conceit of the Tetragammatron of self, including every vile and disaffected insincerity that would pass for passion, but never much for poise, or grace. In other words, it appeared over time that Jupiter, and the rigors of the work in space, had driven some among them mad, or irrational about their decisions and choices, and motivations.

Menima Pearl was the top-nominated lunatic candidate for her supposed connection to 'alien minds' that all the rest would fear, thinking there was truth to the rumours, and then choosing. Program manager Dr. Menudo Wu was runner-up.

Amakmid knew what Doctor Wu, base-commander, was suggesting, as far as his work for the Large Globe Object Music Society. Or, he had the general idea from Wu's earlier request. Wu was sly, he didn't propose a thing lightly, or even directly, but instead simply requested information, like the space-bureaucrat he really was. The LGOMS (Large Globe Object Music Society) was strictly public, as far as the Jupiter Program was concerned. It was an art project, basically, financed by Earth sources, purely academic, the attempt to create a musical symphony from scans of the planet Jupiter, up-close and personal. It had taken many years to prepare for their visit, at the asteroid. So, as a guest, he would see what he could do as far as Wu's odd request.

True enough, for the Alpha-Base guest of that season named Amakmid Beautiful-Truth, the topic was factored by things like body-heat, hair-styling, body-weight or density, eye-color and overall build, fingerprints and Human-Genome Personal Mapping, blood-oxygen absorption rate and blood-fat, and other respiratory details, gender, and sexual-orientation, and general vigor. The reason was, he knew that the Large Globe Music Society could create a tonal-transfer map, of the human body for a living person, the same way they now had almost-completed a similar map of Jupiter herself, and the resulting musical composition and recordings.

This group had come a long, long ways, to attain the vibrations of the main planet, with sophisticated macro-analysis and microbe-level spectrum-scans, translated to the most astounding beautiful music, sounds and tones, projected into a synthesizer-matrix, with assigned tonal channels and methodically-building and then releasing musical ladders. Yet Jupiter had no microbes, no life at all. But, they were all very thankful that Jupiter had learned to sing such a fantasy for them all.

"She still sings, whether you leave here or not. She sings forever," Amakmid said to Deveroux, later.

"It's only music if everyone is playing the same tune," Daniel replied.

"Not if its a planet, an entire world. Perhaps its the ears you're hearing with. You need new ears."

Deveroux paused. How could it be, that his most assured personal confidence in his good looks, was failing to attract an answer to what he was now facing? Maybe the lack of women in his life? Beautiful Truth was not actually a particularly handsome man, nor athletic, nor very tall, too short for basketball or too slow for tennis, nor of very dark skin, nor of an unusual visage or countenance. But he was the LGOMS leader. His mind and awareness had developed planetary symphonic musical transfer-composition.

Old world composer, Gustav Holst ('The Planets') would perhaps have envied the work. And there were always, so seldom to be observed, the twin ghosts of Copernicus and Galileo, wandering around the bases and Jupiter program ports-of-call. Yes, dead souls, like Halloween, apparently seeking audience or to spend time with Sir Isaac Newton.

"My grand-father was an off-worlder, you know, way back," Deveroux replied. "From a distant star-planet. So maybe that is in the DNA, for my hearing. My ears."

"Yes? This is true?" Amakmid said. "Deveroux, my dear friend, I am the same! My ancestors about two generations, even before!"

Deveroux sharpened. "What? What star-system? Cygnus or Hydra?"

"We are the same! Hydra, of course!" Then Beautiful Truth suddenly hugged him, joyous for an instant.

This had happened in Deveroux's life before, and it was not a falsehood. Three father's-father's back, Earth-powers had some kind of inter-world bodily commute, resulting in actual off-world residents in various places on Earth, for many years. Quite a few, but not so many as an invasion. And yes, humanoid, identical to the basic Human Genome, yet 'not from around here'. Male and female, they married, enjoyed life, loved and lost, loved again, lived, and passed away. Sometimes the off-spring or children were not even informed of this, and the information was almost always concealed or suppressed. Indigo Children, Mahanwe, meaning 'rejected ones'. Aliens, ET's, not fully human. Deveroux was also moved. When these types ever found each other, it was a sort of joke.

"Cid Bixi Mimim, Beautiful Truth," Deveroux said. Then they laughed.

"You can't go wrong with an Elton John song," Amakmid said.

The scans used by the LGOMS were designed to capture and record the various vibration realms associated with the planet, the main form of Jupiter herself. They could assign scanned data to musical output channels fairly easily, and then work to create a harmonic musical composition, an amazing ' tune' that could go on and on, for even days at a time. The scans could electronically assimilate all levels of activity happening in real-time at this very dynamic world: Jupiter. Density levels, rotating orbit, speed and path, liquid gasses and solids distribution and contents or chemical components, overall gravity-well readings, reflected light, ambient light, sourced-light, shadow and Solar positional relativity. Heat and cold readings. X-ray and Gamma-ray emissions. Storm fluctuations. And many other factors.

"So what did this Doctor Wu want from you, Amakmid? A personal performance, maybe, for he and his parties?"

"Sort of," Beautiful Truth said. "He wants me to configure the scan-and-compose set up for his own body."

"Oh." His co-worker said, rather blandly.

So, how is it my good looks fail to attract a perfect answer to what I'm facing as far as my deportation back to Earth from this space-station at Jupiter?, Deveroux pondered. And this was only a few hours away, *tick-tock, tick-tock*. They had been delivered their official orders for deportation; they had attempted to solicit a refusal from Earth on the decision, but time ran out quickly. The knowledge of the crimes at Alpha-base, so incredibly distant from bottom-line Earth authorities, were not confirmed, not official. This meant the Jupiter Program rats could easily slip through those cracks, and create mayhem from within, albeit supposedly to the betterment of all.

He had to wonder if the so-called 'deep-sleep transfer' was in any way misunderstood. Would he and Angelo ever even truly wake up again? He had already died before, and it's unpleasant, he knew from experience. It may as well have been an execution. And now, this one, somehow linking his flesh-blood-and-bone to a hidden 'enemy', men from Hydra, long ago. That is, aliens, E.T.'s. *Sound familiar? Don't try this at home.*

Home. The place where when you finally show up, they have to let you in.

The Large Globe Music Project at Jupiter took time to set-up. A concert was planned, at the Alpha-Base, for all visitors and staff, but Deveroux and Angel-Face would miss it. Too bad, that sort of thing was rare. The creations were recorded, however, and would be released later to music-lovers back home and elsewhere.

Daniel wondered aloud about the process, it seemed amazing. To analyze an entire planet, at its overwhelming size, with a streaming data-flow of various components, and convert all that into music. Rather like the song a rock or stone would make, or the music of a mountain, which no one can ever hear or know, until the key to unlock those songs was applied. Or the music of the mind of man, or the body of a cat or dog, even.

He would have spent more time with Amakmid Beautiful Truth, he seemed like a splendid companion or new friend. But, he and his partner Mendoza now were prepping their hastened dishonorable departure. On the second day prior to shipping out, more information was available.

"Oh, you big coward," Vanessa Signo, his PPP case-manager, explained. Her voice was thin and distant, the radio-link seemed off a bit somehow, it was a long way off, 700-million miles.

"They've done the same thing a thousand times, its standard. It's called The Popsicle Express, informally. Officially, the Space-

Authority calls it On-Demand Unconscious Personnel Transport. And that's really all it is."

"But what about body-functions? What if I wake up in the middle of the whole thing? What kind of stress or medical consequences are there, when they bring me back?"

"Well, you'll just be a little groggy, I guess," 'V' said. "There are catheters and diapers for body fluids. They keep you alive, obviously."

"I'm sick to death of being dead already," Deveroux quipped.

"I never really cared for the agency's decision about Seattle, you know. It's like social suicide. All of a sudden, I don't exist. No more friends, no more daily routine or work, no more home, no social life."

"Sorry," she answered, the distance seeming to grow longer or the cold-empty Abyss between them somehow tainting the meaning.

Deveroux paused. He had a handy ODUPT Quick-Start Guide. "Well, according to this, we report to the launch-lifter facility docking port here at the base, tomorrow afternoon. We'll be stripped naked and out-fitted with medical-type apparel, I guess like hospital gowns. Then we'll be disinfected in a clean-room with chemical spray. At some point we are escorted to a room where we're hooked up for life-sustain or life-support, in high-tech coffins for each of us. Very comfortable, I guess, thick nylon pillows, clean-cool air-flow, lighted and dark alternately. Somewhat like a comatose paralysis day-bed temporary hospice. First class medical coffins for the half-dead, all the way. Let's see, uh, oh, yes, here it is. We're fitted for tubes and fluids and then a mask over the face and nose. Then they gas us, not sure the chemicals here, pretty sophisticated. Out like a light, total unconsciousness, projected ahead for the entire eight month trip home. Then I guess they just load us like cargo into the ship. Looks like they are sending us back on the 'Aerotica'. Sound like fun? It couldn't have ended worse, Vanessa, seriously."

"You knew the job was dangerous when you took it, Dev'," she replied.

They talked more, but it seemed fruitless, morbid. Their mission was a failure. Very little that was not already known had been learned, and what they felt they had found out was dubiously false or esoteric to a degree of uncertainty that made any actively helpful conclusions worthless. Lies, rumors and half-truths. *And half-death.*

Other aspects of work at the Jupiter Program orbiting bases and platforms continued in with typically flawless technological mastery. Flawless functionality was essential, for any work in deep-space. If anything, what was bringing down the J-Program was corruption and sleaze from within. And if anything, a program like the one at Jupiter, would never really 'go down'; instead, things would likely fade into a monstrous affair of conflicted interests and greedy, subversive global Earth powers, hungry for more and more and more. And this was very costly, and somewhat hazardous, which was why the PPP was brought in.

The whale-hauler 'Down' had been busy on her schedule to Ganymede, where she loaded very large contents of standard helium gases and frozen liquids, into her huge holding tanks. They also loaded very large quantities of common frozen H20, and then selected batches of the rare helium-3 substance. Commander Brandeis and his Second, Menima Pearl, kept watch over the process, there was no crash or near-miss as they docked for loading (unlike the ill-fated 'Ferrous-2', by then several years in the past).

Pearl, in her typical formal fashion, confessed to Brandeis, who she at least called friend, that the Quantuum-Mind Overlords were no longer still or content, and that the channeled 'messages' had grown dark, with ominous warnings and confusing ideas about Jupiter and the people of the planet Earth. Commander Brandeis regarded his second-in-command suspiciously. Crazy, stupid or both? Commander of a deep-space vessel? Not the best.

"You will please to keep your religious superstitions to yourself," he told her without pity.

"But Commander," she would complain, there on the navigation-helm deck of the Down, leaving Ganymede behind them at high speed, following necessary command maneuvers to leave orbit. "Is it really my fault? Can I help it if my consciousness is doing all this? If it's not under control, you will be the first to know, I can assure you."

"Woman," Brandeis told her calmly. "I have a job to do. You and your space-religion are not helping matters. If the warnings and visions or Super-Mind space-garbage you embrace are telling program leadership your stories and tales of alien-cultures with some plan or design that would somehow effect us here, and they believe you, how could any such thing ever be confirmed except by another disaster of some sort? Are you thinking yourself a witch or psychic-medium?"

"But Commander! How can you say that?"

"I can have you removed from service at any time, dear, and I'll pop you into deep-sleep transport home to a proper lunatic asylum at the first sign of incompetence. You are an officer on this ship, and your responsibilities are critical to the survival of many lives."

"I have performed my duties flawlessly, sir," she argued weakly.

"If you become lost in your dreaming of all this, I pity you, but don't doubt for a minute what will become of you, if you betray us in favor of your ancient masters or whatever you are thinking. I'm not the least afraid of what you've been up to. You are dismissed to your duties. Don't bother me with this again. Thank you."

Pearl would again consult her antique 'minook' idol, the artsy statue with various technical features. It was thought a harmless vanity, like a toy or souvenir one might keep for show. The little Buddha-like thing was actually lifeless and dead, with no mind or brain or heart. With her skilled input and manipulations, the 'minook' seemed to spring to life, full of music, talk and chatter, images, and a sort of odd-ball assortment of meditations and esoteric knowledge, all stored conveniently in its hidden computer memory. Second Commander Pearl kept hers on a shelf in her private quarters, with candles, gemstones and sacred objects, for her private time.

"Am I a tool for something evil happening? Yes or no?" she asked the thing, in a hushed tone. Her work-shift was over, and she had been drinking some of the ambrosia-beverage, with mild narcotics.

The idol seemed to move and bend, a holographic effect took over, bathing the thing in the illusion of life-like motion. It was mostly golden, like a fat little man, bald-headed, with rings and robes, always smiling. A strange tone lifted from the base, a series of concentric circles on which it stood forever, or as long as it would exist in physical reality, no more than a mystical toy, yet alluring and seductive, when the depths of deep-space travel gnawed and bitched at the workers, their hearts and minds, like an aching for something better, such as normal life back home on Earth.

"Who am I? What am I? What am I here for? What am I doing? Where am I going? Where did I come from? Why is all this here? What is left behind if I should leave for a long time? Where is the sunshine I once loved so much? Where is my mother and father? Who am I? Who are you? What is this? When did all this first start? What is my part? Who do I have to please? Who do I have to fight? What are the rules of my existence? Am I being used by something more powerful than myself alone? What is that which uses me? How will it effect me? Will something bad ever happen to me or others as a result? Yes or no, yes or no, yes or no, yes or no??"

The little minook idol was also singing, it was a nonsense song. 'Careful how you answer," Pearl told herself calmly.

CHAPTER 39: Sleeping Homewards

Doctor Wu certainly had his evil ways, it was true. The man was too long in power, too deeply entrenched with heavily vested Earth-powers, mostly economic and military. He cared less about the actual work in space and goals of the program, then he did for the industrial interests that financed entire kingdoms of great wealth back home. The Chinese Envoy alone was worth hundreds of years of general prosperity for those people and lands. The South American Envoy held the fate of many millions of people, in terms of their overall personal provisions and lifestyles, even for their whole lives. Respected family empires and leadership groups depended not so much on the helium or hydrogen or H20 or rare substances from Jupiter, as they did from revenue and income based on the space-work in the J-Program that they participated in, 'back home'. And were these not good works? He knew they were, which only made matters worse.

True enough, as programs such as the raw-materials operations at Jupiter and her moons were developed over many years of technology and dangerous labor, planners found it was sometimes necessary to transport human beings as unconscious bodies in suspended states of livingness. Without the hassle of any fear, upset, resistance, or odd-ball notions of self-importance that could endanger others. Planetary Proficiency Program researchers Dan Deveroux and his partner-associate Angel-Face (Angelo) Mendoza, had been declared 'persona non-grata', that is, unwelcome, at the Alpha-base, even at the top-office level. The reason was perhaps clear to only a few insiders: they knew too much. As base-Commander Wu plotted with far more powerful Earth interests, he could not afford the prick of conscience, and criminal charges, that would follow the PPP's investigation. So it was a deft decision, on his part, to deep-freeze the two men, as a sort of example to others, and send them home. They would of course be perfectly safe.

When the time finally came, a day later, Menodza and Deveroux were helpless except to comply meekly with all terms. If they resisted, things would only be worse, and it made no sense anyway. Vanessa Signo's attempts to change the decision to have them deported from the Alpha-base in this way, yielded no expedited reversal, and would not help them in time. So they found themselves together again, conscious and aware at least for a while, to be prepped for the procedure. It was all much like a medical process. They both had been stripped and 'disinfected', and were naked except for 'robes' or 'hospital gowns', barefoot on the cold floors of the Prep Room, still there at the asteroid base.

"Talk about humiliating," Angelo said. "It wouldn't be much worse if they hog-tied us and packed into pickle barrels for nine months."

"We'll be aboard the 'Aerotica', the ship you arrived on," Deveroux said. "Not in regular rooms. They have still-life cells or deep-sleep berths we use."

"Shit! Nine months? Unconscious? Damn. What if there's a problem?"

Deveroux laughed. "Shit! Yeah, problems. Like shitting. Ha-ha! Good one!"

"Well, that all just depends, doesn't it?" Mendoza joked.

They grumbled more, waiting, then waited, then waiting even longer. The deep-sleep transit process was monitored by medical staff, during the voyage, of course. All standard. Soon, there were nurses and medical staff on hand, and the two men were walked into a boarding-area. There would be a shuttle-lifter up to the 'Aerotica', which would then dock. Once aboard the ship, they would be gently placed into their long-term sleep-cells, and then 'gassed'. It wasn't really that simple, their sleeping bodies had to be fitted for tubes with essential fluids and nutrition, and oxygen, and monitoring devices that would read their life-signs, heart-rate and so on. Angelo was left to consider exactly how his personal bodily functions would be dealt with; they had both been under orders to cease any regular meals for three days.

Rolf Denueri, the Commander and pilot of the 'Aerotica', was fully informed about his guests. The fast-ship, or 'needle-craft', was his pride-and-joy, and his responsibility, in flight, concerning every detail. The vessel was parked in a stable orbit-path that matched the Rosebud Stone, the giant reddish-green rock where the Amalthea-base had been built, as long ago as 80 years or more, at that time. Rolf's second-pilot worked with the ship's navigator and engine-thruster operator, on their pathway home.

The planets had moved millions of miles in their normal orbits since the ship had reached Jupiter. New navigations were needed almost constantly, for the 'Aerotica' or any other deep-space journey, to successfully plot her course through the Abyss, and reach Earth again. The two men who would be traveling in unconsciousness, or ODUPT ('On Demand Unconscious Personal Transport'), really only made his work easier, though he mostly avoided the process if he could. They all did. But Commander Wu had the power to order such a thing, and Commander Denueri was under his orders. It was also his sense of things that there was nothing really wrong or evil with this method. Space-travel had its demands on them all.

Like some kind of space-creature or spider from Mars no one ever knew, or seldom met or encountered in person, Menuda Wu was a mystery. And he liked it that way. The Amalthea-base had become his personal paradise, and he was secretive and seductive in its many pleasures. Why not?, he asked himself. Here I am stuck here for even years of my life, the successful achievement of a lifetime of study and training. My efforts bless millions. Why not enjoy myself?

"Life's not fair," were the last words Deveroux was able to share with Mendoza, after the docking-shuttle had mated with the 'Aerotica', and they had been escorted by staff to their on-board stations for the voyage home. The view of the Rosebud Stone and the Amalthea-base dwindled away until it was unseen except by telemetry devices. Jupiter was as large as Earth's moon even to the naked eye, from there, and so much larger from positions at Ganymede or Europa, and the orbiting 'miner' platforms, as to be literally overwhelming, her colored candy-stripes giving lie to the whirling maelstrom she truly was. The silver Eagle, Denueri's fast ship home, drifted and floated freely and weightless, like a massive feather of sustenance and hope, in the dark Indigo nothingness, then the docking shuttle finally disengaged the air-lock, and rolled backwards, dropping away to return to Rosebud.

Within only a few hours, Deveroux and Mendoza were unconscious, resting peacefully in their hi-tech coffins aboard the 'Aerotica'. All thoughts of the Jupiter investigation, the criminal charges and accusations, the deaths or even murders, the great cost and vast wealth and tangled web of conflicting interests back home, the stewing mystery deep within Jupiter, never really proven or dis-proven at all, all these troubles and challenges, all the technology and science and epic-proportion discoveries, the people and places, the pilots and astronaut workers, all these things now faded into the dream of bodies at rest. Dreaming men, sleeping, sleeping long, sleeping still, trusting one another for their very lives, and sure enough to be cared for.

"They'll be fine," said one of the ship's staffers, peeking in at her work through an observation window where she could view either man, peaceful there within on soft pillows, plugged into various machines. From outside, a computer monitor-screen tracked every detail of their life-signals.

"We have two others for the unconscious transport," the other nurse added. "What did they do, anyway? Very unusual."

""I'm not sure, and I don't care," the staffer said. "Let's just confirm the others, inform the command-deck, and we'll be on our way soon. Aerotica only has six sleep-cells anyway, guess we can be glad they didn't have more."

Their status was logged and reported to various Command stations, the ship's crew, and the hierarchy back at Alpha-base, and then at stations on Earth. Even had they been wide-awake during the voyage home, all details about their passage were recorded and tracked. Somewhere in New York, Vanessa Signo, the PPP case-manager, heaved a distended sigh and moaned lightly to herself, as the news reached her desk. It would be nine months before she would speak with Deveroux again, and they could begin to rebuild the investigation, and their so-called relationship. She was confident they would arrive home, and be revived from the deep-sleep safely. No one really wanted to hurt them, or so it seemed.

The distant stars burned and glowed, cold and wise, eternal, silent.

Jupiter was like a many-colored fat cow, a hideous vast orb of power and energy, there in the Sol-system, beyond even the idea of ever truly being known, nor wishing nor wanting, a gigantic globe-blob of matter and energy, stunning in both size and beauty, utterly inhospitable to human life. The larger ship, the 'Down' under Commander Martin Brandeis, was also starting her maneuvers, following a series of up-loaded goods and substances: helium, hydrogen, H20, helium-3, methane, ammonia, and other rare materials in large quantities, for this was her function.

Two other large whale-hauler transports were active then in the Jupiter planet-arena, along with three small-fast ships like the 'Aerotica', and many smaller service vessels and lifters or loading craft, ship-tugs, etc. And then back home, more ships and bases, at other planets like Mars, as well.

Did men of Earth have neighbors, like ourselves, out there somewhere, far beyond, in the Milky Way Galaxy? Had it ever truly been demonstrated and revealed so all could agree? There was certainly room for error. But second Commander Menima Pearl, of the giant transport piloted by Brandeis, was among those who felt quite certain it was so.

And the reason was, the water.

Hydrogen-Dioxide, two oxygen atoms mated to one hydrogen atom, and so life (as we know it), then possible. Carbon-based DNA-formed mammals, also insects, fish, and microbes, among other types, all totally depending on the availability of water.

Cool, clean water.

"You're good to go, navigator," Commander Rolf Denueri said into the Aerotica's inter-com radio, speaking to his main engine's navigator, two decks below the flight-deck helm.

"Thank you, Commander. Main engine sequence to speed from your mark, to thruster-array, at directional computer telemetry coordinates. Please stand-by"

"Thank you, engines. Let go home."

A few moments, then everyone within the ship could feel the main engines ignite, low and rumbling, a long, deep vibration, with the exception of Deveroux and Mendoza and a few others. Then the silver Eagle quickly gained speed away from the asteroid belt and Jupiter, towards Earth, a mere speck of light, like a star herself, far, far away.

It really was the water.

Not the rivers and lakes and oceans of Earth. Water elsewhere.

Water in amounts equal ton and in excess of, Earth's many oceans and sources. Fresh water, clean and pure H20, frozen in timeless and ageless local proximity to Jupiter, at the moons Io, Europa, and others, in almost limitless supply.

Second Commander Pearl, of the 'Down', was no dummy, despite the so-called enlightenments of the Omni-Mind space religions. Her duties as Second in Command of the giant transport were routine. But what duty did she owe herself? What duty did she owe the people back home? What duty did she owe the so-called Quantuum Mind, and the Quantuum Mind Fellowship? What duty did she owe her beloved 'minook', the clever toy worship idol? Any at all? None?

It was the water, at the large-economy-size planetoids or moons of Jupiter. There were other known moons at the Earth's Sol-system of planets (Saturn, Uranus, Venus, the others), that were known to have H20 sources. Various elemental chemicals were found at all the planets, some useful, some not. Even as the Jupiter program was created and built, and all the effort involved, none of the early planners disputed that the water found at the frozen moons of Jupiter, was very useful and valued.

As astronomers and star-gazers studied the skies and the distant stars, light-spectrum analysis of shimmering rays from those stellar objects, were tuned and refined and studied to supposedly show other whether or not they also had any water, 'out there'.

Daniel Deveroux was dreaming. Only Daniel Deveroux experienced his dreaming, but as a fairly normal person, he dreamed as anyone would. His somewhat Eastern-looking face was serene and blissful, the nurses peeking in through the observation window, now and then (on a schedule). He breathed in the treated air that flowed gently into the chamber. He was dreaming. He didn't know he was dreaming, yet there he was, heading home. The little sleep-chamber was darkened to a comforting coolness, the survival comforts arranged neatly and with great care.

In his dream, Deveroux believed he saw the planet Jupiter, again, for he had seen it before. To gaze upon it, his feet firmly moving him from place to place within the Amalthea-base, at the view-ports and telescopic video screens (the only way it could be done); to gaze at Jupiter from such proximity, it was worshipful, awesome, and scary. No eye had seen it, and no eye ever would. Mother Nature had provided, in a strange way, such that no living witness could ever truly 'see', any planetary object, at all, whatsoever, of the naked eye. The reason being the deadly vacuum of space, producing instant death, for a foolish person to remove his protective space-helmet and view of Jupiter, or another world, as if to see it for himself.

If seeing is believing, than it was also death, in the case of planetary worlds seen from space. But telescopes and video-screens and other means made it possible, or so it seemed, maybe all an illusion anyway.

Then in his dream, he thought he saw shapes and forms, like beings, even like men, but not quite like men, much bigger, expanding and growing and rising up out of Jupiter, as if her bowels had somehow split open and was releasing strange souls into the darkness of space.

They looked like giants, as if stretching and pulling towards a strong force, straining somehow, but effortless, and of different types. Deveroux dreamed, in his sleeping transport, and no one saw his dream but himself. Gigantic shapes and forms like alien robots, lifting away from within Jupiter, spinning into orbit, growing and growing, silent, ominous.

Looking for water.

CHAPTER 40: From the Bowels of Jupiter

"Something's happening," Space-Authority Security Officer Le Van Ho, at the Jupiter Alpha-base, 2,415 AD/CE

If dreams can come true, then also perhaps nightmares. The 'Down' was moving through her circle, longer than half a mile, like a giant fat whale, dark blue, an object at great speed, bringing home her bounty of elemental raw-materials. If Jupiter were 'East of the Sun', then Earth was 'South of the Moon'. For a common man without astro-navigation knowledge, it only meant, she moves, she floats, she flies, among the stars and skies, falling forever, motionless to the perceiver, but faster than the eye might even witness, or the mind truly imagine, to reach Earth, from Jupiter, in only a few months.

There was a poetry to it. At times, from the Command Deck or Flight-Helm of such a ship, at this era of history, a pilot or captain like Martin Brandeis, was welcome to his personal vision of the romantic glory of all that was involved, a stunning and powerful chore for a human being, thrilling. At other times, the same seductive lure of the very idea space-travel, was a bane and curse to be avoided and tossed out, out of his thoughts, rejected with serious prejudice, for the work to be done calling so hard upon them, or to somehow fail, and die. A Commander like Brandeis knew, dreaming, at his work, might bring disaster. Yet it was all a dream, and not unpleasant at all, in general.

"Steady as she goes," was his word-at-the-hour, and so it was. The vessel was trim and flawless to her task, her commander and crew like a symbiotic permanent part of the machine she truly was, designed to sustain every life she carried, a mother with many sons and daughters, there inside her comforting high-tech belly.

Brandeis was sharing about his horses, back in Oklahoma, with one of the communications men, just making conversation.

"If you don't know horses, you might not appreciate it," he was saying. "I'm very proud of them, we have about 30 head. Ten of them are rare purebred Arabians, all stallions or studs, for breeding. Beautiful animals, we treat them better than half the population of Earth, in the regions underserved of wealth, anyway."

"And that doesn't bother you?" the radio-man asked him casually.

"I can't feed the hungry in the waterless African wilderness, by abandoning my horses, now can I?" Brandeis said. "I'm not the king of the world with gifts for everyone, sorry. Someone must also care for these magnificent animals. So why not me? It's a hobby, really, it relieves the stress of space-work, and is very different than here on-deck moving this metal mountain through the Abyss safely to get your sorry ass home in one piece. I'm an animal lover. It's a virtuous thing, you know."

"Well, I'm an animal myself, Commander," the radio-man replied. "I meant no offense. So, you're cloning the horses, at your ranch, aren't you?"

They talked more like this. Conversation was a frequent activity to remedy the long, boring hours, as the ship tracked her path home like a razor's edge, and even the top-command making no decisions or choices at all for days and days. Only monitoring, watching over the engines, watching over the life-sustain, watching over the holding tanks, watching over the batteries and solar-energy collector arrays, watching over the movement of the planets, routine, regular, consistent, understood, necessary and patterned as hard-wire computerized programs to the Down's mechanical innards, so that her qualified perfection would not fail. By this period in space-travel work, men like Brandeis were very confident in their survival. The ships were trust-worthy, and likewise the people and crews, the staff and technicians. Each depended on the other.

The Command deck for this kind of ship was at the forward position, on top of the 'nose-cone'. It was as big as a small super-market back home, all by itself. Spacious for the workers, with various kiosks and task-related stations, three levels overall. Wisely, there was no one single all-containing push-button dashboard of controls, that ran the entire ship. The 'Down' was much too big and far too complicated for that (although a smaller ship like the 'Aerotica' did much more resemble this sort of flight-control style design). So when present on the command deck here, one could view seven basic command-function areas, each tethered to a million lines of simple 'changes', that performed all the needed functions to run the thing overall, from place to place and month to month and year to year. And each of these command-deck platforms was staffed on a rotating schedule, by skilled technicians who were experts in their disciplines.

A Commander like Brandeis could move between them all and oversee things more generally, with any choices or difficulties or questions directed to himself before actionable button-pushing would create this-or-that effect on the ship and her essential operations. The Command Flight-Deck had the appearances of a high-tech mall or control-tower maze, full of large view-screens, computer arrays, gauges and meters, communications gear, and specialized devices for various needed analysis and readings, specialized to an infinite degree of scientific veracity. Just like the ill-fated 'Ferrous-2', in almost every detail.

Doctor Wu was a music-lover, after all, and a collector of rare antique comic-books. So the Large Globe Object Music Society could at least entertain his little idea, it was really nothing, after all. He laughed to himself, the idea was too precocious, just the kind of perverse approach he liked, for that kind of thing. As if to lose himself in a realm of his own making, to escape the painful truths of his life and duties, Wu became more and more obsessed with the technology of the LGOMS.

If they could scan a whole planet, and change the signal into electronic music, why couldn't they scan a human body? And of course, he knew it could be done. Whether or not anyone had ever done such a thing was maybe besides the point, for him. And all he was really interested in was showing off the end-product with the women he kept. Always trying to please the ladies, even if it meant nothing. Worth a laugh, and a spasm or two of bliss. He chuckled again to himself at the idea, then continued what he was doing. Music to my ears, he thought, music to my ears

Then one of the station-monitors attracted the Commander's attention. "Commander Brandeis, a moment, please, sir," he said, moving across the 60-percent gravity floor-room. Brandeis was looking at a computer-screen with details on the positional movement of the Earth and moon, projected month's ahead to the time of their arrival.

"Yes, go ahead," he said.

"Something's happening, sir, at planet Jupiter," the other man said.

"What the hell does that mean?"

Jupiter was by then far behind them, but not so far as to no longer be of interest. And of course, nothing ever was 'happening' at the planetary level, out of the ordinary rotational spinning.

"Look at this," the man said. He had a hand-held PC 'tablet'-type device that linked to other instruments and inputs. It had a view-screen and controls. The man was one of the lesser-level navigators, his only job was to track asteroids, actually, as the most unpredictable factor in their traveling back-and-forth to Jupiter. And in so-doing he of course would track Jupiter as well, and her moons. He was a thick-headed, husky looking Asian man. Brandeis knew all the crew by name. The man then showed him the PC tablet-device.

"This connects directly to my kiosk monitors that view local-region activity in the asteroid belt," he said. "You can see it like a radar map. Here's the Alpha base, at the Amelthea Rosebud Stone, bigger than the others. The computer places the ID tags at the position of all known objects, the scanner footprint radiates out from the ship, backwards towards the planet."

"Yes, I'm familiar with it, so what?"

"See these little dots? There around Jupiter. 24 hours ago, they weren't there, at all. They didn't exist. Those aren't moons."

Brandeis paused. "Get communications to hook-up to Alpha-base and ask for their navigations-monitors, for the same positions. Maybe just rogue meteors or out-of-position rocks from the moons? Why is this important to me? We're 50-million miles from Jupiter at this position and another 500-million miles home. How does this matter?"

"I'm not sure, sir. It seemed unusual, so I thought I would inform you. I've already moved a message to the Alpha-base about it."

"What did they say?"

"They don't know either, but they have the same basic info readings," the man said. "I'll place a data-menu order request as you've directed, sir, on the same positions. It is very odd."

Brandeis was easy about the matter. Whatever it was, it was far enough away from his ship, that it seemed safe not to panic, usually best. 'New' objects? Ships of unknown origin? Giant rogue blocks of ice? Faulty monitoring equipment programmers misinterpreting readings that would otherwise be normal? There were no 'new' objects that could have possibly appeared in the regions around Jupiter, it was not possible. Some kind of conspiracy hoax, maybe?

"Let's see this on your macro-scope tele-viewer screens, that's our best," Brandeis said. So they moved again together to one of the other stations. By this time, other staffers on-deck at the time had taken an interest in the topic, so they worked together as they would any other curiosity or science-puzzle they might encounter, all quite routine.

The macro-scope was the ship's most powerful telescope or external viewer. Perhaps comparable to the early 'Hubble' orbiting Earth telescope, it had tremendous magnification and range. The current image could be routed to a large view-screen at one of the flight-deck stations. Brandeis and the others first needed to work with the operator, to target the view they wanted: backwards from the ship's motion at Jupiter. This took some time, twenty minutes or so. They chatted more. The Commander was among those 'in the know' about the theory of an inner-world ancient secret at Jupiter, of some kind, the rumors that were of such concern to the PPP, and the 'sleepers' aboard the 'Aerotica', Deveroux and Mendoza.

Are we not all dreaming? But he kept his opinions to himself, generally, of such things. Nothing much had ever really been confirmed at all. Two men dead, two men in stable unconscious transport, one sleazy musician and his clone in and out of jail at the Amalthea-base, and not much more about any of it, aside from talk and discussion and conspiracy theories. Related? He was sincere in his private hopes that there was no connection to their current review at the macro-scope.

"There are three views. I'm bringing up number one and number three now," said the operator. "These are high-magnification, we'll get a nice look here. Just a second now."

Second Commander Menima Pearl had by this time joined them on her scheduled work shift. She looked bright and perky in her flight-deck uniform, the common one-piece pull-over jump-suit they used when working. Brandeis looked on skeptically as four or five other technicians gather around the macro-view kiosk. They were all now quite curious.

The two screens flickered to life, blinking a bit as the electronic signals connected. Apparently a third screen-view was delayed.

"What is it?" Pearl asked.

"Some unusual appearances of some kind at Jupiter. Not sure, really," said the asteroid-belt monitor who had alerted Brandeis earlier. Almost an hour had passed as they looked over what data they had.

"All right!" the macro-operator said. "There! Check it out!"

Now they could all see very crisp, clear views in real-time of planet Jupiter. The image filled one screen completely, even overlapping the edges of the screen. The other screen-view was farther out, so the planet moons were seen as well, or some of them. At the same time, the third view-screen also began to flicker and blink, and another view came up, an extreme close-up on the largest moon, Ganymede.

"Looks okay to me," said one of the technicians, producing a short laugh among them. And there she was in all her glory, the mysterious 'red storm' eye showing from that position as well, along with the rippling, flowing colored stripes and layers of chemical substances and fluid ice-gasses. Jupiter's orbit is so fast, that even after a few minutes viewing in real-time, one could witness some actual motion. Her moons were seen as well, somewhat, though not all sixty-three, from the macro-view telescope screens, like twinkling-glowing diamonds or balls, hanging in the empty nothingness around their mother.

"Okay, fine, compare to the unusual data you've got on the planetoid monitoring," Brandeis said. "And whatever they have from Alpha-base, if available."

The asteroid monitor reviewed his PC-tablet type device again, the same one he had brought to Brandeis an hour previously. "Can you program the scope to infra-red spectrum view?" he requested after a moment.

"Infra-red? Sure, it will take a minute. Why?"

"That's how they think they saw something, on infra-red. They were using that filter."

"Anything worth seeing would show up just as well on a regular light spectrum view," the telescope operator said.

"Well, just go ahead," Brandeis told him. "Not a big deal. Whatever they thought they saw, we're way out of reach now if there's any problem."

"Probably chunks of ice that hit something and rolled into orbit in a million pieces," another technician said.

The macro-scope operator was busy at his controls. Changing view to see the same images in infra-red spectrum meant he had to re-direct the external telescope programming itself, to operate its normal functions at that range of input. So, 'it will take a minute' meant what he had said, even at their advanced level of technical mastery, by that era of space-travel, things worked the way they were supposed to only when intelligent people 'knew what they were doing'. Then the same views he already had for them to see, shifted and changed, a 'red shift', and their own eyes could now 'see' Jupiter at the a different level of the normal spectrum of light, there on the large electronic view-screens. What they thought they saw was quite different then they expected.

"Oh my god!" Second Commander Pearl exclaimed.

As bizarre as it sounds, from this view, Jupiter seemed now to be surrounded by huge dark birds, large dragonish creatures, gigantic and stunningly beautiful animals or beasts, and Angelic-looking 'beings', great moving plates and forms, wheels that spun together in strange synchronicity, as if intelligent to themselves, floating 'heads' with human faces and animal faces, angry-looking or with angry mouths, somehow ancient or archetypical, even creatures with four heads on one body, and creatures with giant, golden wings. It was all rather much like a scene from the Biblical book of Ezekiel, now available in 'infra-red' spectrum telescopic view of planet Jupiter, viewed from a large in-transit space-ship, circa 2,415 AD/CE. The figures moved, shifted, stretched, dissolved, formed and re-formed, like a peculiar dance, in the darkness. Jupiter seemed to have somehow at last disgorged her secrets, from within.

There was a long pause, the group at the viewing station for the macro-scope drank it all in.

"Look at that one!" one of the women navigators said. "It looks like a hippopotamus!!"

Then some laughter relieved the tension. Commander Brandeis considered his options.

CHAPTER 41: The Water-Logic War Begins

"Reservoir 22 roll-over team, please advise," heard that day on the inter-comm they all used.

"22 Rollover here, ready. Go ahead," was the reply.

The team of three men had set out in one of the wing-set riders, rolling almost as if in flight, along the mechanical bridgework tressles, in the dome encasing Reservoir 22 from above. Below, a deep pool of pure water, shimmering like silver, lighted only as essential, with rays of artificial light from the wing-set rider, glowing downward. It went down hundreds of feet. Their only job at the moment, to perform a spectrogram. The wing-set hummed, gliding ghostly, three men inside.

Then there was a terrible sound, like steel-on-steel, the wingset gyro-stabilizers shuddered. A rumbling, almost like an earth-quake.

"Rollover? Rollover, are you there?"

The men inside the wingset rider were terrified, as the disturbance knocked loose one of the thing's rail-locks, and it began to slow, and then droop over, and then began to disconnect from their only way back to safety.

"Rollover 22, please respond?" was heard from the comm-link.

The arrangement of events, at Jupiter, at this time, was a confusion of actual real occurrences (even of the life-or-death variety), and bizarre illusions and a sort of hallucinogenic fugue. It may not have been until much later, that whatever 'enemy' there was, in its capacity to create confusion, was recognized by military and security forces as adopting illusion or 'maya', as a weapon, and a very effective one, for both fear-factor, and general dissipation.

"Wow," said one of the Almathea base telescope operators. "What the heck? Look at all this!"

The Jupiter Program overall, including the Amalthea-base and the orbiting moon-stations and ships, communications, telescopes, staff and all that was involved, as far away as Home-Earth, had now somehow descended into a general confusion. And it all began, with the same observations the crew and Commander of the giant transport 'Down' had seen, along with others, even that hour. That being, an un-holy array of curious and mystical manifestations, appearing in the planetary-region of Jupiter herself.

Or so it seemed.

A true Pandora's Box of deadly circumstances followed over the next hours and days. Observation monitors at the mainstay Alpha-base confirmed the same events, as the macro-scope team aboard the 'Down' had. Stranger still, those handling the crisis, at times could not agree on what they thought they saw.

One would say it was a hideous monster of heat and flame, attacking an orbital base; another would witness a solar-flare from out-of-nowhere, as if they were the same.

One would track and monitor a genuine UFO-type space-ship in orbit around Jupiter, then it would seem to vanish, then reappear, witnessed a second time as an ancient ocean-going ship of Earth's salty waves, a 'tall ship' with sails and wooden hull and prow, like a crystal vision. The same? Both yes and no, same space-navigation, same physicality or object-presence, same speed and trajectory. Yet ages and ages apart, both equally impossible.

"This is to be regarded as an attack," Commander Wu declared, as Alpha-base and general program leader for space-work at Jupiter, as things heated up. The 'Down', and also the 'Aerotica', and other ships-in-transit, were in agreement. The orbital stations, such as that of Charlie Benway's lift-and-load stations and ships at Europa and Ganymede, also agreed, as far as the science-witness was concerned.

Doctor Wu was quickly in consultation with the task-force assignment leadership of the clandestine Operation Odyssey, the 'OO', the secret operation that had been in motion, or planning and preparation stages, for as long as 20 years. The 'OO' on-site Commander was Pacabello Retendare, a huge Italian man. He was not a military commander, his work at the Alpha-base was only to organize Wu's previous commitment to begin early stage out-fitting of the 'Solo-Men' mutant-clone operated fighter ships, designed and intended for 'just such a thing'. Clearly, this so-called 'event' was a surprise to them all.

They met in secret, there within Wu's above-it-all command center. "Paca, please," Wu told the man. "We knew about this, we didn't know the day and date, but we knew. At the moment, without information, we can only assume that some foreign source has unleashed some bizarre sort of attack. They probably want the water, or the bases, but for all we know they may be taking slaves for their alien home-world. The OO must respond, and they must respond quickly. Your fighters must engage immediately."

"I don't command the launch of the solo-ships. Only half or somewhat more are even ready," Pacabello said.

"How many is that? Your agency needs my base here to store and prep the fighters. I know you had 2,000 or more, I was involved with the orders. So only 1,000 are ready? Why? Who gives the orders to engage, damn it?? We need defense!"

"The orders originate with the Galactic Event Posture Committee, in Geneva, back home. 1,800 solo's are ready. 600 more need fueling-gear and we lack for the mafu-clones. But 1,800 of these fighters can fly at any time, it takes hours, and the pilots need their orders and navigations. We have never done this before, Wu."

"Then from my office to yours, you have three-hours to contact the Galaxy Posture Committee and move your fighters out against these bat-wing freaks, understand?? Get it done now, sir, or we may not have another chance, ever!!"

Meanwhile, the problem with the Alpha-base water-reservoir system, deep within the Rosebud Stone asteroid underneath the hermetic, air-sealed base itself, with its 2,000 or 3,000 residents and workers, was now in full-stage crisis. True enough, following the collapse of the wing-set industrial inner-railway repair module, and the men inside, fallen into the deep silvery waters of Reservoir-22, it was learned by those responsible, that some some sort of external floating object, had attached itself without notice, beneath the underside of the Alpha-base asteroid, and was apparently sucking out their water-supply like a creepy Alabama backwoods teenager siphoning gas through a garden hose. How they had escaped the awareness of the base-monitor and space-region scans, was a real mystery.

The young water-quality Astro-Biology staffer, Michael Hesidom, found himself on-site handling his part of the crisis. The 30 giant Reservoir Tanks, were all connected by water-filled shafts and pipes. The Alpha-base of course needed the water, most of this was not headed back home to Earth. The giant H20 ice-worlds of Io, Europa and to some extent Callisto, now seemed to somehow be the goal of some sort of Galactic-level plan from some far-off world or species. And why not? Among the staff and crews at the Jupiter Program stations, word was that they were experiencing a 1,000-year old plan by some real off-world alien species of sentient type, to provide vast quantities of quality H20 for themselves.

The deep-interior mechanical forms they had secretly known about for 80 years or more, inside high-pressure Jupiter, had been unleashed, as part of this goal. Or so they thought.

AB Hesidom worked with a Reservoir System Task Force. At the bottom of one of the main tunnel-shafts that connected ten of the giant tanks together, another strange battle was also in progress.

"There's something down there," one of the task-force Commanders said dryly. "I really don't care about the fucking water, they can have it. But if their ship or whatever has a tube up our ass, we could breach life-sustain, lose our air. We could all die within hours."

"Base perimeter structural scans had this for five days!" Hesidom complained. "They knew it was stuck to our butt like a

barnacle for five days! And now all of a sudden they want a
solution!?? Stupid! Stupid!!"

"We didn't know what it was, Hesidom," the man replied.
"We thought it was a space-slug or some bizarre fungus, that was
a natural formation of the Almathea planetoid. Or an infection, a
virus."

They worked together in the long corridor of tech-monitors
and controls that ran the entire reservoir system, 30 tanks the size
of football stadiums, each as deep as 400 feet, filled with pristine
water from the moons at Jupiter. The sub-control rooms were on
top of the reservoir system, still beneath the main base. But this
battle was slower than paint drying on a door no one could open.

"There is a fungus," another one of the task-force workers
pointed out. "You realize that, of course. It's not as much like a
fungus as it is some sort of Loch Ness Monster. The diver-team
that rescued the wingset rig that fell into 22 encountered it. They
were dragging out the men from 400 feet down. Like a fucking
tentacle or giant water snake. It crushed the survival pod from the
wingset, as it fell down after."

"Like a piece of tin-foil, yeah," Hesidom said. "Also may have
diseases that effect water-quality."

A long pause. Then the Commander told them all. "If it
doesn't exist, kill it anyway. That's an order," he said.

"That which does not exist is not real," Hesidom quipped.

On the topic of extra-terrestrial life, or aliens, most of the
space-workers and science-tech people within the program, were
not convinced. They all knew, that the Milky Way Galaxy
supposedly held at least some other worlds similar to Earth,
complete with people and societies. And research showed that
some of these were quite advanced and could move bodily from
star-to-star. More specifically, it was thought that they moved and
operated remotely, from a distance.

But no one really believed it. The distances were much too
great, human space-ships and human life could never sustain such
travel, it would take hundreds of years to reach even the nearest
star-system at Cygnus or Hydra, or Andromeda. But the notion

that the Sol-system's water-rich worlds, such as those at Jupiter, may have some attraction, at least made sense. For a very advanced star-travel capable sentient species, to plan ahead 1,000-years to harvest the same water (and other substances), from Jupiter's moons, (assuming water as the basis of all life), was perhaps not too incredible to imagine.

The local channel for their remote-control attack was none other than Second Commander Pearl, of the cargo-ship 'Down'. And even she didn't realize how she was being used, through her mostly harmless infatuation with the comforts of the space-religion 'Quantuum Mind' fellowship. Yet she sensed it, fearing deeply herself, in her mind and even her soul, to have become a traitor to them all. The little Buddha-like 'minook' idol laughed, in its robotic Moshai-like wisdom, all pre-programmed yet to great effect.

"Who am I? What am I? What am I here for? What am I doing? Where am I going? Where did I come from? Why is all this here? What is left behind if I should leave for a long time? Where is the sunshine I once loved so much? Where is my mother and father? Who am I? Who are you? What is this? When did all this first start? What is my part? Who do I have to please? Who do I have to fight? What are the rules of my existence? Am I being used by something more powerful than myself alone? What is that which uses me? How will it effect me? Will something bad ever happen to me or others as a result? Yes or no, yes or no, yes or no, yes or no??"

From a simple logistical standpoint, they determined that a long sprite had shot forth from within Jupiter, like a half-million mile water-spout, directly out of the legendary Red Eye Storm. This reached a zenith or apogee, and then spread like an umbrella, water-like, but more controlled, and indeed it was.

The ancient machines hidden within Jupiter were now activated, and the huge crystalline fountain maintained itself like a glowing stream or flow of intense energy. From this, the previous archetypical forms moved into stationary sentinel positions, like standing giants, each at a pre-determined point in the Jupiter system. They were, in effect, giant robots, or remotely operated machines. But they did not operate like normal space-vessels at all, further complicating their defeat or efforts to defend the program stations and bases and ships.

The Solo-Men Operation Odyssey fighters began to launch a counter-attack about 18 hours later. They had been prepared, and the Galactic Event Posture Committee was at least ready enough to mount some sort of defense. The OO had various Commanders and strategic planners on-site at the Amalthea-base, and elsewhere in the J-Program system. Pacabello Retendare was only one of these, not really a military man at all. The OO had anticipated something like this for even 20 years. Now, suddenly, a most curious type of deep-space Apocalypse, there in the Abyss of Earth's solar-system by Jupiter, seemed to have unexpectedly begun.

The new fighter-ships were each about the size of the 'Aerotica', perhaps smaller, and of course fighters. Earth had never really had any use for space-ship fighters, (as compared to perhaps jet-fighters in the skies back home over some war-zone). There was no reason, there was no enemy. Each of these 1,800 ships, however, resembled terrifying birds-of-prey, and were operated by the senseless and totally expendable clone-men, created for no other purpose.

If there was a battle to be fought, it would be waged between ghost-robots operated from distant alien minds, and mutant-clone piloted death-ships from Earth. In theory, not one of them was an organic original human nature, or even needed to be harmed, or killed at all. Yet, by virtue of some kind of sympathy or compassion for the human-looking and acting mafu-men, the clone pilots were kept in a state of suspended deep-sleep until needed. It was a horrid and immoral affair, to the disgust and distress of everyone who knew about it. They had created these half-human men, only to be sent into battle and die.

"What's wrong with this picture?" Marciel Peniuer, the scape-goat musician accused of the murders of the Old Men in the program who supposedly knew about the early research about

Jupiter's interior, asked his identical twin, Marciel-2, or 'Marcus'. These two were still at the Alpha-base, in their quarters, going about their business as best they could. The entire base was now on Red Alert status, in a panic of activity to survive.

Penieur wasn't someone who knew anything about all this. But he was a savvy guy, and rumors floated among them all. And the Red Alert Status warnings had included the basic information: there was now a dangerous threat external to the Jupiter Program system, probably of alien origin, and they were handling the crisis as best they could. But the fate of the OO ships and clones was particularly painful to Marciel, and Marciel-2, for obvious reasons.

"I may not be fully human, but I have feelings, too," the guitar-player's cloned 'mafu form' double answered.

"I know you do, Marcus," his father-double replied. "I know."

"I also need a drink of water every now and then," the clone said. He was again working on repairs and polishing one of the wooden acoustic guitars they used in their shows.

CHAPTER 42: Illusions Attack

"Well, it would be my usual vengeance, nothing new," Zuess, King of the Gods, from Olympus Mount, in an ageless realm, concerning suitable punishments for trespassings against what he felt was his planet (Jupiter).

"It's no good, Commander DeNueri," said his Second, aboard the fast-ship, 'Aerotica', now on its way home. "The entire ship is being used as a lightning rod for the third moon's sentinel image. The fucker has locked onto us!!"

That type of chatter in space-work was another long-established 'no-no'. For a pilot in charge of his ship, ship's crew and passengers, it meant they didn't really know what was going on out there, at all, but it looked bad enough to possibly panic about, and it involved something strange enough that they were all just as likely delusional about understanding any real threat. Convenient for an enemy, true.

"Please use standard English," DeNueri replied. The two men were in charge of the successful voyage home aboard the ship, moving very fast over a distance of about 700-million miles, in six months, back to Earth. "Sentinel image doesn't translate well when making decisions."

Their ship was now facing back towards Jupiter. It was a very maneuverable vessel. When they had started their voyage, Jupiter-main was behind them, only less than 48-hours total flight-time on-the-road-again, from departure point at Alpha-base, where Deveroux and Mendoza had been uploaded like dead men. With the emergency, system-wide for all J-Program activated to alert, DeNueri had made the standard choice to choose to park his ship into a routine planetary orbit, until things settled down, and they received more clear instructions.

It wasn't difficult, calling for a fancy sort of gigantic loop-de-loop, and then re-entering their normal stationary orbit as if they were only then arriving. The maneuver was called a "Circus-Drop", and moved them through literally millions of miles, into place again around Jupiter.

His Second Commander, at this point recruited from staff at the base, (not his regular) tried to reconsider. There was much more going on, that they and others had to deal with. Of course the areas around Jupiter are vast, so much larger than Earth's mass-density regional solar-orbit position, there really was no comparison. But, for space-flight work, anything resembling a conflict, an enemy, or ships-at-war, was highly unusual, even by that year, 2415AD/CE.

Knowing ahead of time, even twenty years ahead, for the Galactic Event Posture group, didn't help at all, for a specific pilot. And so it may have been for the so-called enemy, by 1,000's of years. *Since when did a planned battle go as planned?*

DeNueri and others soon realized that the situation was seriously complicated by a sort of hallucinatory effect. There was by then no real doubt that some kind of significant event had transformed the scene at Jupiter. With her 60 moons, Jupiter was now truly at war, at least ostensibly. Ships like the 'Aerotica' could view things on their own telemetry gear.

Groups of the new fighter ships, the 'Solo Men' space ships, were being released, from the bowels of the Amalthea-base. This increased traffic tremendously, but again, the regional space was vast. With magnification, they could see the squads of small, faster fighter ships, heading out to destroy the so-called sentinels, little more seen or actually known than ghostly phantom shapes and an array of dazzling images.

"Dead men fighting robot ghosts," Denueri's Second commented dryly.

It was much more like a swarm of bees attacking the enflamed shadows of giant windmills. The shadow giants could (and did) attack them back, they learned, and end their flights, destroying ships and the clone pilots. They also shifted and changed shape, size, and appearance. The illusion effect made counter-measures very difficult. For Rolf, and his Second, it all only meant they would have to go very slowly, or make no decisive navigations, unless they had to. The ships were in motion anyway. The 'Aerotica' was a transport, not a fighter.

"My meaning, sir, without being a raging idiot about it," the Second finally said, "Our ship and everything in it, is at this point enveloped by the attraction and pull-power of whatever that sentinle monster thing is, that they've stationed at the third moon. Here, on your view-board, look for yourself. This is us. This is it, 250,000 miles out. It has this sticky-shit enveloping energy form coming from it, magnetic, or solid-object gravity-based. And we're inside THAT!"

"Whatever that thing is? Know your enemy," DeNeuri said. "It just gets better and better. If the Alpha-base is under attack, why does the enemy keep his positions only at the various moons? Pretty slow method, I'd say. Why don't they just directly attack the base? Take it over by force, or even destroy it?"

"Maybe they're just looking farther down the road. A few hundred years or what have you," his Second answered dryly. "But for now, if we want to maneuver, we have to demystify the sticky-shit hold it has on us. It's clearly a physical object."

"Please, my friend, don't say it. Just don't say it. I've heard the term before. Giant robots?"

"Well, for lack of a better term. Maybe more specifically a 'magic giant robot designed to suit the intentions of an alien species at a far off world'. If that's any comfort. Sorry, just slipped out that way."

"I don't trust magic," DeNeuri said. "Robots are cool, though."

"Then please explain why your ship can't maneuver?"

He paused to consider the question. "Because something's wrong with it, that's why."

For seven days, or day-long periods, 24-hour cycles, chaos reigned, for those involved. More easily understood as a war, it was better considered as a Grand Theft, to be turned away by the ordinarily presumed rights-holder, logically property of the people of the planet Earth. Big deal, it's just water. The salty seas of Earth still had plenty. Epic regards, far off worlds and peoples who may want a drink, from even very long in the past, even prior to the development of Earth and society. The nebulous-amorphous attacking guardian robots, were hidden deep within Jupiter for ages. Water, at the moons, had a greater value than the Jupiter Program planners had considered.

"Give the fuckers a drink and they'll go home," said Deveroux, seated aboard the 'Aerotica' as well, with his pal Mendoza. They both were wearing the hospital gowns they had been disassociated in, for the trip.

Angel-Face was enjoying some real food. They were in a medical lobby-area, among the holding-cells for the long-term voyage in 'sleep-state'. Daniel had a coffee. One might have thought they had never lost consciousness at all, and were still being prepped. Neither of them cared, they were stressed of many other factors then as well. As the J-system alert came online through all the J-Program stations and outposts, both men were awakened as a matter of course.

This was a good outcome for them, though they had no idea why, or what was going on. They were groggy, then given some food and drink, and did their personal habits. So for a moment they were just sitting and eating, relaxing, assuming together that they could, or would, be dropped back into suspended-state, at some point. Rather like a vacation from a long sleep, then to sleep again.

"But Deveroux, they drink down the moitsure of entire planets. Or, the moons, I mean. Big gulp, yeah," Mendoza added with a smile.

"Hmmm, yeah," he said. "Feels good to move around. God! How long were we out? Oh yeah, my monitor. Let me check."

On his wrist is a personal computer device, that tracks his sleep-state while on-board. "What?" he said. "Less then 48 hours? Shit! It feels like a million years in that box! This says only two days!"

"Well, they're okay," Mendoza stated simply. "Better than a coffin."

"I wonder if they could keep a person alive that long. A million years?"

"I gave up on eternal life, concentrate on here and now. You're still under the drugs."

"We all have limits," Deveroux said. They were eating muffins with juice and coffee, then rested and chatted more. They both felt very much defeated.

"Whatever we didn't learn, it's okay, the PPP will retire us gracefully," Mendoza said.

"We blew it royally, Angelo," he said. "Months of work. The Empire Jupiter Club is too powerful, too many layers."

"Then they'll have us shot, I guess," Mendoza joked. This was not the case.

The pilots of the 'Aerotica', along with their technicians, had their predicament under review. They felt immobilized, but it wasn't exactly true. The Europa moon guardian was targeting their ship, it was closer than others at that time, easy-prey after the ship rolled into parking-orbit. This involved some kind of technology that scanned or viewed the magnetic resonance of the 'Aerotica', and also created an immobility, freezing some ship's functions.

This was certainly a panic, for the Commanders in charge, because the ship and its occupants could go off course. It was frustrating, all they could really do was sit and wait, and see if the sentinel would strike. As a cobra hypnotizes a snake-charmer who hypnotizes him right back again, DeNeuri consulted with the Alpha-base, and rigged a response.

"Blow the fucker to bits!" he told Pacabello Retendarde, the on-site OO ship-managemer and clone-handler.

"No, no good," he said. "Sorry. We are handling this, we have fighters."

"What's it made of? At Europa?"

"Uncertain."

"Have the other ones attacked?"

"Yes, but we met them earlier close to their stationary roles, so they only attacked the fighters, that were attacking them. So for all main stations, so far, so good."

"My ship is trapped, Paca," DeNeuri answered harshly. "I can hold orbit on the long circle only so long, in terms of my people here, understand? You need to provide coverage with your Solo-Men ships. I need back-up."

"I can't blow them to bits," Retendare said. "They are immaterial, non-metallic. They source to the energy stream, from the planet. They act like solid-forms, then change. So no, Commander, I can't blow them to bits. I will inform flight-command and have your position on the schedule for the fighters ASAP. From our best current location for orbit pathway and Solo-Ship fighter groups, maybe ten hours. You won't even notice them, they are going to Ganymede, not your position."

A few hours went by. Deveroux and Mendoza were now re-introduced to the general population of the 'Aerotica', which was a pleasure. This needle-craft ship was only populated by two dozen people, tops. In an emergency, all hands would want to be fully awake and alert, if possible, including the 'sleepers'. A rude awakening may be better than none at all, they had some catching up to do.

Then without warning, the entire ship's crew, that hour, spiraled down into sudden disaster, or so it appeared. A high-pitched alarm sounded everywhere: *BING! BING! BING!*

Then a Voice from the flight-deck, the news wasn't good: *Attention, please, ship-board staff, crew and passengers. This is a General Stage-One Life-Sustain Breach Alert. Ship's oxygen and vacuum-seal to external space has been breached by hostile action. Please find your personal life-sustain oxygen and vacuum seal products immediately! This is not a test, this is not a warning. We have approximately 20-minutes to total interior life-sustain failure. Attention please, all-hands find and activate your life-sustain products for vacuum breach immediately. May god save us all.*

BING! BING! BING! BING! Then the message repeated, over and over, as they all would then scramble.

"Aw, fuck," Deveroux said. "Just shoot me. God. We're all going to die. Again."

"This way, please," on of the medical-staff told them. "Quickly, we don't know how bad the breach may be. Hurry now, suit-up!"

The flight-deck was now at maximum panic-level. According to their controls and monitors, a lower-deck port-structure had breached to the vacuum of space. Breathable essential air was leaking out into the vacuum. If it wasn't halted or dealt with effectively, and soon, it could cost them their lives. That part of the ship, would be airless within minutes. And if other parts of the ship didn't achieve perfect air-tight seal against the lost parts, that air would be sucked out into the nothingness as well.

"Sealing second hull-chamber external loading port 2 and 3," said one of the flight-deck technicians. "Also shutting down live-operating systems for those port decks."

"Gimme' a minute to clear that area for personnel," said DeNeuri, on the flight-deck helm. "Then seal the entire section at the inter-deck platforms, the deck gateways. They have air-seals, on each one. Find the controls on those now, please, activate to isolation containment atmosphere."

"Yes, sir," the man replied.

"How the hell did we breach! This ship doesn't breach! It's impossible!!"

As all this was going on, the flight-deck crew also had to slip into their air-suits and helmets and activate the inter-comm radios that worked between each suited team-member. They really were not the same as full-suits for external work (for example); these were only emergency life-sustain helmets and basic pull-over protection. They would only sustain life without any other support from the ship, for a matter of a few dozen hours.

The ship's inter-linked radio sets chattered with various commands and orders, needed info for various decks and crew functions. Passengers, medical, hospitality staff, all also were prepped to survive in their suits, including the two researchers for Planetary Program Proficiency.

In a few minutes, Deveroux and Angel-Face were easily resting again, only now with their air-helmets in place, and the basic pull-over protection suits. A trained man could get into his own suit in less than 120 seconds, executed properly.

The alarm-system was still sounding, like a freaked-out Paul Revere: *BING! BING! BING! BING!*

"*Attention, please. This is a General Stage-One Life-Sustain Breach Alert. Ship's oxygen and vacuum-seal has been breached. Repeating, we have a breach to life-sustain, please remain calm and follow directions. Please find your personal life-sustain oxygen and vacuum seal products...*"

Deveroux clicked the top of his helmet, a plexi-glass kind of bowl, against the top of Angelo's, where they sat near one another, waiting calmly. The radio-link that connected them all did the rest. "So, is it good to be alive, Angelo?"

"You know, Dev', you have something hanging from you chin. I guess that's your tongue inside there that you talk with. And then words coming out, but I can't make sense of what you're saying. Good to be alive? Yeah, it's good, sure. Good to be alive."

"Glad that's settled, then," Deveroux replied.

CHAPTER 43: Just Checking

"No, it's not, it's not that, nothing of the sort. I never heard of it anyway, I don't know what you're talking about, and I don't want to hear any more about it, or anything similar to itself, if that's what that is, because I never heard of it, and I don't know anything about it, and I don't want to. Thank you." --Vanessa Signo, Planetary Proficiency Program case-manager, responding to a public inquiry about the Jupiter Program, later wired to broadcast television and entertainment outlets.

Vanessa Signo was walking back into the ivy-covered campus-like main building that comprised the semi-secret upper New York Planetary Program Proficiency headquarters. She had forgotten to bring a selected box of electronic paper-work from her office-cubby, with detailed facts and research about yet another PPP assignment: they had been asked by Global Earth Council to look into a long-term scheme to acquire anti-gravity technology from extra-terrestrial sources, based on the secret and horrific genocidal elimination of regional indigenous peoples in a small South-Pacific island community. What the connection was, no one could really fathom. But, as Vanessa often found herself thinking, *'work is work'*.

Without a doubt, Vanessa was an appealing and attractive woman. She was at an age where the fullness of life can blossom in a woman, into a vibrant health and physical beauty and poise, that one might truly fall in love at first sight. But she was also a woman with a purpose, a cause or clarity of intentions. More than just a secretary or desk-jockey manager.

The PPP teams and crews maintained a jocular certainty, a fraternity of simple understanding, that they were 'truth-tellers', or 'fact-finders', in this new and complex world. And the value to those seeking answers, even very important or critical aspects of problems that affected many lives, could be much. The agency had been born and developed, by world authorities, for a very specific function, as the world turned so much older, by 2415-16AD (Common Era). It was something Vanessa felt good about, she was proud of her work and labor, the results they could get, and many successful inquiries.

The outer entryway to the PPP offices opened wide, as she passed inside. Other workers greeted her, and smiled. Vanessa was dressed in a peasant-style long colored-pattern skirt, and a tight-fitting top, under a loose cotton shirt, with wooden-and-cloth fabric sandals, her hair pulled back to a tie. "Hi Vanessa," one man said, as she passed (an older man with the program). "How's Jupiter going?"

"Not good," she answered shortly. "I think it blew up."

"The whole planet? Jupiter?"

A small laugh there, then they passed by to what they each were doing. It was no secret, at least among those concerned with such things. Reports over the past four or five days from the Alpha-base and transit ships, were confusing. *What the hell is going on up there?*, Vanessa wondered. And she was not alone in the quandary of it all. *Did Daniel and Angelo even survive?*

Even at the moment, she couldn't be sure. Without seeing him with her own eyes, and confirming his aliveness and healthy condition herself, she had to acknowledge that space-travel meant that her part-time lover Deveroux could blink out into eternity and death, at any time between point-A and point-B, gone forever.

"Hope he's all right," she told herself. She moved easily back to her work-area through another doorway and then down a hall.

Because the Jupiter Program was a typical space-based operation, the links to Earthside authority were similar to any launch-site or rocket industry-based facilities. These had become much more sophisticated, by that era, but some aspects never really changed much.

There was a great deal of hard-wire high-tech communications gear, towering structures with satellite-dishes and high-energy power-sources to project vast amounts of information and computer-data, to people working 'out there', many millions of miles away. There were the launch-pads and landing-strips, similar to regular air-ports for jet-liners. They didn't actually do that much new launching, with the big rockets, but there was a regular flow of traffic in space-planes that could lift passengers and crew or technicians into orbit, to dock with deep-space ships for longer voyages. Other features at these places included military, construction and building or fabrication, like hangars and materials handling, high-tech areas for solving technical issues and testing. They also had medical and people-services.

So it was at one of these sites that the news of events at Jupiter started arriving back home. The US-Western site at Puerto Rico was one of the strongest for rock-solid communications to the Amalthea-base. It took a few days for operators to make sense of what they were hearing. Even those in-the-know regarding the Operation Odyssey program and the 20-year secret design-and-build efforts for the small space-fighters and clone-men pilots, were somewhat shocked.

It was one of those moments when something planners had anticipated, even for a long time, was happening now under their noses, and seemed un-real to them, if only for all the years they had created their own ideas and notions about it all. The picture didn't match the picture, and somewhere in-between, was the truth, essential to their survival.

And the word went forth:

Staff-base Update/US-regional Space-Authority, Puerto Rico: CLASSIFIED ONLY, please do not distribute or release. Jupiter Program Orbit-Dated 8.20.2416.3, Certified transmission. Alert Level TEN: Deep-space observation reports significant disturbance at planetary Jupiter of an unknown nature involving qualified unconfirmed outer-system intrusion by proxy, attack-oriented remote robotic galactic-origin servers. Motivation: unknown. Alpha-base at Rosebud Stone has responded with pre-planned OO fighter-craft and defensive weapons. Enemy forms create illusion and hypnosis; 13 dead among tech-staff, workers and pilots. Loss includes three moon-base systems ice-harvesters and machinery, lesser critical value. Semi-thesis, circumstance theory: consult with GEPG, OO management, PPP or Space Authority. Situation regarded as abnormal, at-risk safety and survival-status, Alarm Level Ten. Rosebud on-site difficulties water-reservoir systems intrusive misunderstood. Widespread systemic failure at many stations. ALARM LEVEL TEN-Update staff-only, Authorized info-data transmission, pathway 2x87Bc.01.2 ONLY.

Vanessa had received the same update. Many of them had, but the general public had not. So, in a sense, no matter what was happening in the lofty realms far above planet Terra, the common soul here at home cared nothing, knew nothing, and had no real interest, other than perhaps curiosity or speculation. If the sky was falling, if it fell on their heads, if the Illuminati knew about it beforehand or even planned it, it meant nothing. And the reason was, the decision-makers and choices lay elsewhere.

All false, all true, all illusion, or all very real, it was meaningless nevertheless. For a person like Vanessa and her associates, because they were of course also part of society and the general public, it was like a commonly held secret they had to deal with, yet also un-real, for being so distant, and so strange.

She settled back into the office-cubby, a pleasant, well-staffed part of the building-operation. PPP was mainly research-only, so much of their computer-networking was oriented around data-sources and current-hour real-time info, from all over. It was a clearing house for useful and condensed content and material about all sorts of things, worldwide and beyond, usually specific to their given task at the time. Sort of like a very flexible and complete library.

They also had management functions, and other needed operations. She searched her desks and cabinets for the box of papers she wants, spread-sheet read-outs on financing and science-contacts for the Horsehead Nebula Arabian Thoroughbred Breeding Farm in Oklahoma, the one owned by Commander Martin Brandeis, of the 'Down'. As she was looking for that, she consulted again with their new Computer-Advisory Analysis System.

"Computer on, please, passcode-authority V. Signo, 'monkey', please set for inquiry. Waiting," she said. This while digging through cabinets full of boxes and looking at the paper sheets for what she wanted. The vid-screen monitor to the Advise-Computer blinked alive, with images and controls.

"Welcome V. Signo. Analysis System active for your inquiry analysis. Please proceed."

"Thank you, computer. Please access all universal data-base and information for query regarding world Jupiter-Program status, deep-space Rosebud Stone at Amalthea, and associated programs and platforms. Waiting."

The thing buzzed and hummed. The machine Voice was now that of a pleasant male-sounding voice. *"Ready, data-base occupied to access. Proceed, please, Vanessa Signo."*

She took a private moment to form a question. Distracted, it was a habit with her, the PPP computer application could help her think about the right path for the work, the thing simply had more information available to its circuitry. "Thank you, computer. Question: what is the current status of the transport called the 'Aerotica'? Waiting."

A moment. *"The fast-ship transport 'Aerotica' is in orbit around planet Jupiter. Ship navigations are scheduled to return to Earth in approximately 6.4 Earth-lunar cycles. Passengers include P-program workers Deveroux and Mendoza, in sleep-state on-demand unconsciousness, and 11 other passengers, also the crew of five, Rolf Deneuri Commander, Second Commander is Philip Deneuri, his brother. The vessel is traveling in standard orbit-path at 200,000 rpm's, measured as miles, in a traverse trajectory. Fueling took place at departure from Rosebud Base, J-Program ..."*

"Silent, please. Hold please."

The thing stopped. Then she seemed to remember something. "Computer, question: does the 'Aerotica' currently suffer from an air-breach or external hull fracture, that threatens the life-sustain system? Thank you."

A longer pause. At this point, the computer would need to access current Earth-side telemetry and navigational scans, and also info-data streams from the 'Aerotica' herself. Other data-streams could arrive from Alpha (sometimes called the 'Rosebud' base). But for Vanessa, it meant so much more.

Then the machine seemed to restore its voice-reply tone-generated English language action:

"*No.*"

Vanessa finally found the box of papers she wanted, the logs and data-base on the horse ranch in Oklahoma. "Oh good!" she said to herself. "No air-breach to space, guess they'll live."

"*Yes,*" the computer now intoned mechanically.

She glanced back at the video bank. This was mostly an audio/spoken word system, so the vid-bank showed only some controls and basic functions. It could also display maps and renderings, still images and video-sourced material, and other audio, such as communications-links.

"Computer please repeat, the 'Aerotica' does NOT have a security breach air-vent fracture of any kind at this moment? Please respond, waiting."

"*No, Vanessa Signo, passcode 'monkey', Program Proficiency case-manager, New York Hosting, PPP Advise-Consulting computer analysis-system, to query: does the current status of deep-space operating vessel 'Aerotica', Rolf DeNeuri commanding, yes-no does this vessel suffer from an external hull breach for life-sustain. The query response is truthful in the negative, or, 'no', this vessel has no air-breach to space at this time. Sources available. Waiting.*"

This was a bit of a surprise. Vanessa of course knew, or thought she knew, that her man, Deveroux, would be a stone-cold slab of dead meat if the ship lost oxygen. She didn't love him, that would only ruin things. But she liked him, far more important to her personally. He was a complex man, and it was obvious to her, (perhaps not others), that he was mostly lonely, and depressed.

Deveroux was a very self-controlled person, he held himself in a strength, perhaps it was his grand-father's off-world blood, a drop or two, anyway. He was sexy and playful, but not extreme or rude, and also good-looking. But beneath all that, his heart was hanging by a black-and-bleeding thread, greasy with years of discovery, that in real life, cheaters win, liars prosper, strong-arm tactics never fail, and weak or frail people are easy prey for every variety of abuse. Not that he was innocent.

But the work, the so-called 'truth', these followed him around for far too long. So V's mothering instinct had incidentally become the ground-floor of their relationship. And her office had received the basic data from the 'Aerotica' as Puerto Rico, in direct contradiction to what the computer was now telling her. Everyone else, the Earth-based telemetry and telescopes, the Alpha-base scanners, and Deneuri, believed that the ship had breached air.

The computer was silent. "Computer, please establish a link to the radio-communications at the 'Aerotica', if possible from your network. Advise, please? Is this possible? Waiting."

The computer paused. Then, *"Thank you, V. Signo. Processing your request. Estimated time to completion, four hours."*

"Four hours! They'll be dead in 15 minutes from right now!!"
"Processing speed upgraded. Please hold."

CHAPTER 44: Truthful Information

"He is an immoral person," Cryolia Linsom Ee, Space Authority Security Force Commander, at the Rosebud Asteroid Alpha-Base, privately to a friend (during the crisis), regarding base-manager Doctor Menuda Wu.

The time of this crisis, concerning the Jupiter Program, roughly 2,415-16AD/CE, later came to be known as the 'Water Logic', perhaps for lack of a better term. Most folks had no idea, they were not concerned, and the details were not available, or hidden. The elapsed time amounted to a few weeks, at the climax of the ruin, confusion and fury that came, a total of about a month or so.

What was known was fairly simple, yet astonishing enough, as to be doubted: some far-off alien people, had placed powerful and mysterious giant machines deep within planet Jupiter, even 1,000's of years ago, to secure for themselves a supply of H20, from Jupiter's moons. When the future finally arrived, Earth's own similar chemicals and substances harvesting program at Jupiter, was by then in place. Working remotely, and at the site, through secretly advanced ships of their own, having traveled very far, and also working through human bodily vessels, such as the mentally seduced Second Commander of the 'Down', Second Commander Pearl, as things came to a head.

The ancient machines became active to their purpose and function, and 'all hell broke loose'. The results among them, good or bad, healthy or not, human or not, well-intended or conspiratorial, looked something like this.

It was happening at once, all at the same time. First, or most impressive, was the fireworks display of bizarre illusions and archetypical forms and images, somehow enlivened or given motion and power, emerging from inside the belly of the giant gas monster-planet, Jupiter. It was quite a show.

The Solo-Fighters with their clone operators were dispatched to fight them, but it was as if they were fighting ghosts and shadows. The Operation Odyssey plan was not fully ready, they had their own dates and schedules, based on approximations about the secret inner workings of the planet Jupiter. Soon there were thousands of the Solo's space-ships moving in-formation, and heavily armed. *But against what?*

"Solo Leader, this is Alpha Formation Attack Command, please respond with your position?" The radio-link, secure to the general population at the base and Jupiter moon stations, buzzed and hummed.

The OO fighters were specially designed and built over 20 years. They looked like somewhat smaller versions of the fast-ships, the 'needle-craft', like the 'Aerotica'. Sleek and fast, about the size of a football stadium, in length, they had special engines, and would never return to Earth at all, or land on any planet or moon. A single pilot could navigate one of these, but they mostly were manned by two clones.

The mafu-form 'men', or simulated men, much like guitar-player Marciel Penieur's double ('Marcus'), looked and acted just like real people, but they were not original DNA human beings at all. If they died, it was thought not to matter, or even if they needed to be destroyed, for whatever reason.

"Attack Command, this is Solo Leader for your formation, confirm please," came the radio-link reply.

"Thank you Attack One," said the dispatcher, or attack commander, (a team, not an individual), keeping track of their movements on a radar-type telemetry, back at Amalthea-base, in the relative safety of the impromptu command center for operations. "Solo One, just read me off your navigations and targeting, please."

The distances were great, it never seemed to match the blips and diagrams and avatars on their computer screens, used to 'watch' what was going on. Ever in motion, Jupiter and her 60 moons could be as far away as a million miles from Alpha base, a distance equal to the distance from the Sun to the Earth. Proper navigations shortened the actual flights, but there was no other way, all the fighters were stored at Alpha-base, mostly under-cover, there was no place else to keep them.

At that very moment, the Solo Leader was at the spearhead of a formation of 250 fighters, moving together at great speed.

"Thank you Alpha," the clone pilot replied, from the hot-seat flight deck of his ship. "Alpha, our formation is headed at 61.235-degrees center-point to Sol, off Alpha-base position at last mark, towards enemy at Ganymede orbital base."

"Copy that, thank you."

"Estimated time to enemy position is roughly 34 hours, to engage. Formation is 252 fighters, oriented in classic diamond formation, over perimeter-area of 10,000 nautical miles. All systems and ships are at full-functional status. Targeting will eliminate, destroy, disable or otherwise engage unknown alien object or figure in the vicinity of Ganymede station. Solo Leader out, please confirm data. Waiting."

Like any of the ships-in-transit, Alpha-base and other telemetry and rocketry tracked their movements and choices. That was the easy part, like watching a bowl of glowing ants moving toward other glowing ants, on their monitors, each ant identified and known as a 'real' object in the spatial-region or vicinity, at a certain speed, in a certain path.

Ants.

So this would go on and on. At Ganymede, and the other moons and positions within the more than 100 year-old Jupiter Materials Program, the 'enemy' had taken up his apparently stationary positions. A mystic or visionary might have wanted to map it all out like some zodiac or eternal calendar, with the figures like gods or deities, *all new, all-powerful'*, but of course the circumstance was not like that at all.

The Elder Podliakov, from the early years, who had worked mapping out Jupiter's mysteries when the program was actually being built, tried to describe what he personally thought the 'enemy' was all about. Following the dismissal of PPP agents Deveroux and Mendoza (without much explanation), and the considerable loss of Penieur's somewhat dual-purpose guitar classes, Podliakov and others on the 'hit list' that seemed to have ended the lives of Montrose de Montrose and Philby, could only withdraw and wait.

This wasn't so bad at all. They were all respected members of the program staff, they had their roles and jobs, and the main base and outer stations were secure enough. Podliakov found himself chatting with Amakmid Beautiful Truth, the Large Globe Object Music Society Director.

"Our project is ruined," Amakmid was saying. "With this all going on and whatever the fuss is about, it's impossible for us to go ahead with the scanning and mapping and our Grand Concerto Event. Too bad. We came a long ways for this. I'm sorry, go ahead, Professor. I interrupted your train of thought. About those...those things out there."

"Your fears may be justified, sir," Podliakov told him. "I'm not an expert, but I know a few things about what they're facing, the so-called enemy. The easiest way to describe them, in terms of their power and destructive capacity is, that they come from us, from our subconscious, or animal consciousness."

"That makes no sense, Podliakov, with all due respect," Beautiful Truth replied.

"Of course not," the older man answered. The two of them had found respite in his own rest-quarters, a room set-aside for him to live in, at Alpha-base, after the trouble with the PPP men. Podliakov was often normally at the station at Ganymede, the one run by Charles Benway, the harvest-director on-site. "They're not like us, my friend."

He may or may not have known that Amakmid was a recognized extra-terrestrial himself, in some small portion of his bloodline. Just like Daniel Deveroux. Podliokov then went on with his theories, as his guest enjoyed a drink, resting in a chair, listening and trying to understand, gazing at an aquarium of colorful fish-creatures.

"The alien mind is very different, and of an intelligence sufficient to cross the galaxy, from star-to-star. That feat alone is stunning, mankind will probably never even come close."

"But why are they here?"

"A long time ago, Jupiter was identified from far away, as having supplies of life-sustaining water, the ice-planets, and other water, huge amounts, really. Of course, we take our water-supply for granted."

"Like these fish," Amakmid said, laughing to himself a bit. "They swim and move in the water in the tank, but they care nothing of it, have no awareness of what it is, and if it is removed, they quickly perish."

"Yes, well, the same mind, or intelligence, created the ghost-machines inside Jupiter. How, I have no idea, I would guess a type of seeded compression, like a jack-in-the-box on a 1,000-year wind-up."

"Fine," Amakmid answered. "I am not without intelligence of some kind myself. Please explain. What is a ghost-machine?"

"This is what I am telling you, I don't really know. This is the enemy. They form like plasma or energy-types, organized very finely, or with a very acute design, and powered by temporary local energy-draws, much like our own solar-panels, or solar-energy cells. But they draw power from everything around them, without much discrimination, except of course the way the energy is used. And you, and me, and all the people involved out here, are among the energy-sources they use. Including our so-called consciousness or awareness. This is why I say, they come from us, in a sense."

Beautiful Truth watched the fish in the tank. Some were bluish, neon types, some bigger, some rather bland, some very slender, and also star-fish, rocks, little plants, and bubbles. "I think I know what you mean," he said. "We use a similar sort of system with my Large Globe Object Music. The scanners draw from everything, at every level."

"The machines the aliens designed can transform from any local energy source, atomic, sub-atomic, solar, motion or kinetic, actual life forms like human beings, and fish too, I guess. In a way, the energy sources are limitless, given they know how to unlock the types. Even very basic matter, something as dull as iron, or the dirt and rocks of Mars, or the stones in the asteroid belt, are not unavailable to their use."

"Yes, yes," his guest said. "And so no matter how hard we attack them, even our biggest bombs or destructive forces, they somehow can turn that energy into their own sources, and even grow stronger."

"Something like that, of course I am not really certain of it all," the old man replied. "But yes. A tar-baby effect. We attack, but it accomplishes nothing, and even makes them stronger. And then of course their actual material purpose, or their actual goals and activities, are concealed or hidden, because they are also drawing upon conscious human brains and awareness, to create the illusion of magnificent angels or amazing visionary archetypes, and incredible frightening giant pink elephants."

"Seems theatrical. Like a circus."

"They don't discriminate, why should they? If it is a material object of some kind, good enough, that will do just fine. A planet, a ship, an asteroid, a huge source of frozen water, a fish in a fish tank, a clone, the Sun. So, for my sense of it, these fighter ships they've sent out will do no good at all. They may as well be shooting their guns and missiles and their hell-fire, at a magic heat-mirror."

Even the space-program's best thinkers and most advanced science-and-astronomy workers, could not genuinely fathom what they seemed to be facing. And Podliakov was among the oldest, and best thinkers they had. So to some extent, it was hopeless, and there seemed to be no chance whatever for success. If 'enemy' was so far advanced, as to reach into their very souls, what could they do to repair the obvious damage, or even save their lives?

As the formations of specially designed fighter-craft found their way to Ganymede, and similar sets of defensive space-craft were dispatched, the battle-lines, (more like circles), were unbalanced. The twin orbital stations at Ganymede, assigned for hauling up materials and processing and loading into the large tanker-ships, were under a bizarre kind of siege. Here, there were about 200 or 300 regular workers. Each of the major moons (Europa, Io, Callisto, others) had come into the territorial oppression of whatever it was that had been released, from within the silent gas giant, Jupiter.

"Diamond Group Leader to Diamond Group Formation, we're approaching orbital status at Ganymede. Setting up for planetary navigations, to be transmitted to all fighters. Please prep to missile-array targeting, and weapons to forward. Instructions to pilots from this position. Confirm to formation-group tracking at this time. Thank you."

The ships were spread-out safely, to 10,000-miles, at this point numbering 250 ships. These were large enough space-craft (they had to be, to cover those distances even that fast), that one may have assumed that the planners of any battles, were intending the ships to deal with an enemy like themselves. But that would have been easy, or at least a more fair fight.

There was more inter-link radio. "Diamond Leader, this is quadrant-three flight-control, Airship-133. Captain, what the heck do we shoot at??"

And to be truthful, the leader didn't know. Earlier scans and tracking had shown Ganymede to be possessed by a more normal vessel. It wasn't a 'space-ship', but had an elongated hard-surface form, seeming unguided by intelligence, or a crew. More like a satellite. But those Alpha-base data-streams changed by the time they arrived, and the planet now seemed shrouded by a dark cloud-formation, very dense and black (not blue-indigo like the Mother-Night of the Abyss). It spread around Ganymede (which is very large, the ninth-largest globe object in the Sol-system), not completely, but of a visceral substance-type or material density. As the Diamond Group Formation executed her navigations, and the planet-form was more obvious, it was not pleasant to see.

"Let me get more information, Airship-133," the Group Leader said. "Truthful information, that is."

The attacks continued.

CHAPTER 45: Wake Up and Enjoy The Crisis

"Mother nature wins every battle. But why, why oh why, did my loved one have to die?"--from a song by musician Marciel Penieur, Jupiter Program Amalthea-base, 2,416AD/CE

Rolf Denueri, Commander of the 'Aerotica', felt helpless, trapped, and somewhat afraid. By this time in Sol-system space travel and labors at Jupiter, the machines and technology were at a very reliable level. Accidents were rare. Likewise, ancient alien transplanted 'water logic' sentinel ghost-machines, suddenly released to do their work, also quite rare. This fact was not lost on those involved, the planners and staff at the mainstay Amalthea-base, the pilots and ships and orbital base loaders and 'miners'. They also felt helpless, trapped.

"It's like waiting for the other shoe to drop, like a ton of lead, right on my pointy head," the Commander commented, still on the flight deck. "I get the idea whatever that thing is, it's enjoying my anxiety. What do you think?"

"We're screwed," the Second Commander said. "It's gonna' win."

There was no air-leak, or hull-vacuum breakage, at the 'Aerotica', as they had thought to be true. The ship's systems, even at the most basic level their science could demonstrate, had indicated this very serious emergency. An air-breach to the vacuum of space, left unattended, could kill them all within only a few hours or even minutes. The ship's emergency alert had sounded loud-and-clear, a signal-flare type SOS radio-message had been quickly dispatched, in the hope of some kind of rescue, were they all to suddenly lose life-sustain capacity, and the staff, crew and passengers had quickly gotten into their temporary life-sustain 'space-suits', mostly only air-helmets and vacuum-sealed 'bags' that would protect the fragile lives Denueri and his Second were responsible for, there in the empty, weightless bliss of nothingness, where the ship had departed on her way home, only to be delayed if not destroyed.

But there was no air-leak. It was an illusion. If what had come out of Jupiter's deep insides would destroy, it would be by illusion of destruction.

It took a few hours to figure this out. And yes, it was a panic, to be sure. Not the kind of thing a serious space-man could ignore. The meters, hull-security indicators, oxygen-level and life-sustain controls, and alert-systems, were trusted impeccably. They had to, they had nothing else to go on. So the emergency proceeded as if quite real.

Yet, after a few hour's worth of near-death fears, waiting patiently to suffocate, as if by a strange mercy extended, perhaps an accidental lack of interest in their death, upon investigation, they would all look around, somewhat dazed, to find they had plenty of air, there was no life-sustain breach, or hull-fracture at all, and there was no danger. In fact, none at all.

The only remaining fear was that the hypnotic effect that had led them into this terrifying delusion, would do so again somehow, and against this there seemed to be no defense whatever.

"What a relief," Deveroux was finally able to share with Angel-Face Mendoza. "It was a only sophisticated illusion."

They and others were now able to remove their personal life-sustain gear, breathing freely. Only hours previously they had both been in the deep-freeze suspended-consciousness transport cells, ordered by Doctor Wu.

"I guess," Mendoza said. "What the hell was that all about?"

"False alarm. You can go ahead and get out of that monkey-suit."

"False alarm? An external hull air-breach of a deep-space ship 500-million miles from home with fifteen lives aboard? False-alarm? Like, who screwed up that little detail?"

"Well, at least they woke us up. My head hurts. Not sure I really care for suspended-sleep state transport at all," Deveroux said. "I need to find out what's up with all this, if possible. We're helpless. But at least maybe we can pitch in with the battle somehow."

They began to pull together their personal items. In the sleep-state cells, both men were treated like patients in a hospital, with semi-nude 'gowns' or robes, IV-tubes, life-sign monitors, fluid-elimination kits, oxygen-masks. As planned, they were to be asleep for the entire voyage back to Earth, a period of at least six months or more. With the crisis, however, they were now both wide-awake, and at least a bit pissed about things in general. They were informed on their status, then busy with other problems.

"That Menuda fucker at the base screwed us blue, Dev'," Mendoza said. "I'm surprised he didn't just have us killed. The two science-guys are both dead. For-real dead, you know? Who was behind that? Him or his friends."

Their consciousness and full awareness was to some extent welcome. Of course they could do nothing of any consequence, there was no button they could push or genius-remedy that only they knew, and the others didn't. But they were capable men, all-hands on-deck, nothing wasted in terms of human resources, a good rule during a crisis. The PPP was respected as a reliable and trustworthy global-Earth resource. And ship's Commander Denueri understood, things were not as they seemed.

Deveroux found his ordinary ship's deck-attire, they were all very similar. Mendoza did the same. Deveroux's basic plan was to learn as much as he could, then help out 'as-able', and see if things improved enough for them to eventually find their way home safely. The pull-over jump-suits resembled togas or tunics.

As guests and passengers, they were updated on the general situation, then left to their own. There was really nothing to be done, for their part. Even just two or three sleep-state days now behind them had also taken a stress-toll on them both.

A bit later, Deveroux was connected by inter-space radio to Vanessa Signo, who had been trying to reach them from the Planetary Program Proficiency HQ in upstate New York. The set-up was in a tech-resource area for passengers aboard the

'Aerotica', with some privacy. Standard communications protocol confirmed the link.

"Just enjoying the crisis as best we can, Vanessa," Deveroux told her. "It's typical space-travel stuff. Nothing ever really happens, and if it does, it's way too late to do anything about it. We're basically trapped at one of the Jupiter moon's. They were shipping us back home as frozen popsicles. Something went wrong, giant robots or some cartoon crap like that. Probably invade Earth and enslave all mankind or something."

"God dammit, Deveroux," Vanessa replied. She was genuinely upset. It had taken hours to connect to the 'Aerotica', and the conflicting reports told her that her agents were either already dead, soon to be dead (within only a few minutes), or the victims of some cruel hoax from a so-called higher-intelligence with vastly superior powers. A cruel hoax might be preferable, given they were all still alive. She did like the guy, after all. "The tracking said the 'Aerotica' had breached air and everyone aboard was dead. I was a little concerned."

"Well, just think of me as a ghost, then. Tell the truth, you were always a little cold about me, V," Deveroux said. "You know?"

"No, Dan, no, that's not true. I mean, I didn't actually know your grand-father a few generations back was from Cygnus or Hydra or whatever planet it was. But it's not like I don't care. I felt really good about San Francisco and that full-body sex toy of yours. Don't get personal. I like you, but..."

Deveroux laughed. He had been in deep-space on assignment for the PPP, for nearly a year, at that point. And another year previous to that, doing the preparatory research back home, hoping to pinpoint the troubles with the Jupiter Program, as they had with other assignments.

"Vanessa, you need to communicate with management and the General Council, concerning our assignment out here. By now the 'water logic wars' out here must be known, at least somewhat. It's not something they can hide. It's totally out of our hands. Our assignment is officially a complete failure, without a doubt. We'll be lucky to get home alive."

"Yeah, I know," Vanessa replied. "As your assignment case-manager, it is disappointing. But it's not your fault, okay? Just get home in one piece. K'? Deveroux? I like you, okay? I mean personally. Really."

It was as near as she could come to telling him that she loved him. He had trouble with the concept himself. Romantic love, marriage, fidelity, kids and family, were not very popular by year 2,416AD/CE. Most people felt sex was a purely recreational or athletic activity. As much as two-thirds of Earth's population was also infertile, and could not even procreate at all. *Love? Surely you jest.*

"As horrifying as my ego and pride can be, I really can't take full responsibility for the current crisis out here, sorry. I mean, horrifying to myself, that is. Pretty arrogant to end up trapped in this tin can out here waiting for something worse, you know?"

The radio-link was stable, but the 'Aerotica' was not going anywhere, at least for a while, though of course in-motion. The Solo-Fighters had arrived at the various points-of-contact with the ghost-ships that had emerged from within Jupiter. Much of the battle was commanded from the Rosebud Stone (Amalthea) base, and some of the efforts were successful. The alien-ships could not totally escape material reality, and once their illusion-making powers were overcome, Solo-Fighter commanders found the core-elements to be much like satellites, or simple un-manned 'ships', placed in orbit at points near Jupiter. So the strategy was such as sending blind men into a dark passage, seeking what had blinded them.

Additionally, the hypnotic effect of the 'mind-control' qualities, was less so where the mafu-form clone pilots were concerned.

"Robots fighting robots," Vanessa answered Deveroux. "I guess that sums it up."

"My report will be more detailed," Deveroux told her. "If I get home, that is, in 'one piece', as you say. It may take some time for me to sort everything out. Clones versus alien mind-altering hypnosis machines. Does that sound right?"

The call went on somewhat longer. Vanessa was glad to confirm that her agents were still alive, as well as all-hands at the 'Aerotica'. There were some other minor official protocol matters between them, then the radio-link went down. Deveroux drew back for a moment.

At Earthside, the distance between them seemed hard to grasp, as if he was really a ghost, or dead. Vanessa remembered they had already declared Deveroux dead, some time ago, to avoid prosecution by the power companies on his previous case. He did good work, he was well-liked at the agency.

Her thoughts were dreamy, somewhat, to visualize the man, drifting there above, like some kind of an angelic-man or flight-born lover, who she may never see again. It was the nature of things, but obviously anyone would wish for complete safety and success.

The 'Jupiter Water Logic' problem was quickly becoming old news, as far as the elite insiders were concerned. As near as anyone could tell, or would tell, it wasn't quite as bad as they thought. The Jupiter Program would probably survive. The materializations had not actually undertaken a great deal of actual destruction.

The bases were not obliterated, and there was no general slaughter of the space-workers, pilots, or mafu-men clone-pilots. Reports and details did show deaths, almost entirely among the Solo-men Fighters. Twelve others had perished, since it all began, under various circumstances, including an attempt to physically approach one of the alien satellite objects.

So it wasn't as if the circumstance was not serious or of a genuine concern. But if the general wisdom of the Galactic rarity of water was behind it all, and the other side was peaceful in its own goals, to acquire water sources, then it made sense to 'let the professionals handle it', and give up on a great deal of fear, back home at least.

Formations deep inside Jupiter had exploded into new orbits around the gas giant planet near Earth, and taken up stationary permanent new positions. Why worry about a thing liket that, after all?

It was also the case on Earth that no matter how horrid things were 'out there', no one really cared, among the general population, being a hidden matter. Life goes on.

In a moment, privately, Vanessa Signo realized that she was probably 'in love' with Daniel Deveroux, and was painfully yearning for his safe return, so they could be together again. And she hated that she loved him.

"Don't try this at home," she said to herself, as she left the radio-link kiosk at the PPP offices, in upstate New York. It was a clear, sunny day, with few clouds. She tried to visualize in her mind's eye how it was that the blue sky of daylight hours, concealed the stars of the dark hours, and those never moving, or seeming not in motion at all, the 'firmament', silent, eternal, still, at peace.

And somehow in her little girl's heart, so was she.

CHAPTER 46: The Galactic Posture

A battle, a Water-Logic War, it was not normal, there in the distant Earth solar-system planetary area near Jupiter. Always in motion, that world and its moons, and the 50-million miles plus distant Amalthea HQ of the main operations management, also in motion, not even actually visibly affected or seemingly disturbed by it all. The scale and size of the planets and spaces in-between, certainly beyond the naked eye to discern any ships or activity, were difficult to actually 'see'.

Earth history would record the period as about 2,416 to 2,418 and even far beyond that date, Anno Dominni (AD), or Common Era (CE). As with most wars, it wasn't so much the battle itself, and the losses or the dead, but which side would control the 'spin' or legacy, as to what had actually happened, down the road.

Probably few would ever really know the truth. In this case it was very easy to invent, because the enemy were determined to be robot ships of highly advanced alien design, the origins of which were completely indeterminate. So, the Earth-system planners could come up with whatever legacy they wanted, and they did, like filling the the blanks of an empty dream. The official communications looked like this:

"In-House Only (classified Wu by penalty): Jup. Resource Command data-log 789B-22b: Re: Operation Odyssey, M. Wu/confirmed; threaded details to all-points (participants): details-info-update on conflict and attack-formation success/destruction as recent internal planetary-object ('alien') orbital-ejected robotic space-transit 'ships' have been under assault and clear-present danger to Jup. staff, stations, property, technology :: OO attack ships and un-manned 'solo' machinery dispatched on rotating schedule or repeat assault to disable, destroy, dismantle, eject or capture 'enemy', now five weeks from Earth-calendar date, 10-21-2415, outcome uncertain: 345 Jup./OO fighters lost or destroyed, all-hands un-manned mafu-pilot/clone

trainees decommissioned, (rested from use). Also Ganymede orbital outer station-decks suffered intense heat damage and radar-signal beacon-dish collapse (more); Null-zero 'enemy' robotic forms captured or destroyed, however, unconfirmed disappearance or unidentified location-navigational on three previously identified stationary-Jup. orbital 'enemy' configurations, no results from scans or trajectory. Earth-targeted movement highly unlikely and under review for Galactic-deep Earth/contact/attack/instrusion scenario, ref. Galactic Posture Event Planning. Science-tech advises 'alien' forms from within Jup./planet are pre-planned epochal local Galaxy distance-remote H20 or other substance collectors and to be regarded as 'illusions', or projection-mental-confusions/off-world non-Earth technology, as 'superior'/no redeeming contact potential or goals at this time. Other damaging results to Jup. regulars and stations. OO/Earth-based programs were miscalculated for epochal inner-planetary Jupiter release (alien forms); seeking advise and Earth-Govt./Global Security, travel-time delays effective cause/effect ratio command necessity --"

The reports and threaded discussion among all-points participants seemed endless. Tired and frightened as well, about it all, and the fearful demons unleashed from the 'illusion-makers' within Jupiter, and notions about off-world Galactic cruelty, the two agents (researchers) for the PPP, were useless to any battling or wars, and wanted very much to be back on solid ground at home. Having been unfrozen after less than three days in the deep-sleep transport, aboard the 'Aerotica', Devereoux and Mendoza were now required by law to cooperate with the emergency.

Weeks passing by, with more fruitless chasing away of phantoms created by the ancient machines from within Jupiter. Menodza dreamed of his Marta, in Belize, and his children. Deveroux, for all cold and calculated reasoning, felt alone, fearful, isolated, indeed very much as if he had died, prior to departure from Earth. The two friends could only gaze together out the

'Aerotica's various portals and communications-links, and vid-screen viewers, to wonder what had really taken place. The so-called 'enemy', mere zodiac-wonders, now monstrous in some bizarrely vigorous semi-stationary anti-Earth activity, were to them much like strange, giant birds, that someone else must deal with. It was much a waiting game, for a useful truth to redeem the circumstance.

"I like the one at the Io moon, don't you?" Deveroux commented. By now they were simply in hurry-up-and-die mode like they all were, still on-board the 'Aerotica'.

"What do you mean you like it?" Mendoza said. "That's the enemy."

"Well, yes, sure," Deveroux answered. "But the one at Io is more like a circling ring of fire type thing and has all those colors, have you looked at the scans? It's just cool, I think, I mean, personally, that's all. Of course it's evil and all that. Obviously."

"Traitor. Judas."

"That's not funny, Angel-Face. You take that back."

They were both much as two angry children, at this point, more than two years into the PPP research on the Jupiter-program. The journey into space had been a waste, in terms of any ordinary research that the Planetary Program Proficiency agency had taken on in the past. For whatever reason, it was a disaster, for them personally as professional global servants, and also as friends and individuals.

"All right, I apologize," Mendoza said later. "We may as well be art critics at war with the colors someone has used to defeat us. It's pointless."

A more specific analysis of the circumstances by which Earth conducted the so-called 'war', showed the enemy as organizing into a semi-permanent sort of 'zodiac' near and around planet Jupiter. Commanding officers counted 60 of the enemy ships, more like robot satellites, but some of these were apparently fading in and out, as if of an invisibility, or ephemeral nature.

What was really worrying global Earth powers, and the Jupiter program management, was the idea that the 'visitors' would never go home, and were commended from afar to be stationed there around Jupiter, as if to remain forever. It was a new situation, if even the Galactic Posture Plan had attempted to foresee such a thing. But it now seemed very serious, concerning the sovereignty and local rule of Earth and her various stations and platforms and outposts, at that time.

If 'they' never went home, or were not overcome or defeated or in some way controlled, the Jupiter Program would fall to a strange new order of power, or end completely. It was to the global Earth like a too-soon introduction to an unknown Galactic family, with unknown rules and logics and means, Mankind now to tippy-toe into participation in a family of worlds and peoples like ourselves, far away in our own Milky Way galaxy, some of them of much greater technology and science and powers.

Equals? Hardly. Step-children of some ancient parentage? Earth cared nothing of it, we still had many of our own problems and our own answers. But surprising and astonishing, many 1,000's of years in the making, this was assured.

"Under new ownership," said the Astro-Biology water-specialist, Hesidom, who worked in the reservoir system. His partner for the current shift was another reservoir task-force team member, a man named Henrique Valdez, who controlled various reservoir pumps and pipes that were used to move the large quantities of water, from off-station transports, or to places needed within the Alpha-base. Valdez was a dark-haired Hispanic, educated to his task, a veteran of work in space. Hesidom's comment was not very funny.

"We can't even see them!" Valdez said. "They're like ghosts! The station-platforms at Ganymede were toast, they said. Melted by searing heat! The fighters had found the core-robot satellite beneath the maya-illusion projection, and were set to blow it up. Once the thing recognized the attack, it fought back and melted half the base at Ganymede. We don't even know how, but it was no illusion. It will take years to repair, if it ever is repaired at all."

"And it's still there, in stationary orbit, right? Is that the intelligence?"

"Apparently," Henrique answered grimly. They were both now again staffing the deep-asteroid water-reservoir tank control room shelf or hallway, a large series of technology and control-monitoring rooms, where any operations concerning the reservoir system were maintained.

Neither of them had much to do with the so-called war regarding the planet Jupiter's new children, the 'visitors'. And they were glad to be out of the loop, in a certain way. But as with all the J-Program workers and stations and ships, everyone worked together. So a certain amount of information was common to them all.

"Under new management, for the entire solar system," Hesidom quipped. "For a thousand years."

"We have other problems, and I probably won't live that long," his co-worker said. "Check this out, I could use a second opinion here. I've been monitoring this for three hours. We still think there's something wrong with the inter-tank pipes and external loading tubes on Reservoirs 22 through 28. You tell me, what does it look like to you? See? This is the tank temperature-control measurement data, for all of those, 22 through 28."

Hesidom had seen the same data-screens, many of them had. While everything else was going on at the same time, with the planetary military objectives and the mystery of the zodiac arrangement of the visitors, the Alpha-base asteroid deep-rock reservoirs were experiencing a series of screw-ups, that were somehow connected. The scientific imagination that needed to recognize and understand the problem was also spinning visions of horrifying invasions and alien take-overs, or bad water, or worse. Even their best science could hardly compete if the invaders were creating mind-altering experiences, like a very clever weapon, that trumped other types, even harmlessly, in a sense, except for the enemy's goals. So Hesidom's job included working with Valdez on whatever was needed, within the reservoir system.

They must have water as well.

He moved from where he worked, to view Henrique's data. "All the water is kept at temperatures that make it possible to use for normal functions, and also to be safely stored. Obviously the deep-space abyss is frozen, below anything serviceable, intense cold. So the reservoirs are supposed to stay at 45-degrees, too cold for a hot bath, but potable. We use the water for everything, only a small fraction. But look at the measurements on 22 through 28, the same group, the big tanks."

Hesidom reviewed what he could understand at a glance. The logs showed a steady 40 to 47 degrees for all the tanks (except those where the water was actually kept frozen, as ice). But Valdez and others had highlighted a week's worth of logged data at Reservoirs 22 through 28, and it was easy to see that the temperatures would spike, for portions of the tanks near the bottom, to as high as 102 or 104 degrees, from the base of 45-degrees. Then after a few hours the temperature would drop back to normal. The strange thing was, none of them could figure out what was causing it.

"It's that THING," Hesidom said. "The goop monster."

"Sure, sure, that 'thing', not exactly very helpful, Hesidom. Let's assume there's something down there, like they crashed the wing-set assembly and it sank to one of the intake pipe shelves, right?" Valdez countered. "And whatever it is, it's crawling around like some goddam virus or giant octopus, and it has a body-heat, or some reason it's warmer than the normal water temperature. That's what? Another illusion? And where the hell did it come from and how the hell do we get it out of our water tanks?"

"Drain the tanks, do a dry inspection," Hesidom said. "If it needs water to survive, it might even die."

"We thought of that. At least six tanks are affected, maybe more. That's like 400,000 cubic-tons of water. The other tanks are full, or nearly full. So there's not enough space. We could off-load to one of the whale-haulers, but with the competition against the invaders or whatever they are, the big tanker-transports don't want to dock. Too risky. They don't want to be melted down to molten metal lava, you know?"

Hesidom paused. "Well, just don't drink the water, I guess." He laughed. "Look, Henry, if we can't find it, but we know it's there, then we're not in the wrong to keep the place safe and under control for the safety of every living soul on this base, right?? Giant fungus, alien robot, water-sucking salt-monster, does it really matter?"

"Not funny. Now you're on my nerves, Hesidom, that's two not funny jokes in a row."

"Sometimes I find sarcastic humor and humorous reflection very insightful. It helps me with new insights as a joyful and harmless process I have applied to hundreds of difficulties all my life," Hesidom said.

It was to laugh, so they did, and the darkness and fear inevitably lifted just a bit.

CHAPTER 47: Sleaze In The Solar Breeze

"In the vibrational realm, where neutrons dance and sounds and noises comfort Man, the ear is not the thing at all, not at all," Amakmid Beautiful Truth, Large Globe Object Music Society, 2,417, AD/CE, from the Sky View Platform Deck, Rosebud Asteroid Observatory.

For the Jupiter Program deep-space operations manager, Doctor Menuda Wu, there at the asteroid-base (the Alpha-Base), his private kingdom, the crisis at Jupiter was rather like a somewhat annoying distraction from his personal enjoyment of life, and little more. Wu was not the sort who doubted that death was the end, and nothing at all beyond, for himself, or for anyone, and this view set the tone for all his decisions and choices, those he made for himself, and those he made for others. It was a sterilized, narcissistic, bitterly hopeless and grandly wicked concept of life: grab all you can, in every known form of sustaining personal acquisition, use, use, use, then surrender at last to the Nothing, like a creature of no real use to the Universe but as a plaything for his own desire, and not much else, and no one else. Greed and selfishness hardly described his temperment.

Wu was Wu, it was all very plain to him, and if others didn't understand, it was their loss, and their weakness, only to be exploited by himself or other creeps who kept the same view. And it was very effective and very seductive, the money, the power, the position.

So, if some alien species of sentient awareness had planned to bother him (personally) from ages past, with a distant remote-controlled revelation of prepared and planted machines that would for whatever reason suddenly emerge and start their strange works, Doctor Wu, no warrior and no great friend of Earth's people, was only left to his tower of comforts and other machinery, to wait until things calmed down enough and he could do what he wanted, instead of what he must. The path of least resistance meant he knew his powers and controls and servants and ships and general goals, and beyond this, the Jupiter Operation he managed was 'just a job'.

Menuda Wu served his masters well in this way, a solid administrator with no heart, and only enough brains to get the essential program tasks organized and in-place, season to season.

"Just get it over with," was his standing command directive to all concerned, as far as the current crisis.

And no warrior, not him. Not a pilot, never a space-walk in a deep-space survival suit. He couldn't deal with the means and strength needed to weld a high-tensile aluminum beam from point-to-point necessary to hold a radar antenna in place, standard to spec and within tolerances. But he had a general idea. Doctor Wu had no real navigation talent, such as plotting the course of ships from Jupiter to back home.

This was a demanding task, to be sure, he avoided such things and left them to others. If base functions for life-sustain oxygen, heat-and-warmth, water, food, electric and communications went down or failed, he could not repair any of them himself. He didn't really like astronauts and space-workers, they were superior in social status and physique, so he avoided them.

Doctor Wu also had no appreciation for self-defense, martial arts, weapons, or the personal protection and safety of others. He could not calibrate measurements for orbiting objects in motion such as other asteroids or Jupiter's moons, even with the help of computers. He had no idea how to deal with air-lock ship-to-ship docking methods. He had no idea how to deal with materials handling and loading to the large whale-hauler transports, or how those ships docked. Doctor Wu had a general knowledge of all of these systems, but he had never actually done any of those operations himself. This was just his management style. He was pretty good with computers, though.

"Just get it over with," he told his subordinates, when they asked him what to do. And somehow, deep within his thoughts, he felt certain the current crisis would never really be 'over with', at all, and that the prospect of an inter-Galactic handshake over Jupiter's resources, was at least amusing, as far as his role.

Plans for the Large Globe Object Music Society's 'Concerto Solaris Jupiterno', were cancelled. For Beautiful Truth, this was a heavy sadness and loss. His group had come all the way to the Jupiter base, completely equipped and staffed for the artwork involved, a distance of 700 million miles and a journey of several years. And prior to that, the same music-society had spent many years, mapping the vibrational dimensions of other celestial globes, the Earth's moon, and Mars and her moons, in particular, and Earth as well. Much time had been spent mapping what they wanted to create at Jupiter.

The technique was 'new'; the science-involved was not unusual, it was merely a matter of connecting the scans and energy profiles they were already using, to musical electronics and sound-output and recording gear, then creating tonal patterns that functioned like music to the human ear. If for no other reason than the cancellation of loss of all his hard work, Amakimid may have resented the alien, or extra-terrestrial, 'action', they all felt they now knew to be taking place at Jupiter. But then, he was also not totally without his own un-Earthly origins, by a few generations anyway.

"It's over, it's lost, all our work, all the plans, the travel and preparations. I may never live to see this opportunity again," he told his composer, a musician from Sydney, Australia, Robert West. West was about 60 years old, and had a white goatee beard, and a childlike appearance of wonderment about almost anything. West was the sort seeming confused when operating an automobile, but actually well ahead of most of his peers, in his area of expertise.

"Even if you die, or the robots kill us, they cannot destroy Jupiter, and the sounds and vibrations of Jupiter will go on forever, my dear friend. So it is never really lost at all. Someone else will build on our work and create even greater music. From the Sun itself! Can you imagine!!" West said, but it was obviously to cheer his co-worker up about the loss of the planned concert.

Amakmid smiled and nodded the affirmative. They had in fact such a plan. The scans and vibrational maps, at various levels of interpretation, were being developed and the revised, all the time. When the Globe Object Music Society began, it was a very simple idea. They scanned Earth's moon, with radar, and gravitational, and motion, and X-ray and infrared, and in other ways. The scans were 'real time' or 'live', and then linked directly to the synthetic music output sources, high-powered synthesizers scaled to standard musical notations, and some not, which were more free form.

Over time, they learned that they could create 'music', directly from the heavenly objects. Or, as the ancients had said, the 'music of the spheres' had now become a reality. But for classical music-lovers and those who enjoyed traditional 'songs', like rock-n-roll or the music of someone like guitarist Marciel Peniuer, the LGOMS music was somewhat boring, and tended to sound like Earth's deep-ocean blue whales or sperm whales, calling out to each other in the depth of the blue waters.

But it was authentic, and composers like West had learned to nuance the scanned vibrations and sound-source output for a wider variety of music forms, some truly delightful. Amakmid compared it to surfing the Hawaiian island Big Kahuna coastal waves on a musical surf-board.

"Wipe out," he told West, who smiled. Australia also had big surf ocean waters.

Who knew the innards of planet Jupiter would send forth their implanted alien-architectural forms, during their visit? As usual, the space-program and various Earth-based space-arena activities were tailored for educational and research learning and students, with many visitors and users who applied for the privilege.

Even as the OO fighters were circling their various targets for obliteration, Menuda Wu and his staff had as many as 100 on-site University students and academic or science-based visitors and guests to deal with. Wu fancied himself quite the charming host, and it wasn't easy to explain some of what was going on. So he didn't even try.

And as some knew, and some didn't, among these clandestine and scandalous affairs, were Wu's 'private parties'. Or, orgies, to be clear, of the sort Deveroux and Mendoza had walked into prior to their deportation. *Yeah, sure, yeah, sure, yeah, sure.*

They were pretty bad, and over time, every sort of debauchery and Bacchanalia was pursued passionately, embraced and encouraged. No sexual pleasure was withheld, short of felony-level offenses. And other pleasures, decadence. Earth-science and pharmacology had by year 2,416, invented some astonishing basic brain-chemistry drugs. Amakmid Beautiful Truth was trying to understand Wu's request, concerning the use of his equipment and skill with the vibrational planetary music scans of the LGOMS.

"You scan the planets and asteroids, correct, and program music from that?" Wu said to him. "But this is for a party with my guests. You scan my body, and some of the girls, the same way. Smaller scale, obviously. You circle the vibrations and scans back to the orgasmic pleasure centers of the brain, using the Circulator. It's harmless, guaranteed. It has colorful soft cloth wraps and a foam gel, with energy-sources to the skin and body parts. We recently had them shipped in, for the base workers and general staff health and well-being."

"The Circulator?" Amakimid replied. They were together in one of the chamber rooms beneath the main operations-center, where Wu had asked him to his office for a consultation. "Yes, yes, I've heard of it. An advanced sex toy of some kind. Wow, yes. Sexy."

"In other words, I am asking you to do your LGOMS scans and musical vibrations, as a sexual experience for a private party. See? We'll have a real thrill, I bet, eh? Uh, by the way, if you care to join us, that is not a problem, maybe with a guest? A man of extra-terrestrial bloodline, you know, and with the girls, I mean. The parties are within my private patio and hot-tub decks, very discrete. What do you think? Eh?"

Amakmid paused. "That is the most perverse and kinky thing I have ever been personally exposed to in my life, sir," he said. "Of course I'd be glad to help any way I can."

They both smiled and shook hands like frat boys visiting their first whore-house. "See what you can do," Doctor Wu said. "Give it a few days. I'll be in touch."

"Sure," Amakmid said. "So, how's the inter-galactic invasion going?"

Wu paused. "Uh, well, we're losing, big time, totally out-matched for technology and science advances on the part of the enemy. Earth has no real chance to defend itself. Even worse, the real 'enemy' are so advanced, they don't even have to show up. Remote-control from who-knows-where, a billion light years away, and they're still kicking our ass. My personal best guess, they will be taking over Earth's Solar-system soon, as a sort of long-distance water source. They could give a crap about Earth or it's population. Just my opinion."

"Oh, I had no idea," Amakmid said. "Do they like music?"

"I have no idea. All I really know, they like water."

Amakmid waited, then started to leave, feeling depressed about what he had heard. "Water and music. All that lives enjoys water, and music, too, yes?"

"Yes. And sex. See you soon. Dismissed, for now."

So he left Wu's office, back to his own rooms and staff. Amakmid didn't really like Menuda, either, he had that sleazy sort of smooth-skinned gloss and perfumed glean, a sort of fleshy-pasty-stringy unhealthiness and overall gloomy mid-range Siddartha-on-the-skids modernity, that others recognized as the fruits of wealth, power and position. For Amakimid, it was not very significant. He usually got along with all sorts of people.

CHAPTER 48: Just Another Orgy

"At this time in our common historical journey as planet Earth's human population, only somewhat more than one third of Mankind can reproduce or have children, due to the final disposition of the Acquired Immune Deficiency Syndrome, 100 years ago or more. As we now believe, the Galaxy we call home is also probably home to other beings and selves or sentient forms very like ourselves, on other worlds too far off for us to ever attend or travel. Do they have similar problems? Are we, and other self-aware species in the Galaxy, heading for extinction? So, as the Jupiter Program we created to supply ourselves with limitless clean fuel and energy from hydrogen and helium and other materials, now reaches a new time in its ample provision for us all, we must agree, and the Council must agree, and consider: does it sustain, or does it drag us into a future ruin that our children must manage? Space is no place for children, for our children, or for birthing babies. Here at home, the Earth does not perhaps deserve any more children, if we cannot treat them well. There is a justice to that. Thank you, and members of the Galactic Event Posture Committee, for your time and attention during this crisis. Cid Bixi Mimim."

Alphonso Carbon III, Global Earth Space Authority leadership, at the Conference on Current Affairs' general assembly, Jupiter Program, Puerto Rico base, Dec. 15, 2,419AD/CE.

Another orgy was again in-process according to J-Program Amalthea-base Manager Doctor Menuda Wu's wishes, there within the confines of his very private upper-deck suite and private playground. Wu was only popular by virtue of his authority and ability to bring together such parties. Otherwise he was mostly disliked. The Alpha-base pretty much ran itself, as it was designed to do. He was the head-honcho to be sure, but much like an unwelcome spare-part in his own machinery, as far as his own day-to-day activities, he was frequently bored and unoccupied.

Yes, he had friends and confidants, people close to him, but the wealth and provisions needed for the Jupiter Program's Amalthea headquarters to function well, also meant that Wu and others lived lives of considerable luxury and accommodation, such that his parties and orgies were common-knowledge. To be invited was a coveted matter. Staff and crew felt they were good for morale and even healthy. This was a different time, more than 400 years beyond the turn-of-the-Century at calendar year 2000 AD/CE.

Visitors arrived at Wu's Wonderland via the same elevator that Deveroux and Angel-Face Mendoza had used, (escorted by guards, prior to their eviction as frozen popsicles from the base in deep-sleep transport). Wu hardly remembered the incident, in his world they were just another trouble, bothersome emissaries from powers back home who would stir his waters, probing into his realm, looking for problems and causing him more work, more meetings, more lengthy and difficult explanations and more grasping, exhaustive apologies.

Of course he knew the program was corrupt, full of problems and malcontents. The Planetary Proficiency Program was an agency global Earth authorities dispatched to hound-dog troubled program hot-spots through focused research and first-hand review. Efficiency experts, connected to very significant planetary power-groups. Dan Deveroux and his Mexican pal may as well have been librarians searching for misplaced Dewey decimal system books on some dusty bookshelf, long forgotten, out-of-sequence. They were no real threat.

But with the inter-Galactic deep-interior Water Logic War, the Jovian explosion of hostilities that had then flowered into the sudden reality of a problem, for Doctor Wu and many others, two more frozen popsicle men, in deep-sleep on their way back home to Earth safely aboard the 'Aerotica', made sense. It wasn't done a lot, but they really were quite safe. Besides, the two researchers weren't much fun at his parties.

A long, low hallway opened from the entranceway elevators that moved guests up from other levels of the base and its wide array of rooms and tech-centers. Wu's party-playground was designed in a sort of Roman Holiday, Italian Modern patio style, with tiles, brickwork, bronze railings, pools of steamy hot water, statues and columns, plant life and flowers or running ivy clinging to walls and lattice-work, and strange fragrances and perfumes. No, there was no ornate throne, as a king or tyrant would have. But there was a stage, such as for music performances or speeches.

Everywhere all around, the high-tech glitter-glaze and glisten of view-screens, entertainment modules, computers, telescopes, and even some of the new-style fortune-telling 'minooks' that Second Commander Menima Pearl of the transport-ship 'Down' enjoyed, here and there. Massage tables were popular, gym-equipment, food-buffets, a locker-room, sleep-areas or private rooms, and showers. Truly a pleasure palace.

"Yes, I have a rather small penis, it's true," Wu was saying to a woman, with him at one of the buffet tables. He was genuinely shameless. They all had cloth wraps and robes, with sandals. There was music from somewhere. The woman was an Asian staff-assistant, about age 28-years, she worked in human resources and staffing.

Wu continued, hardly even humiliated by the topic of his lesser manhood. "About, well, you know, when the flag is down I'd say a solid three or four inches, loose. Some of the ethnic men have much better equipment in that sense. I understand these days I can have it elongated for greater pleasure, I may try it some day. The surgery, I mean."

She giggled, trying out some oily pickles from a silver tray. "I always wanted one."

"A penis?"

"Yes!!"

They laughed. "We can share, dear, we can share." Wu commented. They took their finger food and moved around the room. Other couples and guests were enjoying the hour, maybe

30 or 40 people. Yes, there was a certain amount of coitus going on, copulation and groping, but most folks were discrete enough. Many people were nude or semi-nude.

Among them were the guitarist Marciel Penieur, and his identical clone, 'Marcus'. Exactly why Wu had invited them was questionable, and they suspected somewhat that status as frozen-popsicles was ahead for them, too, much as Deveroux and Angel-Face. But for the moment, things seemed to be working out. They were nude as well, with appropriate robes.

Marciel had his guitar gear and was pondering if he should do any music performances, or maybe have some kind of sexual experiences, like others were.

"More than 500 clones!! Gone, dead, toast, obliterated, never alive, if they were even legally ever alive, technically, or official persons, that is," said the duplicate, 'Marcus', commenting idly on the recent battles in space. They watched the room from one side of the large patio. Marcus was in fact a clone himself, he and Penieur were much like brothers, and also best friends. It was hard to tell them apart, both in appearance about middle-age, skinny or slender build, Caucasian with dark skin, rustic.

"Look at that asshole, Wu," Penieur joked beneath his breath. "He actually thinks the women like him!! Ha! What a joke!"

Wu was now entering a pool of tepid water, with the Asian gal (her name was Mika), another male, and two other attractive ladies, one was ethnic-African, the other Caucasian. The other man with them, as it turned out, was the Large Globe Object Music Society composer, Amakmid Beautiful Truth. Amakmid had an Eastern look, like a Hindi man, somewhat.

"An original human being, Grade-A, farm-bred, corn-fed hambone DNA, yeah, true," Marcus answered. "Everyone has an asshole, Marciel. But I think this tyrant has two."

"I suppose," said his 'father', the original Penieur. He smirked a bit, chuckling. "So what, the clones in the airplanes? I mean the fighters, the space-ships? Yeah, true, more than 500 gone, I guess you're right. Tough deal, they gave up their short lives to save us. Think positive, at least as my clone, you have it a little easier. With the guitar shows and traveling around and all. I can get you a legal identity, Marcus, if that's bothering you. You can pass. All it takes is a certain amount of..."

"Money, power, and corrupt friends," Marcus added, completing his thought.

"Right. Authentic Earth-born human DNA helps, too."

"I got that. Second-hand, but I got that."

"I know you do."

They watched, they waited. Amakmid Beautiful Truth was helping Wu and his guests out with setting up the 'super sex-toy' they called 'the Circulator'. The thing was basically a set of pumps that moved a special magnetic gelatin-fluid stimulating chemical, through some colorful boa wet-wraps, that lovers could enjoy together, and then was charged with a very mild energy, that stimulated the epidermal layer of skin and sexual organs in a thrilling way, or so they said.

"Like your blood is on fire," the advertisements reported in bold print. *"Get circulated!!"*

"Gawd, damn, what an idiot," Marcus (the clone) added. "Look at him! He's putting on a show!!"

They were all inebriated, or drunk, with all kinds of elixirs and substances. The current favorite was a beer or ale-type drink, with an ambrosia that included a variation on an ages-old 'beverage of the damned' with mild narcotics, once called absinthe. True enough, the technology made their sexy romp awkward, nothing new there. Naked and oiled flesh pressed and stroked for passionate release, as Beautiful Truth rigged several of the Circulator devices.

By applying what they used for the Large Globe Object Music concerts and recordings, with electrode-scans and sonograms, and also electromagnetic image-resonation scans, and others, for all three hot bodies, (and Doctor wu), there in the steamy pool where they played their erotic joys together, lovingly enough, Amakmid orchestrated the stellar sex like a true Mephisto. Other orgiasts had gathered around to encourage them and cheer them on. It was quite a show. Music filled the halls.

"Go! Go! Go!" they cheered, men and women in robes or nude, also sexually aroused, drunk, or just plain debauched. Wu, in all his fleshy glory, was the center of attention as the two women delighted him in various poses, slobbery, gross, ravenous. All three of them were draped in the multi-color fabric boa material, with the oozing gels and Circulator stimulation, on high setting. Electronic music beat tempo: *boom-boom-boom!*

Marciel yawned. He and his clone were distant. "Let me get this straight," Marciel said blandly. Marcus (his identical twin) had by this time found for himself a plain old beer, another Orange Buddha, from an ice-chest. "The scans on their bodies are in a feedback loop to the guy's rig, right? So he gets like their body vibrations and cellular level activity, and muscle-movement, body temperature, blood flow, like that, all scanned into his computers."

"Mmmmm, ya, that's what I got from it," the clone said. He sipped the ice-cold brew.

"Then the signals go back into the sex-toy machine thing? And it gets them like a real crazy buzz while they do it, is that it? Sort of like an orgasm pleasure-amplifier?"

"Mmmmm, I dunno, ya, I guess."

"Why not just do pleasure-stimulating implants directly into their fucking brains, for Chris'sake!!"

"With that son-a-bitch, he probably would. Maybe more intrusive."

The party group watching Wu and his playmates were still cheering, or now going at it themselves, on either side. The scene resembled a carnival dunk-tank or maybe mud-wrestling at some bar back home, such as at the Hotel Caligasta where Ralphie-Boy 'Hector' White, the disgraced captain of the doomed 'Ferrous-2', was still hiding out, on a hot summer day in the West near Reno, Nevada, wiping the salty sweat from his forehead in 105-degree heat. But no one ever called him any more, at all. It was over. *"Go! Go! Do it! Get it! Woooooo!!!"*

Just then two very attractive women approached coyly and joined Marciel and Marcus. The guitar player and his clone noticed immediately: the women were identical twins, half-naked from the waist up, reddish brown hair, green eyes, pale skin with freckles, and two sets very healthy-looking mammaries. Of course Marcus and Penieur appeared to be identical twins as well, though this was not exactly true. The men turned towards the women, smiling. Even their freckles seemed perfectly identical.

"Well, hello, hello," Marciel said.

"Hi there," one of the gals said. "Quite a show over there. Geez. What pigs!!"

They gazed again un-believing at the scene, across the patio, things had now reached a tribal fever, it was a genuine gang-sex moment. Beautiful Truth had somehow given them the ride of their lives, a thrill they would not forget, the couples sweated and panted, as if exhausted, their eyes rolling back in ecstasy, moaning, heaving, throbbing, stroking.

Hot water and perfumed steam vapor rising, others having sexual moments as well around them, others dancing, many nude or in colorful outfits and painted flesh, wild and outlandish, a crazy scene, a demented bliss, without malice, an innocence, living lust and writhing rapturous pleasure pit.

But then...

Without warning, there was suddenly a sort of unusual jolt or rocking motion, throughout the whole structure, like the entire Alpha-base had moved or jerked a fractional amount, all at once, from some outer force, unseen.

BbboooooooooMMMmmmmmmmmm!

There was no accompanying noise or sound. They all were shaken, no one failed to notice. It was like the Rosebud Stone at Amalthea, at about the size of the smallest of planet Mar's moons, had hit a bump, which was impossible.

A cry went up from the crowd. All 38 of the naughty guests in Wu's pleasure palace were thrown to one side, some grabbing at rails or walls, a few falling over, depending how drunk they were. A food-table buffet toppled over with a loud crash. One of the ornate Cupid statues fell over and broke in two pieces. Water in the hot-tub pools splashed over its rims.

The whole group let out a cry: *Whooooaaaahhh!!*

The guitar-player and his clone, and the two women they had only just then laid eyes on, eagerly dreaming of their own loves, were also tossed to one side briefly. Then the earthquake-like motion stopped.

BbbooooooooMMMmmmmmmmmm!

"What the hell was that??!!" Penieur cried out loudly, holding steady onto a brickwork wall.

Things settled down, there was a hush and dread. They were all experienced space-travelers, and knew, almost nothing they could predict would jolt the entire base and the Amalthea planetoid itself like that. Maybe some sort of rouge comet or unobserved meteor out of the Sun had hit them. Dazed, some of the partiers sensed things were not as they should be, and calmed down enough to check it out, serious and even grim.

Then an alert-sound on the in-house public address: *Beep! Beep! Beep! Your attention please, attention please. This is an emergency alert for all base staff and crew, please find your service-task technical posts for emergency status. Emergency alert. This is not a drill. Emergency alert, all base staff and crew, please find your service-task work posts for emergency status immediately. This is not a drill."*

Maybe the party really was over.

"Dammit," Marcus, the clone complained, getting up from the floor where the mysterious jolt had knocked him over. "Twins!! And now this!! I hate space-travel!!"

CHAPTER 49: Anarchy At Amalthea

"Please, a moment, sir, Mister Brandeis?" said Second Commander Menima Pearl, from her station at the spacious and grandly technologized flight deck or helm of the huge whale hauler, the Jupiter Program space-transport called the 'Down'. Pearl was fully able to perform all her usual duties as Second Commander, during the now month's long crisis. Their ship, along with any of the others caught up in the bizarre circumstance of the extra-terrestrial manifestations from with the huge gas giant planet, had been directed to a stable parking orbit path, which was very large or wide in circumference, but not taking them home yet.

Powers handling things didn't necessarily want a lot of traffic back to Earth, for fear of wild tales and unfounded rumors, or even truthful reports, given the program was in jeopardy.

Pearl had been alerted to scans of the source of the rocking jolt, or earthquake type energy-shock, that had interrupted Doctor Wu's orgy. Navigators aboard the Down, monitoring all corridors and pathways more or less constantly, also held views of the base, the nearest and most dependable deep-space resource for any of the ships-in-transit. The Alpha-base was well-equipped, well-staffed, well-maintained, and could deal with almost anything that might happen as the ships and men dealt with the on-going task of harvesting and delivering the huge volumes of helium, helium-3, hydrogen, H20, and various valued non-life materials, mineral-fluids, and so on.

Brandeis, the whistle-blower who had spent a year or so gathering details and data about the failures of the Jupiter Program, himself a long-time space-ship commander (at the 'Down'), had grown less fond of his Second Commander than he usually was, over the course of their current voyage. It wasn't her fault, she was a pleasant enough woman, sane enough, and very skilled and efficient at her work. They even enjoyed laughing and chatting about things at times, socially.

"Yes, commander, what is it?" he answered her.

They were now in the middle of their duty-cycle schedule, in their jump-suit pullover uniforms. Brandeis was deep into a review of a plan to get them home, escaping the monstrous illusions of the robot-conflict. Pearl was at her ordinary station, where she mostly kept track of command-function ship movements and pre-planned changes, hour-by-hour, often bored to tears.

Brandeis stepped over towards her, releasing a staffer helping him, and some computer machines they were using. The 'Down' would only be going home once base navigators and those waging war against the mysterious shadow-machines from within Jupiter found a way to move them safely out of danger. The ship was a long ways from the base, relatively, at that moment.

"Well, it's all part of this whole thing I guess, but I'm reading just now from these scans, the base has been attacked, or, I guess you say, attached, by some kind of ship, one of the robot-machines from the planet. See for yourself."

She paused. Now of course Brandeis was interested, but he also knew how to remain calm and not accept such a thing as truthful without confirmation, there in the cold indigo Abyss.

"There was also an SOS from base-communications to all ships," Pearl added.

"Just within this past hour?" Brandeis asked her. "You seem rather bland about it."

"Yes sir, only just 45-minutes have passed as I confirmed what I thought I was seeing. Of course I always remain calm at my duties, sir. There have been many such distress calls."

The 'Down' Commander quickly eye-balled her telescopic monitoring screens, knowing what he was looking at well. There were three mains, one on infra-red, another as a common-view macroscope, and a third as a type of advanced radar. As they had learned, the infra-red level scans revealed the illusion-making robot ships much more plainly.

"Show me," Brandeis told her curtly.

So Pearl did as she was asked. The Alpha-base was configured as an image tagged with an identifying icon, in real-time and relative distance to other objects, marked with numbers that showed transit times (more efficient than miles or kilometers for their navigations). The infra-red scanning telescope settings could be overlaid against the other types, so the elusive forms they were fighting also appeared on her view.

Many of these had not been identified yet. Some of the enemy forms seemed more powerful than others, or had specific functions. Remaining squadrons of the Operation Odyssey Solo-Fighters (the specially prepared clone-operated fighter vessels from Earth, stored and launched from the base), also could be viewed as flickering specks. The 'Aerotica', the Jupiter moon orbiting harvesting platforms, and other large ships like the 'Down', also held positions. Their ship was well-equipped, Menima Pearl's station-view could bring up an accurate map of the entire region of the Abyss they were concerned with.

"Magnify Amalthea-base view, Sector 14.2, times 500 enlarge, please," Pearl spoke to the voice-activated machine. Now the tiny dot that indicated the home-base at the Rosebud Stone filled the screen, so they could see it more clearly. These were powerful telescopes and viewing scans. Brandeis leaned closer.

"The base reported an emergency, they were not certain what happened, or what caused the effect, but the entire facility was shaken by a sudden jolt, like an earthquake," Pearl informed him. "All levels were effected, all we received so far was the basic data-log, and the emergency beacon hailing. It's not good, sir, this is very serious for them I think."

"Show me, Ms. Pearl," Brandeis said.

"But at this view here, see that shape? It looks like an external form that has attached itself to the underside of the base, like a machine, or one of the robot ships from the main planet. It's not one of ours, I can tell you that much. The base does not have a docking facility on that side of the rock itself."

Brandeis considered what seemed to be yet another in a long series of bizarre effects associated with the revelation of the Jupiter off-world machines, or 'alien robots', that had suddenly catapulted them all info chaos. *God, what now?, he thought.*

He didn't believe in God as an active participant in human affairs, generally. *Ask the goddamn Quantuum Mind, woman*

"How bad was the shockwave or jolt? Damages? Do we know?"

"We can only really read some of the scans that track mass-density vibrations, there has not been a report from the base," she answered him dryly. "But my guess from those, it was a rough ride. Like an earthquake, the planetoid is as large as a small moon, so, I'm guessing a 7.0-Richter scale event throughout the entire mass. But the base is still functional for basic life-sustain, I can tell that much, from these, and the streaming data-logs."

"Fatalities?"

"Only the clones in the fighters, that I know of. Not sure. There were other human deaths, though, I believe at Ganymede, they got roasted. The base population has not reported deaths from the sudden shock or jolt, the earthquake, at all, yet."

Brandeis took over some of the view-screen kiosk controls, froze the image and filed it for analysis with other specialists at their own stations elsewhere aboard the 'Down', with an expedited command-request from himself personally. But he could see what she saw, as well as anyone: a mechanical looking form, about the size of the 'Aerotica', was very near or actually attached to the bottom-side of the irregular shaped moon where the base had been built as long ago as 80 years or more, even over decades of work.

To lose the Amalthea-base for some reason was unthinkable. It was not one of their ships at all, they would have had an ID beacon reading, and also as Pearl had pointed out, Alpha had no docking at that spot on her underside (with no real 'up' or 'down' direction in space anyway, for navigational purposes). So it was an attack of some kind, or would be interpreted that way regardless. It was also extremely rare to even consider such a thing, in all his years as a pilot or astronaut. *They must be ape-shit over there, he thought.*

"Your opinion, Commander?" Pearl asked him coldly. Calm hardly described her demeanor. Bitterly determined to survive was maybe a better description. Her hunky female largeness and womanly strength seemed to firm up, like a bulky athlete ready for a contest she would never really engage.

"Disaster, total disaster," he replied. "That kind of a jolt? Across all levels for the entire asteroid, including all facilities? They could have lost power, systems down, life-sustain, communications, antenna and sunlight loss, even gravity-generators off. Not to mention whatever that thing is, or whatever is inside it, or what its purpose is. You know, fucking giant alien monsters could be boarding and taking over the base command-center even this moment, for all we know. Disaster. Amazing."

"I doubt that, sir," Pearl responded. "The distances involved with star-travel for any living thing prohibit any actual movement of life forms from world to world, even within our own galaxy. With all due respect, it's not possible."

"Yes, I keep hearing that phrase over and over during this evil crap we're experiencing now for the past 40 days," Brandeis said.

"Fifty-eight Earth-day 24 hour cycles since the first manifestations, Commander," she corrected him.

They paused, considering options as usual. Their training and background as space-workers told them instinctively to take their time. Other members of the flight-deck crew and staff had gathered around where Pearl was stationed. There was a certain amount of troubling chatter and talk back-and-forth among them, as the news spread among them. *The Alpha-base? Lost to the monsters? They would never get home!!*

"What about one of our shuttles, reposition the transport and dispatch as a rescue effort?" one of the communications operators suggested.

"We have ten of those, but they are very small, only maybe 20 passengers max," Pearl added. "And we couldn't get anywhere near the base, short of a complete re-calculated navigation and a three day maneuver just to position this ship for the movement of the shuttles. The shuttles couldn't even launch at all, anywhere short of that. We can't reach Alpha in less than eight or ten days, sir."

"Does Earth's Space Authority know anything?"

"Find out," Brandeis answered. "And keep a constant on-going hailing-request back to Alpha until they answer. Starting right now."

"Yes, sir." The radio-operator hurried off to see what could be done.

"What ship do we have that is closest to Alpha right now, then, Pearl? Bring up that information, please," Brandeis said.

The whale-hauler's Second Commander skillfully worked her boards, various items related to over-view and ship-to-ship ID. It only took her a few moments. For just a moment, the cruel allegations against her seemed to fade.

"Here we go," she said. They all listened closely. "For distance-capable vessels, we have the 'Utilitarian' at Europa; travel-time is two weeks; the 'Cosmic Mother' and sister-ship 'Brokerage', both at Ganymede. Those are transports like this one. Travel time at least 30 days, no good. Three squadrons of the OO fighter-craft, very little use unless they want to blow the thing up or shoot it down. One group is two days out, the other five days. Each formation has about 120 vessels. But those are clones and they have to be told what to do by electronic remote. And they are very small, no real passenger space at all."

"From Alpha-base," said Brandeis. "I mean, the remote command aspect."

"Yes sir, so, obviously no help there," Pearl continued. "The mafu-form clones they are using can't actually think for themselves at all. They could only be directed by other human operators by special command-controls."

Another long pause. Pearl listed more of the available ships: "There's a science-explorer ship, the 'Bethlehem Star', near Io moon, on the dark side of Jupiter, the far side. No good, too small and even longer travel time, 60 days. The planets have moved and so has the base. We're talking 90 million miles for some of these.

We have the 'Bolton', a fast ship, but also maybe 20 days, or I guess faster, ten days out. Small ship."

"What about the 'Aerotica', the one they placed our friends from Program Proficiency on in deep-sleep? I wonder how they made out? Where is she?"

More keystrokes and voice-commands, the ship's computers promptly responded. "Keep in mind, Commander," Pearl added as she worked. "Many of these ships are in no position to help any effort at the base, if needed. The so-called alien presence is still keeping them fully occupied with their own troubles. What if the situation at Alpha-base is just another advanced illusion created to confuse us? Could be true, we know they are doing that now. And it works, doesn't it?"

Commander Brandeis was at a loss. They seemed helpless, yet again. Communications back to base were being attempted.

The 'Down's' full array of scans and messages back to Space-Authority Earth were applied. A strange space-ship or star-ship had whacked into the Alpha-base with the force of ten 7.0-Richter scale earthquakes back home (given the effect was throughout the entire Rosebud Stone and also the base facilities). The same vessel or machine, its intentions and capacities or endowments totally unknown, was now attached by some means to the base, home of almost 3,000 people from Earth and elsewhere.

"See if you can spot the 'Aerotica'," he told Pearl. "And if we have a chance, let's go for one of your Quantuum Mind channeling sessions. See if can contact the All-Knowing and clear some of this up."

Pearl smiled at the idea. She knew it was not a very serious option. "Yes, sir," she said.

CHAPTER 50: Star Ship

"I have to say, you just never know about rock-n-roll. You know?"-Space Authority Security Officer Cryolia Linsom-Ee, to Security Officer Le Van Ho, at the Jupiter Program Alpha-base

What they had observed, from the navigation and command deck of the long-delayed in-transit 'Down', was now in play there at the Alpha-base, millions of miles off from their position in the dark-blue skies. There was of course a vast supply of planetary water, good old H20, at some of the Jupiter moons (notably Io and Europa, where frozen pure water was available in supplies as large as entire oceans of salt-water back home on Earth). The Alpha-base also needed her own water supply, and stored millions of cubic-tons in a reservoir system, beneath the main facilities and buildings. This was simply good practice when the base was designed, they knew it was needed, and had many uses, including life-sustain.

By year 2,418-19AD/CE, Earth technology could desalinate her ocean's briny salt waters for drinkable H20, at a much more efficient pace than in previous periods. But to extract clean-burning hydrogen fuel from water was not working out, and valuable helium, and helium-3, was also rare, on Earth. So the Jupiter Program's 100 year mission was not to gather water, exclusively, at all.

What the program's designers had not perhaps anticipated, was that the age of the Milky Way Galaxy, perhaps billions of years old, and considered at least potentially populated elsewhere among her trillions of stars, by sentient and self-aware beings like ourselves, capable of inter-stellar travel--somewhere out there, water, the basis of life, was much more needed and rare than here in our own humble Solar-system.

So what Commander Brandeis and his Second Menima Pearl thought they saw, through the 'Down's telescopes, following the unexpected 'earthquake-like jolt' that disturbed Doctor Wu's party-guests at his orgy, began to play out in this way, up-close and personal:

The heartlessly annoying alert on the Alpha-base's in-house public address continued to drone: "*Beep! Beep! Beep! Your attention please, attention please. This is an emergency alert for all base staff and crew, please find your service-task technical posts for emergency status. Emergency alert. This is not a drill. Emergency alert, all base staff and crew, please find your service-task work posts for emergency status immediately. This is not a drill.*"

It didn't take them long to figure out what had happened. Taking action as some sort of reasoned response was more difficult, and in a sense, would occupy Earth's space agencies for even hundreds of years after that. From the Alpha-base's own navigation and space-traffic center, or command deck, one of the helmsmen quickly reviewed their external sensor-scans as his habit of duty, and reported to the on-deck Commander.

"It's a ship, sir," he said. "Not one of ours. It must be from among the alien forms that appeared out of the main planet. It's basically nudged up against our lower deck and underside, and has attached itself to the main asteroid, just below the water-tank system."

The Duty Commander winced. For him, his job was also often very boring, he only had to keep track of any in-coming guests or docking ships that needed rest-and-recreation services for the crews and space-workers, or maybe log and review navigations among their numerous materials-harvesting platforms at Jupiter's moons. It might not have been so bad if the ship was operating normally, following some sort of protocol, even if they were from 'out there'.

But an alien ship? From another world? Unannounced? Death didn't seem as harsh compared to his xenophobic dreams and paranoia concerning non-human sentient star-travel enabled life. He was a tall, dark-skinned man of African descent, well-trained like them all, confident, assured in his work, and currently terrified.

"I wouldn't call that a nudge, helmsman, would you?" the Duty Commander said. "How confident are you in your assessment?"

"I can show you images from our external cameras, Commander," he answered dryly. "It's a ship, and it's NOT one of ours, and it's hugging the bottom of this base. Judge for yourself."

"I will. Bring up all your proofs and scans, do it now," the Commander ordered.

"Yes sir, I have them here."

The Duty-Commander started shouting. "Communications! Link to the water tank crews and monitoring staff right away, that's where it's connected to. Life-sustain and systems operators, please! Find out damages in the reservoirs, the water tanks underneath. Also get your teams on any and all base damage, every fucking inch of this rock. If there's a micron's worth of an air-leak anywhere on Alpha, I want to know about it in the next 10 minutes!!"

The regular on-site crew at the base command-deck for that shift, seemed numb for just a moment, stunned as they all were. Up until then, any hostilities or danger were 'out there', even millions of miles away, to be dealt with by the disposable mafu-clones and robot-piloted OO ships.

"That's an order, people! Job one is life-sustain and system's damage assessment! Don't worry about any goddam aliens! It's pointless, base-security is for that. Find me an air-leak, a crack in a wall, or battery damage, a broken window or leaking toilet, and you'll save all our asses if they broke anything when they attached to us. Go! You know your jobs!"

They all started moving to their stations and tech gear kiosks or controls and communications. There started up now a din and chatter of info-links and radio-talk, all the essential base-operations were reporting in, a wave of information and data, some cries for help, requests for orders, confusion and dread. Electric and power functions didn't seem disturbed, they had power, lights, and basic life-functions for air-seal, heat-and-cool, mobility from deck-to-deck. Most if not all of the command-control computers still worked fine. These were of course a very real comfort.

"Beep! Beep! Beep! Your attention please, attention please. This is an emergency alert for all base staff and crew, please find your service-task technical posts for emergency status. Emergency alert. This is not a drill. Emergency alert, all base staff and crew, please find your service-task work posts for emergency status immediately. This is not a drill."

The Reservoir System Control Decks were almost half a mile beneath the base central-command helm where all of this information and fear was being processed. Astro-Biologist Michael Hesidom, and his superior officers, did most of their weekly routines deep within the Rosebud Stone itself, keeping track of the water-tank systems, the water-quality, water-temperature, water-uses, loading water from external ships (not that often at all), and then dealing with the pumps and connectors needed for using the water on out-going ships.

There were other tasks they needed to keep up with for things to work right: the water was purified or processed for chemical-perfection, and any unwanted substances. Sometimes the water arrived in huge frozen blocks, such as the one that had broken free and floated away dangerously at the Ganymede station, prior to the crash of the 'Ferrous-2', so those types had to be handled differently, (melted). They also had to maintain the tanks and holding reservoirs, if there were any flaws, replacement parts, broken seals or malfunctions, all carefully monitored.

The Reservoir Systems Control Decks were a series of long hallways and upper-level transit-corridors connected by elevators and lifts, built into the reservoir tanks themselves, on 'top'. All their necessary computer-controls and scanning-monitors, level-detectors, pump-diagrams and operator-controls, were housed here.

As with elsewhere at Alpha-base, immediately following the earthquake-type event, workers at the Reservoir System were in a panic. They had already assumed 'something was wrong' with their tanks, given the mysterious disappearance of some of their water. A lot of their water was somehow leaking away, and also the mysterious temperature variations they had noticed. And then there was the incident with the wingset-rig, that had dropped down to the bottom of Reservoir 22, and what the dive-crew felt they had viewed 'down there': basically some kind of a monster or glowing alien fungus or giant octopus thing from another planet, moving through their pipes, and wreaking havoc.

"Probably just wants a drink," Hesidom said, with atypical good-cheer. "You know. Thirsty space-creature? Like in the movies? Just exhausted from the long journey, maybe only wants some rest-and-recreation, looking for a good time. Ended up here."

What they now thought they knew was straight from the base command-deck, and the Duty Commander's team: a UFO, so-to-speak, had attached itself beneath the base, being an unidentified space-ship, in a time when within the constraints of their Jupiter Program methods, such a thing was strictly excluded, and an unidentified anything, for obvious safety reasons, was always a problem, and very unwelcome. An external camera-video scan image had quickly circulated to many of the base positions. Hesidom and five other Reservoir systems-operators could now take a look, on their own computers. All this had happened within about an hour after the initial jolt.

For safety reasons, almost any area anywhere outside the base and at the entire Rosebud Stone, had some accessibility by view-screen cameras and video. Above, below, top, bottom, sides, many buildings, external landing pads, loading areas, gates, view ports, antenna arrays, solar energy collectors, external docking rigs, pumps and lines, lights, jets and storage, maneuvering: if they could see it, even remotely, the tech guys could figure out what it was doing or not doing, and assign work or tasks to repair, re-align, rebuild, replace parts, etc. The only major difference in this case was that they were looking at what was supposedly an alien ship from an off-world planet or home.

It looked like a very dark-colored, hard-surface or metallic wafer or long, flat, semi-convex, palm-shaped wedge or angular gigantic leaf, thicker towards the center, and tinged at the back edges with very bright colors, as though hot or heated, or connected to some power-source, like thruster rockets, only much different, such as for inter-stellar travel at high speeds, perhaps much faster than any of their ships in the Jupiter program, in order to reach the distant stars within even many normal human life-spans.

"Wow," one of the Reservoir Operators said. "Super shit from deep-space. Ugly fucker, huh?"

"Destroy it!!" said another man (named Ballard).

The group hushed fearfully at the thought. "How? A bomb? One of the Solo mutant fighter ships, or even a whole fighter formation? And what about us? That wouldn't put base hull integrity at risk? No thanks, Ballard, no thanks!!"

"Negotiate, then," Hesidom offered.

"Hardy fucking har-har, Hesidom," Ballard replied scornfully. "You speak alien-monster languages?"

There was a pause. "Look, our orders from the top-deck right at the moment are to check for damages, secure all systems for operational functions, and re-establish essential systems for any further activity," Ballard told them, somewhat taking charge. He was a bald-headed white man with a cranky disposition. "We're not really supposed to inter-act with any beings or creatures, or negotiate any bargains, or destroy any ships. Okay? Not that I personally wouldn't want to learn..."

Just about then, from down a few desks at where they were looking at pictures of the 'ship', there was a secondary local computer alert-sound: *Boop! Boop! Boop!* A bright reddish glow from the computer-viewer washed over the operator at the post, blinking off and on, as the alert continued: *Boop! Boop! Boop!*

"Holy shit! Oh my god! Captain,uh, Commander! We got a problem! Holy shit!" said the operator at the station.
Ballard, Hesidom and some of the others attended the man's upset exclamations. "Calm down!" Ballard said. "What? What happened? Just tell us!"

"Reservoir 22, 23, all the way through 30 and 32, are losing water, uh, really fast! The tanks are draining, like, totally! According to these meters, that's like, 400 million cubic tons of pure ice-water, sir! We spring a leak! A big one!"

Hesidom leaned into Ballard's shoulder in the reddish glow. "See, Ballard? They're thirsty. They're getting a drink, that's all."

The alert and reddish blinking PC screen glow continued in an agony of helpless watching and waiting: *Boop! Boop! Boop! Boop!*

"Relay the data to command deck," Ballard said. "Get a mini-sub crew into the water and get a dive-team down to the main pipe-way that goes between the tanks and to the pumps for off-loading."

"They're dry, we can't...the tanks would be empty."

"Why? Why not just let them do their thing? Maybe they'll go away?" the young Astro-Biologist Hesidom said.

Ballard paused. "Because it would open to the external abyss, asshole, and the vacuum would suck us all right out into the nothing like a bunch of deep-frozen fucking ants, that's why? Sound good? Not to mention other life-sustain. Yeah, they want a drink, very funny. They want a drink, maybe they don't care if we all die in here for lack of oxygen for them to get their H20. If the water-suckers have a way to drain us, after the water, next the vacuum pressure would collapse the pipes and tanks, creating a huge main-base external hull breach. From the tanks. Everybody dead in about four or five hours."

"Oh, I see," Hesidom answered blandly. "Yeah, you're right."
Boop!Boop!Boop!Boop!

"Acknowledge the damn alert and turn that stupid thing off. Relay the water drain data and situation to command deck immediately. Get your maintenance mini-sub ready with a crew. We have to get down there and cut them off. If the remaining tanks are dry or there's not enough water depth to launch, go down another way."

"Yes, sir."

CHAPTER 51: Thirsty

"We're trapped, there's no way out," Astro-Biologist Hesidom said, in a panic.

Hesidom had by then, 30-minutes or so later, found himself among those recruited to go down into the depths of Reservoir 22, with a dive team quickly assembled to deal with the crisis. Reservoir 22 was a central tank that connected to others by a long, deep shaft, that acted like a hub between other tanks and pumps. There were other reservoirs in the system that did the same thing, it was designed so that one in every six tanks had large water pipes, like spokes in a wheel, at the very bottom. The main shaft was large enough for the rescue-sub, (maintenance water-submersibles), to move between, about the size of a small road, or tunnel, back home. In this part of the system there was still significant depth of water.

The water was crystal clear, all around the small sub. Five crew were within, the pilot, the engines person (a woman), two external swimmers ready with dive-suits and tools, and a person handling life-sustain and communications-computers-electronics (Hesidom). It was very crowded inside, these were small underwater vehicles for repairs only.

The deeper parts of the water-tank system had rows of small lights, dim and tiny, seldom needed or even turned on at all. The submersible was tubular and longish, like a van or truck, on a small road, or tunnel, back home. It was a silver-blue color, and had portal windows, and moved by jet-streams of high-pressure water. The submersible was only intended to be used for maintenance on the tanks, from within. The jet-stream water engines hummed and rumbled beneath where the five-man crew were now beginning to perceive what Hesidom meant: they were indeed trapped.

The reason was that the alien vessel or ship, that had 'bumped' into the main Alpha-base framework beneath the reservoir areas on the Rosebud Stone, was sucking out their water at a fast rate. It also seemed to be fetching home some sort of giant slug, that had somehow been hiding away within the tank system, even long before the alien ship showed up on their long-distance scans at all.

Ahead of them, somewhere, in the tunnel, was the giant slug. Behind them, or in the opposite direction, was where the deep-shaft pipe-works were pumping out water, to what they now felt was the strange alien visitor's ship, external to the base stronghold, open to the Abyss.

The best guess now was that the slug was an alien water-doggy they were using like a canary-in-a-coal mine, to see if the water was wet enough for them, or truly H20 at all.

For space-workers, interesting and clever of perhaps historically unusual ways to die were sometimes hot topics of conversation, during rest periods or time off. In a way, with a view to the giant slug in the water tanks there at the Alpha-base, that small crew in that mini-sub team, might have reasonably thought themselves to be among the very first human beings in to truly see and witness a living creature from another world. Or this particular one, in any case. And this was 400 years past the year 2,012-18AD/CE. It wasn't as if planet Earth wasn't well-known among them to have hosted an alien or two, of sentient or humanoid variety, over the ages.

But as Hesidom now peered out the small submersible's front view-screens or ports, with the rest of them, it was fairly obvious. The thing was blackish, moldy-looking, slow but muscled, or powerful, much like a slug or snail without a shell. And big, too, filling the tunnel-way almost entirely, writhing around, tendrils of a prehensile form and snakelike motion. It must have grown quickly, maybe over many months.

What if someone had planted the thing there? Maybe one of the guests in the upper-levels, one of the travelers, the guitar-player clone, he was a strange one!! Or the science-students, with the planetary music scans, or the efficiency experts from PPP, maybe one of them was an alien spy, transplanted by off-world forces to seed the tanks with a tiny microbe that later became this thing, growing and growing, proving somehow to its masters that there was plenty of the wet stuff here. A wet-worm monster. Clever? Or just easy? When you've come 50-million light years to steal large lots of H20 from a populated star-system, simplicity gets the job done.

"We're trapped, there's no way! That fucker is going to crush this sub like a seagull crunching down a crawdad!! And if we pull back, we'll be sucked into the whirl-pool!!" Hesidom again cried out, for the rest of the team to calmly consider.

"What a way to go," the pilot said. "I'm going to crank the water-jets and ram him! Maybe it will back away and we can pass."

"Extend the bow winch we use for towing, on front beneath the maneuvering jets. Get some speed and pierce the thing's hide, bloody its ass or even kill it!" one of the divers said.

The mini-sub was rocking back and forth, they were at a depth of some 400 feet, within Reservoir 22. The plan was originally to find a way to disconnect from whatever ship or space-vessel had attached itself to, at the loading pump shaft. The ship they had identified after the big jolt was real, there was no doubt, they now knew it was really there, a certain danger, even if perhaps not hostile.

The dive-crew and sub would move directly to the connecting point and close down the water-flow, or cut off the pipe-flow, or even blow up the portal it was using, or seal it with a hard-forming instant plastic cement. If the enemy wanted to destroy them, it would probably already have done so. On the other hand, accidents do happen.

The pilot and engine's operator worked together to bring the small vessel's water-jets to highest speed or pressure. Even at high-speed it was only a small submersible, and would only maybe reach a few hundred feet-per-second. The internal electric motors made a high-pitched sound. They could also see that the movement in the waters surrounding the small ship were turgid with a flowing dark crud from the Wet Worm thing, a sort of goopy darkness that seeped and flowed outward around where it hid in the path through the tunnel between Reservoir 22 and 23. Hesidom could now clearly see, their water supply here was badly polluted, even vile.

The mini-sub was a good distance off, maybe half across the bottom of the tank, each tank being roughly the size of a football field. As the water-thruster jets kicked in at top rev's, the contraption lurched forward towards the monster. The pilot was not joking. He was going to ram it. The mini-sub gained speed.

Obviously, as dynamic circumstances may unfold, the Alpha-base Commander and Program Manager, Doctor Menuda Wu, was over-ruled, regarding his decisions and choices for management at the base. Mostly this was when his area of science-tech knowledge was insufficient, and also when decisions were perhaps too big in scope, such as the Odyssey Operations Solo-fighter craft and clone-men pilots, in anticipation of these galactic events. So at this point, with an alien space-ship possibly ready to board the base in a hostile manner, the Space Authority was really in charge.

This suited Wu just fine, from one of the recovery bed's in the Emergency Medical Services hospital rooms, where he was taken immediately following the orgy, and the upsetting jolt from the alien space-craft. The physicians who reviewed his medical condition were fairly certain he wouldn't live much beyond a few more hours longer as far as his biological passage as a human being.

Wu had suffered a massive heart attack, during the bizarre sexual encounter and clever application of the 'Circulator' sex toy, combined with the Large Globe Music Society's vibration/amplification scan-energy loop, enfolding he and his partners in a few minutes of extremely passionate sexual pleasure.

"He's not going to make it," one of the M.D.'s said to a nurse in charge of Wu's care. "I'm sorry."

She looked at him, sternly for a moment, and then they both broke out laughing, sort of a chortle.

"Sorry? Why?"

Everyone hated Wu and blamed him for the current crisis, it was just the fact of the matter. The women he had been intimately embraced of with his wild party were also happy to see him on his death-bed.

The nurse smiled. "This may be the only medical case I have ever heard of, in my entire career, of an adult man killed by a sex toy!"

The M.D. on-site shook his head. "No, not really, dear, not at all. Rather more common than you'd think."

What this meant for the safety of the Alpha-base overall, however, was that the Security teams, under Commander Cyrolia Linsom Ee, and her Second in Command, Le Van Ho, were immediately responsible for a coordinated effort to overcome the supposed invaders. Should they blow it up with bombs of their own? Send out space-walking astronaut soldiers to board the thing or pry it off the underside of the base with levers and machines? Or should they try to communicate with its operators, who may well be light-years away controlling things remotely, just like the other satellite-ships still in place around Jupiter at her moons?

"*Bith, dan vow tam beyon, ko tu withl mafieta, bin rat rat tom gyon vo tutalen, mic frommage tartarag, bi van wit, kat tan tan borfofa, borfofi, nona nan, vatomic, ya?*"

This was Van Ho's way of asking Cyrolia, if it might be best to somehow appease the intruders. But how?

"Le, they maybe only want water, you know? We have plenty, we have even entire planets with plenty of water, like on Io, or Europa, giant lakes of frozen H20, not even salty, lakes bigger than the skies over your mother's grave!! Why should our people die when we have plenty to share??"

Van Ho seemed to agree, but they were not necessarily available to the time needed to explain it that way, and respond accordingly. The main idea from their point of view, was to start shooting anything and everything that wanted inside the base and didn't happen to be a regular employee, or on two legs. *"Vattam bittin eedenit vittum bin vartabo, gitta nobnob fatyum? Gin vab not, fortum fit fit din gan narda bituinit fifintyunnuuunin, ya?"*

So, this settled it, and there was no dispute, and she agreed totally. It might save all their lives if they handled things less aggressively. If it was possible, they would appease the visitors, at least from a security point-of-view.

"Cid Bixi Mimim, Second Commander Le Van Ho," Linsom-Ee said. He smiled, but they knew, a crisis is a crisis is a crisis, and work in the Abyss was always only an instant away from certain death, given a certain level of mistakes happening.

The mini-sub diver-team submersible failed completely to interest the very large and annoyingly deadly Water-Worm Monster, even at 400 feet-per-second, as the winch-hoist assembly on the sub's prow more or less crumbled like a tooth-pick against its side, rather than actually draw any green or purple alien blood, as they had hoped.

It hardly mattered to AB Hesidom and the sub crew if the Command-Control teams in charge now had found a peaceful way to introduce a remedy or solution. Because, as they now learned the hard way, the powerful whirl-pool drainage of the reservoirs' main water supply levels, was dragging the sub, end-over-end, toward a certain conclusion of some kind, into the hub-shaft and pumps to the external loader-docker lines.

"That's it! Suits to lines! Get out if you can!" the pilot told them.

There was a certain amount of screaming as all this was happening, from inside the small underwater tool-kit submarine vessel they worked from. End over end, over end, over end, the silent waters cold and flowing, yet not nearly so deadly-sure as the frozen empty and vast, space itself, not the final frontier at all.

"Bith, dan vow tam beyon, ko tu withl mafieta, bin rat rat tom gyon vo tutalen, mic frommage tartarag, bi van wit, kat tan tan

borfofa, borfofi, nona nan, vatomic, ya?" the Second Commanding Security officer repeated himself, even as Cyrolia Linsome E listened to him again. She rather enjoyed the lilting, curious language he would sometimes use. ""*Vattam bittin eedenit vittum bin vartabo, gitta nobnob fatyum? Gin vab not, fortum fit fit din gan narda bituinit fifintyunnuuunin, ya?"*

Michael Hesidom, the youthful and courageous astro-biologist, was soon dead from a head-concussion during this effort, in the mini-sub, at Reservoir-22. Doctor Menuda Wu was later also dead from a cardiac event of uncertain type as well, at Level Four Emergency, soon after that.

"Cid Bixi Mimim," added Van Ho. *"Cid Bixi Mimim."*

CHAPTER 52: The Aerotica Rescues

"A chance to speak out, a moment to come forward in clear view of the general community, transparent, without penalty, bias, attack, or any need for immunity, just basic useful self-expression, to improve the individual, for the benefit of all. Is that it?"

"Yeah, that's it. That's it."

-- Vanessa 'V' Signo, Planetary Program Proficiency assignment desk, to Daniel Deveroux, research agent assigned to failures within the Jupiter program.

A Level Four General Emergency at the Amalthea-base was every bit as rare and unusual as the crash of the 'Ferrous-2'. It first meant they could all die or become seriously disabled for normal operations, due to some unforseen circumstance that was taking place. It meant secondly that there was some kind of essential system failure going on, either the power system, the life-sustain, critical communications, etc. Thirdly it was an indication that immediate action was necessary, whether it was organized or not, whether it was approved or not, whether there was time to study things further or not. The fourth level involved orderly evacuation plans in place at every base in the system for all persons. So, yeah, it was a big deal.

Angel Face Medoza, Daniel Deveroux's partner, had found his way into the pilot's deck, or flight-deck, aboard the 'Aerotica', which had been delayed in their voyage home like the other nearby ships.

Rolf DeNeuri, the commander of this 'fast ship', had taken a liking to Mendoza, for whatever reason. He was a very likable man.

"So what do you think, Hispanola? Que?" he said, joking somewhat, also grimly determined to survive.

As complex as the Jupiter system of planets and moons was, DeNeuri's navigations had eventually moved his ship into orbit around Jupiter itself, in a wide angular ovoid, with a very high apogee. As the gas giant herself roared in her orbit every ten hours to a full circle, his ship passed around her like a wheel on fire, very high in a circle, then narrow, then very high again, turning back every 18 hours. At the top of his orbit, the view back towards the Rosebud base, by high-powered viewer, was possible.

So about that time, nearing his high-point, he and others could review what they thought was happening, much as Commander Brandeis was able to do earlier. Of course the distances from point-to-point were vast.

The fast ship's flight-deck was smaller that that of the 'Down'. The 'Aerotica' was only a fraction the size of the giant transport. Her flight deck was fully comfortable, and fully equipped, connected to very similar gear and technology. Only five men could work in the space allowed at one time, but DeNueri's ship had sub-stations at other decks, like they all did. Angel-Face found he could share some insights with the Commander on a personal basis, 'off-the-record'.

"Not liking that," Mendoza said. "They got an enemy ship stuck on their ass. Not liking that. Probably killing them, or they're already dead from something else. You think?"

"Not according to our tracking and monitoring. All the bases are in contact around-the-clock, without a gap, there's no point at which any astronaut is not connected to the entire system, if he has the tracking," DeNeuri said. "It's a standard safety measure for a long time. According to our monitoring, there are presently 3,516 life-forms at the Rosebud base. And that's more than their full number."

"Does that include the duplicates? The clone-men?"

"Yes, it does," DeNueri answered.

"And you can track the life-forms over there by number, in other words, any living creature, by count?"

"Yeah, it's standard. Not 'any living creature', actually. Not bacteria or a worm in an experiment. But people, yeah, in general. In the last six months the same calculation had dropped by 42. But many of those were departures during the conflict, and others were deaths among the clones, or re-assignments, like you and your partner Deveroux."

"Okay, fine," Angel Face said. "What about the so-called aliens, if you will? Do they show?"

"Uh, no, they don't," DeNeuri answered. "The reason is, the figures are not based only on a simple biology scan, like body-heat or motion, or breath. The scan is based on electronic tags in the clothing. Most of the space-workers don't even know we use them. And most of them would be very glad we do, too, at least at a Level Four General Emergency, like today. I guess some of the malcontents feel it's spying. So, we can't figure on any alien beings or life-forms, like bacteria or somebody's gold-fish, over there. They don't scan, at least not in standard. But we can see their ship, the one that hooked onto them. That's what started all this."

Mendoza chuckled. "Sure, without a doubt. What does it look like, anyway? I mean, it's real, then?"

DeNueri turned and worked some of his controls, connecting to a sub-station. The 'Aerotica' was not in any danger, neither was the 'Down'. But they weren't going home, yet. The illusion forms from within Jupiter's high-pressure depths, released they now believed by some far-distant remote clockwork, were now like semi-stable clouds or small glowing white or dark star-shapes, some of them in motion, some out-putting odd energy and strange, unfamiliar signals, an alien zodiac arrangement of powerful and mysterious 'things'.

"Depends what you mean by 'real'," DeNueri answered him.

The previous ecstatic display for general confusion and 'water logic warfare' had settled down. But Jupiter program leadership had other fears. They had begun to feel that the enemy were not going home, the far-distant powers seemed to have intended to place them at Jupiter on a long-term, or even permanent, basis.

And this meant that Earth had new neighbors, and had entered an entirely different historic era, when available H20 was either to be shared with a thirsty off-world people, or not.

If the illusion form sentinels did not retreat, or could not be dealt with reasonably, Earth had lost the water-logic war.

"Viewer magnification, please, this is flight deck. Hello?"

A moment. "Yes sir, telescope and scanning. Go ahead."

"Bring me up here to my viewer on the bottom of the Rosebud, the unidentified ship or object we're looking at. Can you get that?"

A moment. "Yes, sir, checking that for you now."

Angel-Face was perhaps merely an observer, but it was obviously a tense situation. He and Deveroux had been un-frozen from their deep-sleep status, and were considered valued crew-members. Both were very experienced in all sorts of situations. DeNeuri held no particular loyalty to the choices of Dr. Wu anyway.

The sub-station telescope and viewer operator came back on the line, his voice piped into the flight-deck. "Commander? I can't get you a live image from Rosebud, at our current position, for another few hours, sir. Sorry. But we viewed the target, uh, oh, I guess 20 hours ago, and we have archived those images on data-base. Would you like to view that file now, instead, sir?"

"Yes, go ahead. Send it up. Thanks."

Within a few moments, DeNeuri and Angel-Face, along with two others on the flight-deck, could again get a good view of the alien 'ship', on his view-screen. as the operator said, the view was from 20 hours ago.

The screen was similar to a computer-output, large enough for them to gather around, but not really dominating the helm at all, as one might expect. It flickered a bit with electronic colors, and lines, and then the image cleared up.

"This was from a previous view-scan, it was not a 'real-time' view," DeNeuri said.

They had of course been keeping track of things at Amalthea-base, to provide whatever help they could. At the Level Four Emergency, DeNeuri did not need any higher authority than his own judgment, to decide what he might want to do. But the 'Aerotica' and other ships had standing orders to remain apart from any danger or 'conflict', and to remain in the Jupiter region of general space. They didn't want them running home right away, because they might be needed. And now they were.

"Commander, what if we reposition her so telescope can get a real-time for you?" the navigation officer suggested.

"Yes, go ahead, talk to the sub-deck for telescope and find out what's needed, maybe we can swing the orbit wider before the turn. Go ahead, then inform me right away if it's possible. Work it. This will do for now."

Mendoza was obsessed for a moment with the archived image of the alien ship. DeNeuri skillfully magnified the viewer, he had already examined it many times. "Damn," Mendoza said. "That's a new one on me. Holy shit. A fucking UFO. Damn. Never saw that before."

Rolf chuckled. "Yeah, cute little machine there, isn't it? I mean, look at that shell, it looks more like a turtle, or even like a living thing itself. Like a giant black shell. It's irregular too, not symmetrical. No external lights or even thruster ports that we've been able to see. No communications antenna or dishes. Not even really any doors."

"But, commander, I don't think I understand," Mendoza said, puzzled, as they all were. "This isn't one of the illusions that arrived when Jupiter opened her guts up, you know what I mean? When the red-eye let loose her insides and we got those orbiters, at least that's what they told us on deck. This isn't one of those?"

"It could be, but it's not clear," Rolf answered.

"It's not the same, sir," the navigator added. "They've reached that conclusion from Ganymede base, pretty specific. The reason is, this thing has passengers of some kind. There's somebody inside that one, they even attempted a signal and it was received. A 'ping' code. They pinged us back, in their own alien way, I mean. That was proof enough. The other one turned out to be machines. No ping. Nobody home."

Now the communications and life-sustain officer on deck also commented. "Yeah, that's right, that was the intelligence," he said. "Ganymede had a full report over almost two weeks. The ones from inside Jupiter were eventually all accounted for, once the fighters cleared up that the illusion-projections were not the real thing. So they could plot their positions, and then try to knock them down."

"Which failed," the navigator said.

"Well, yeah, okay, the whole fucking deal, Jarvis, sure, nobody knows what's going on out there right now, sure as hell not us here on this ship. They were 'real', but they were designed to apply illusions. And they're all still right where they left them in orbit. We're just watching the parade. But that turtle-shell thing on Rosebud approached from another direction, way the hell off course from Jupiter-orbit. Like a comet would, cross-section to the regular orbit paths. Then the thing just joined up. But Ganymede tracked it, routine. Their info is solid."

"And now it's sucking water from Amalthea-base and they got a Level Four?"

"You got it. Sir, if I may. That thing is probably a star-ship, sir. In my opinion."

Angel-Face was obsessed with the archived image file. Magnified, they could see it clearly. Almathea was the size of a small planet, like one of the moons of Mars. The 'up' side, or the 'top' was relative, but the view on the vid-screen at the 'Aerotica's flight deck showed a strange, dark, shell-type form, about equal in size to the 'Aerotica; herself, that had made a very clumsy 'docking' at the underside of the Alpha-base, where any other ship would have never connected at all.

"Ugly fucker," Mendoza commented.

"Well, that's alien star-ships for you."

"Maybe not going for cosmetic appeal."

They laughed a bit at that idea.

There wasn't much else to see. Analysts from various positions were studying the star-ship in much greater detail, as the staff and crew at the asteroid-base dealt with things in whatever manner they had to. They all took turns checking it out, trying to put their heads together for answers or action. Nothing seemed to gel into a coherent plan.

After a bit, the man called Jarvis, a navigator, spoke up. "Commander? We can move the 'Aerotic' in another 30-minutes, by taking a sooner apogee on reducing the circle-path of our current orbit. No problem. The apogee isn't for a while, but if we move in 27.3-minutes from now, at thruster-driver mains only, to a sooner turn, you'll have a live view of Alpha for about, uh, about three or four hours sooner, I guess, from my figures."

DeNeuri smiled. "No shit? Good job, Jarvis. Yeah, get it ready. Move her in 27 minutes so we get a view. And also, Jarvis, after that, when you have a chance, get me navigation to move 'Aerotica' over to Amalthea base. Like we would for docking. Just the nav's. We've got to help out somehow. Lifeboat."

"From here? We can't even begin to move to Alpha-base short of six days or so, sir. Too far."

"I can move this bird at .05 par-sec's in 30 hours maximum acceleration, mister. Don't bother being conservative. Get me a navigation to Alpha-base at top-speed. Then I'll have an option. They need help. Just do it. Just the nav's."

Mendoza seemed to snap out of the gripping, almost hypnotic view of the alien star-ship, from the archive file. It was something to see, for anyone, a real one-in-a-million. "I never really had any idea, how you operate on the flight-deck of a ship like this. Outstanding."

"Well, it has to be done properly, but we know how," said Commander DeNeuri dryly. "I need options, and so do they. No worry. We'll be heading home, eventually. It's a much simpler business, in some way, probably."

"I hope so, Commander Rolf," Mendoza said.

"So do I, sir. So do I."

CHAPTER 53: An Alien Zodiac

"They are not visitors, they are not aliens from another world. The current crisis at Jupiter is not an invasion, it is better thought of as a reunion. They are not aliens. Another way to think of it, they have always been here, even all our lives. And now we know it, and that is the only change." --Martin Courio Ne-Airlo, planetary advise panel to the Space Authority laison with the Galactic Posture task force, at an emergency session to the World Nations Assembly, Geneva, Switzerland, June, 2416AD/CE (regarding the Jupiter Program situation at the time).

After a time, the courageous task-force and defense crews at Amalthea-base and far away, such as at the helm of the 'Down', and other ships, and the various sub-station Jupiter moon raw-materials processing bases, realized that the bizarre alien 'star-ship thing' that had attached itself like a massive barnacle to the underside of the Rosebud Stone, was leaving.

"Good-bye, weirdness," someone there was glad to say.

Even with all their frantic attempts to defend themselves, or to destroy it, negotiate, change its path or create a favorable scenario, including several deaths, when things at last changed, it was by an initiative of the 'thing' itself. This was a great relief to be sure, they still really didn't know what it was. *A star-ship. Well, what is that?*

Commander DeNueri's ship had managed to arrive fairly quickly, over a distance of millions of miles, in the functional vicinity of the main base. This may be difficult to visualize. The 'Aerotica' was a so-called 'needle-craft', designed for only few passengers or materials, and fastest-possible speed, given the great distances. The relative position of planet Jupiter and her many moons, were always in-transit, and so were all the asterids in the 'belt', a known pattern of helter-skelter rocks, flying in a flattened circle over an impossible distance and journey. Earth, then Mars, then the asteroids, then Jupiter, circle and circle and circle.

Rosebud was calculated ages ago when the Jupiter system was built, based on her proximity to Jupiter, as more often closer together, than further away. But this changed and changed again, another pattern of more frequent togetherness of celestial objects in motion. The only other option for designers was a main headquarters or management base at one of the Jupiter moons, such as the giant Ganymede. But they wanted a resource closer to home, so the travel of ships would have a resting place to restore fuels and services, and then onward to Jupiter. Amalthea, a smaller planetoid, was a good answer for this.

The Alien Zodiac Crisis, or Jupiter Water-Logic War situation was unsatifactory in any case, but it was only inevitable. The other monitoring-tracking services, including those on Earth, and the other Solar system space-based operations at Mars, Earth's moon, and other orbital positions, assumed a watchful waiting philosophy, even throughout the earlier appearance of the illusion forms or invaders. And those had not departed, the things had stabilized from the previous panic, such as notions that *giant alien robots from within Jupiter were attacking Earth*.

It wasn't like that at all. They were instead very old technological marvels installed within Jupiter many thousands of years in the past by inter-stellar persons who were planning ahead for their future water-needs.

It was sort of sad, in a way. The Rosebud Stone base at Amalthea had temporarily played host to a genuine off-world visitor of some kind. People had been killed, but not as a hostile act. The Alpha-base was a ruin in many ways, but most life-sustain systems were in-tact. The strange 'bacteria monster' that had somehow been discovered in their water-system reservoirs by people like Michael Hesidom (deceased), and their attempt to destroy it for their own sakes, or learn what it actually was, soon had shriveled up into a clump of organic matter like a large sponge that had collapsed in on itself, then like a dead chunk of alien fecal-material, lain lifeless at the bottom of one of the connecting shaft-tunnels between the huge resevoir tanks, where some of them had died. When the giant black barnacle shell spacecraft departed, the monster died, a single-task extra-terrestrial turd, apprently intended only to 'test the waters'.

"Bye! See you next time!" Commander DeNueri's brother, his second pilot aboard the 'Aerotica', commented, with his usual dry humor. "Fuckers!!"

They had moved their ship closer to Rosebud, over a matter of about about 72 hours. The idea was to assist if there was a necessary evacuation. And also so they could get a good look at what was going on, as the dark, wet-looking, external 'shell' type alien space-craft, what they thought was a star-ship, locked onto the Alpha-base from beneath, then detached methodically, and drifted off, as any space-ship would, for departure. The 'Aerotica' was still in flight, on her way. The ships were always in motion.

"Maybe they're just selfish, or biased, you know, don't be so harsh," DeNueri said. His brother, also a French-European, felt him foolish to be so cavalier. But everyone was pleased the matter seemed to have resolved. *Peacefully? A matter of opinion there.*

Deveroux and his partner Angel-Face, on-board the 'Aerotica', learned whatever they could, like the others. Angel was assigned the boring task of handling the ship's modest need for Earthside tracking and monitoring logs, filling in for the regular technician who was needed on other tasks during the crisis. In constant contact, without any disconnect, the information both ways was mostly technical (ships' position, motion-trajectory and navigations, life-signals or crew-status, power-engines and communications, observations and celestial objects), but the normal data-stream logs were a mess while the regular worker was called up for crisis-duty on other tasks the Commander needed right away.

So this tall Mexican, good with tractors and hayfields, with his snow-white facial hair and oddly boyish grin, his olive skin and his jokes, could be found seated restfully at a secondary deck computer-station, clicking off data-bundles to confirm their receipt and accuracy, or completed transmission. Everyone liked Mendoza once they got to know him, with the possible exception of lifeless, brainless machines and computers.

"Does that thing even work?" Deveroux was saying, standing nearby with his friend, flustered and upset. "Computers! They run everything!"

"Yes, it works. Fully functional data-stream link to launch-command and space-authority back home," Mendoza answered him. "Have your coffee, Dan. You're a wreck. Relax. I think this whole crisis is over. No winners, no losers. Stale-mate. Some stupid inter-galactic matter, they took what they wanted and set up their permanent robots. It will take a hundred years to straighten it all out. You and I will be long gone, that's for sure. So relax!"

But Deveroux was not ready to relax. It had been by then nearly two years in deep-space. As expected, he hated it. The original investigation had exploded into an entirely out-of-hand epic inter-planetary, inter-stellar exchange. How could he or anyone truthfully investigate something like that? It was very frustrating. And his pal was absolutely right.

Both of them, the PPP, and the entire spectrum of space-workers occupied with the Jupiter Program, were mere specks of life on a Galactic chessboard, as unlikely to really influence things or make real decisions for the good of mankind, as one of the 'temps', the mafu-men clones, whose entire lives were created to substitute for natural-born humans, under painful or demanding circumstances where anyone might be hurt or killed.

And many of them were. The Operation-Odyssey fighter ships and battle-plan formations were like swarms of bees sent out to attack a raging bonfire. They had no real hope of defeating the 'enemy'. The OO-planners didn't know this, until things started to unravel, and the day-date-year-and-hour of the alien appearance came, with the illusion-robots to take up their stations around ancient Jupiter, the fat, multi-colored mother of so many moons and mysteries.

The Jovian OO forces, 20 or 30 years in the making, could seriously claim that 'no one was killed', when it was all over, despite thousands of 'de-commissioned' clones. A costly exercise in how Earth might approach a planetary defense, one fine day, and also a heartless and cruel Gestapo-Nazi method as far as any 'almost human' beings such as guitar-player Marciel Penieur's body-double, Marcus.

"So if it was me, and they sent me out in one of those space-airplanes, to fight or destroy the robots, and I was killed, you would just have another one of me made up, then, is that it?" Marcus said, to Marciel, as things were settling out at Alpha-base. He was only trying to understand.

As a clone, life for Marcus was rather strange, he had never actually been 'born', he had no mother or father that he knew of, and really no childhood. Many clones were never even informed that they were 'not human', at all, and lived and died believing themselves as normal as anyone they met or encountered (who also thought them 'human').

But Marciel Janus Penieur, in his music-career, had chosen to deal with his personal clone as a family-member, thinking it compassionate, which anyone would debate as even a horrible mistake. Marciel often felt guilty and loathesome towards his clone, but there was a bizarre affection between them as well, like an identical father and son, exactly the same age. There was a magic-mirror component, walking and talking and eating and drinking, playing old guitar tunes, there with him, a prop, a gimmick, alive and very human, yet not-so at all.

"You're one of a kind, Marcus," Penieur replied. "You really can't be replaced. Sad but true. It's also very expensive. Do you have any idea how much it cost to have myself cloned in the first place?"

"Don't even tell me," the clone answered. "I couldn't bear it."

At this point, affairs at Alpha-base were at least somewhat back-to-normal. Once it was confirmed that the 'black barnacle' star-ship had disconnected and seemed to be leaving, the red-alert emergency status for all-hands was lifted, and then reduced to a mop-up effort. The Space-Authority Security Forces were basically in control of everything, once the death of Menuda Wu was confirmed as well.

The guitar-player and his clone had only to stick-with-the-program and follow the rules for a while, like the rest of the un-assigned residents. This only meant they could hurry up and wait, to see if they all would suddenly die or be killed for some reason, and that had not happened. For them, at that time, they were enjoying their rations at a base cafeteria again, along with others: chicken tacos, coffee, cookies, and some calming sedatives they all were issued during the excitement, as a precaution against any further panic. There was also vegetable soup, and a very sweet cake with syrup-icing.

"Well, it's worth it, believe me," the guitar player said. They both had plates of food. Marcus wanted tea instead of coffee, which the felt was poor quality. "By having myself cloned, I'm the world's first double-duty guitar singer-songwriter act. Or, at least, I've never heard of another one."

"There was that gymnast who had herself cloned," Marcus commented. "The gal from Ukraine. God she was cute."

Marciel laughed a bit. "Sure, I remember her, that was ten or 12 years ago," he said. "Milana Otravello, or something like that. She did the floor-routine in anti-gravity. But the guy who cloned her, it was sexual. He wanted to have two of her in bed. I don't blame him, she was hot. Her manager or something."

"They never did get the anti-gravity Olympics off the ground, though," Marcus said, somehow sad. Then they looked at each other and laughed out loud at the pun.

Other residents having something to eat nearby, and most people who knew Marciel and Marciel-2 ('Marcus'), may have raised an eyebrow. Most people assumed they were identical twins.

"Well, as mirror-images go, you're the best, Marcus," the guitar-player said. "If I wanted a female Olympic gymnast, I'd look elsewhere. So, if you croak, or died, or were killed, or drafted for some stupid battle, I'd be sad. I probably wouldn't have another duplicate created. One is plenty for a lifetime, if you ask me. But I'd miss you, I really would."

Marcus took a moment and hugged him. He was strangely affectionate, as if he knew his time were short to learn anything about 'love', even if he could. 'Fraternity' may have been a better word for it. They had all been through hell at the Alpha-base for a while anyway.

They was busy all around with repairs, re-booted systems both essential and non-essential, recovery from various distasters such as the situation in the deep-rock water reservoirs, the lost men and machines, the shock and trauma. But everyone co-operated for the best interests of the survival of them all, and this was not misunderstood.

"Have your coffee, Marcus," Marciel said.

"Having the tea, instead, thanks."

CHAPTER 54: Useful To My Enemy

Ralphie (Hector) White, the Commander of the 'Ferrous-2', had by this time heard about what was going on with the space-program that had canned him, by then as long as four years previously. Sour grapes that White had been eating steadily since then, flavored his interpretation of events, along with ignorance, and a certain bias typical of space-workers at that time, and Commanders in particular.

Superiority in retreat, or the defeated exclusive elite, and Ralphie-Boy was superior to many types who may have dreamed of commanding a giant space-ship, such as the 'Ferrous-2'. To crash one was unthinkable, but he was long past any actual self-hurting guilt. He never really blamed himself, he didn't feel it was necessary.

"It was the horses, the ones they were doing experiments on, not Brandeis, but there were other animals," White told his friend Billy, there at the Hotel Caligasta, 'home for lost or neglected astronauts', one day about that time, at lunch. "DNA experiments to create super-human astronaut labor."

Billy's robot-controlled cargo-trucks were doing well, and White had been able to help him somewhat. The Reno, Nevada-area and endless night of the desert wilderness in the American West, were ideal for White's hobbyist star-gazing and astronomy. He hadn't neglected to view Jupiter and her moons over the period between 2014 and 2017, during the Water Logic Wars. But there was not much to see, from Nevada, anyway.

"What horses, Ralph? What experiments? When are you ever just going to let it go??" Billy said. It was understood between them that White was obsessed with the past. He had a good excuse, it seemed. But it grew old and cold far faster than it ever would end, as a conversational topic.

"The horses, Martin Brandeis, his farm in Carolina. Not his specifically, but they had been doing the DNA tests and then cloning. And the clones were what they used to pilot the new space-fighters, at Jupiter. That whole Nazi program, Operation Obliteration, or whatever. This past year."

White was working on a sour-dough toast and coffee for breakfast. Billy had a day's work ahead repairing a remote-control receiver on one of his trucks (he only had two, but he kept busy and made good money that way).

"So fucking what, Ralphie? They do what they want."

"No! I figured it out! My ship, the Ferrous-2, my entire career! The 'Ferrous-2' had been loaded with a forbidden technical item, without my knowledge, on one of the storage decks. The men who were killed at the orbiting loader-dock, died in an attempt to secretly bring altered horse-DNA results from equine experimentation, to the Amalthea-base scientists from Astro-Health Corporation, working on creating super-mutant space-men for on-going labor at the moon-units. And the Unions hated the whole thing, obviously. So they crashed my fucking ship and blamed me!! Mystery solved!! Fuckers!!"

Billy laughed. "Pegasus, then, right? The space-horse?"

"Horse's ass, my man. Horse-shit, for want of a nail. But yeah, Pegasus, sure. Red horse. Horse-head Nebula. It's taken me years, but I have proof. It wasn't my fault and the real culprits had real motives. Look at the storage logs on the 'Ferrous-2' for that voyage. There were medical shipments."

Commander White would have plenty of years to think about it all, in any case. He knew he would never see justice, personally, so he comforted himself with a bitter-sweet knowingness, that sometimes permitted a laugh. Alien robots at Jupiter? Sure, sure. The Earth's moon gazed down in her icy-cold and barren whiteness, like an ugly, gigantic egg, from the dark indigo as seen from Nevada, that evening, near Reno. White was okay with things, personally, by then, in a 'not-okay' kind of way.

Hobbyist astronomy and telescope star-gazing during the Water Logic Wars had become a small cottage industry all its own, back on Earth. Many amateur telescope users had been able to spot the new formations around Jupiter, which was quite a stir in the sciences.

Others involved in the crisis and scandal were also either 'okay' or 'not-so okay', as current Jupiter Program events and emergencies folded up like a bizarre road-show, or dried up like a waterless flower in the wilderness of some forbidden world where flowers did not do so well.

Commander Brandeis of the 'Down' also felt he would never see justice, especially regarding the long-list of detailed and mostly true or perfectly legitimate complaints against the J-Program leadership. He had lost all faith in the Planetary Proficiency Program, and Daniel Deveroux and Angelo Mendoza. By the time of the Water-Logic Revelations, the new shit hitting the old fan had begun to stink at such a lofty level, that Brandeis knew the Astronaut's Union-Labor disputes meant very little in the grand scheme of things. And his other complaints as well.

So, like a good space-ship commander, he buckled under once again and only did his duties, flawlessly as usual. He took to re-reading an old hardback copy of 'Watership Down', by Douglass Adams, about bunny-rabbits and rabbit-holes in the English countryside, circa 1970, his ship's namesake. His nerves would eventually calm, and in-between, it was the same, the same again, the same endlessly it seemed, then the same, but different. So, Brandeis was 'okay', according to himself.

His Second Commander, Menima Pearl, was looking to be seriously 'not okay', as things were winding down. True enough, she had maintained her post and duties as second in command aboard the 'Down'. And their ship was finally able to depart from the Jupiter-system orbit-paths, heading home to Earth, with much relief. But because she had fallen into the role of 'channeling' an Uber-Mind or Galactic Overlord Consciousness, 'god' or such-ever as she may have felt to be the case personally, as a well-meaning priestess/leader for the Quantuum-Mind Fellowship, the result was that she was slated to be de-commissioned as a spaceflight officer, similar to what had become of Ralphie-Boy White.

For being useful to an enemy, though never intending it that way.

Second Commander Pearl already had her official dismissal. She was being canned. The Quantuum-Mind Fellowship was intended only as a spiritually uplifting and positive or happy religious-guidance center for sometimes stressed-out space-workers. Instead, it had nearly driven her mad, and she was personally suspected as a link to truly sinister powers behind thousands of years of planning what was going on within planet Jupiter. A dupe, a pawn, used by far greater powers, even without her knowledge.

But the space-program could not tolerate that scenario among their officers, or even the suggestion.

Alone during off-duty hours in her private living-quarters, she had given up on consultation with her antique idol, the 'minook'. The bronze or golden figurine would come to life with a spoken word from her, offering random wisdom, Buddhist koans, poetry, and even personal guidance, all based on 'divine mathematics' within its circuitry. *"When in doubt, punt,"* she told herself.

So she had it unceremoniously destroyed in one of the ship's incinerators, like a gollum's magic ring, there in the darkness of the Abyss by Jupiter. The charming idolatrous high-tech minook melted into a warm glowing mass of red molten metal and other substances, and then was atomized, the ashes and smoke ejected into space-itself, with many other objectionable materials the 'Down' would routinely reject. Garbage.

"I never moved any information or orders or commands from my own consciousness that had any effect whatsoever on the outcome of things at Jupiter," she told a friend, privately. "I was not being used by some Galactic person or persons, to influence events here, like moving ships around or triggering the Jupiter robots to do anything. Never."

"But how can you be certain?" her friend said.

"How do I know? Because nothing was done, by me, towards a foreign or alien goal, or tasks, or ends. Not even once. And I wouldn't have allowed it. My conscience would not allow it."

"You mean even if there was some sort of channeling, you or others never took action based on those ideas, so it was a self-canceled dream."

"I would have been a traitor to everyone here if I had permitted myself to be used in that way. And the Quantuum-Mind isn't organized that way. It's really just part of ourselves, our own higher selves, or higher awareness. Love-consciousness, you might say, maybe under stressful or frightening circumstances. And love doesn't do that. So it never happened."

"I believe you, Pearl. I do."

"But I don't think some people believe me about it. So I'm soon out of a job, by the time this big blue whale hauls us home again. Oh well. Big deal, I've had a good career."

"You have, Pearl," her friend said. "You've done a lot of good. You're a good person."

So they smiled and hugged. But Pearl would continue to dwell on the idea that she had fallen to the level of some sort of psychic whoredom, a tool, even to her own harm. So the 'not okay' pattern may certainly have applied. And her peers avoided her, she was 'mahanwe', Afrikaans for 'rejected'. Later on in life she suffered from a type dementia, as she grew much older.

The surviving 'old men', the five or six Jupiter scientists from the early years of the program who did essential early research, dispersed now to live among the moon-bases at year 2417-18CE/AD, had by this time faded again into a sort of comforting obscurity. Their only real claim to fame, or position as players in the unfolding drama, then folded, was that they were clued-in early to the 'secret' about planet Jupiter, anticipated for many decades or even centuries.

The murders, as a result, of Philby, and Montrose, were not ignored at all. They 'knew too much', and all they had thought to be possible, regarding the gas giant planet, the 'secret', then suddenly exploded into a six-month crisis scenario, extended from then into the far, far future.

But the surviving 'old men' themselves, with Doctor Wu now deceased, were not heroes. Yet among the program regulars, there was a chummy sense of victorious 'I told you so' thankfulness, that things had not been worse. Legal remedies and court-room charges followed.

It was widely assumed that Wu had initiated the slayings based on orders from the corrupt global powers back home, whose vast wealth was at risk, even far into the future, given that events at Jupiter would turn this way or that. Loyal and true, innocent of any wrong-doing, highly trained, valued men, and also human beings with lives and dreams of their own, then were dead, gone, expendable, lost. The loyalties of the dead or lost were elsewhere. A sad, historic descent in the Earth-born space program, known for 'the right stuff', and then to find itself one day, murderously corrupt. A comforting obscurity, for all the wrong reasons.

"What about the body?" the Elder Podliakov asked, speaking to one the academics, the same man Deveroux had bunked with temporarily aboard the 'Down', a mature scholar and poet. Podliakov felt the man might know what had been done for Menuda Wu's funeral or memorial. But there they were, at Ganymede station, sometimes called 'Station X', Charles Benway's position, mostly up-loading bulk helium and ice. The same scholar was now living there, his Sabbatical journey intended to include first-hand views of Jupiter.

"The program manager? Wu? They had a memorial and sifted his ashes into space above the moon at Copernica, I was told."

"I never liked him," Podliakov replied. "The man had a personality like a plate of raw eggs with nothing on 'em."

"Ha-ha, yeah, a real egg-head." The academic turned away back to his books. The deaths were no one's happiness, but Menuda Wu's replacement was a new start, much-needed, they all felt. Wu may not have been the cause-and-effect origin of many troubles and problems with the Jupiter Program, but his administration was widely felt to be vaguely sinister, if not downright evil. Perhaps too many long years of work in space, or too much power, or too much secrecy.

"It was Wu behind the murders, you know," said Podliakov. "He knew what was going on. Those men were my friends."

The academic frowned, scowling, also angry and hurt. His work was disconnected from all that was happening, but with the mess that had unfolded during his visit, he really could not accomplish his goals in peace.

It was the same with the Large Globe Object Music Society, and their leader, Amakmid Beautiful Truth. The research and projects had to wait, even for a very, very long time. "Well, murder," he offered in a hushed tone. "Perhaps the manager, and many others, somehow, behind it all. But with the overall revelation at Jupiter now, it's doubtful even a very high-level commander knew anything about it, really. I guess because it's so old, or so...alien. Is that the word? It is not a human thing, is it, to kill another person like that, for gain, or convenience?"

"To conceal a wrong," Podliakov said. "Alien minds from other worlds may perhaps be excused, it's foreign territory, after all. All they wanted was water, 1,000 year's worth, I guess, hard to say. But they didn't kill to get what they wanted."

"There is no death, Mister Podliakov," the academic said. "Didn't you know that?"

They laughed a bit together at the idea. *Maybe so, maybe so. Maybe not. Maybe so.*

CHAPTER 55: Ladies and Gentlemen

Even 400 years from today, it's hard to predict just how difficult, or easy and simple it might be, to assemble the powers and governments of the Earth, to represent, plan, attend, or deal with, anything at all that might take place at the planetary or galactic level, involving our Mother, Gaia, Earth, Urantia, home forever.

Oddly enough, the very same types of events, (for example, a world-destroying carnivorous monstrosity from god, or a solar-stellar nova that would burn us to a crispy cinder, or negotiations with other Galaxy Folks, on whatever terms, or outward bound colonization of distant worlds, and many kinds of scenarios), these had been the topic and passion of countless cults, religions, prophets, gurus, scientists, world leaders and space-exploration sources, and predictions of big changes that would effect us all and our corporate future as 'humanity'.

Fears, uncooperative groups with vested agendas even centuries old, ignorance, insurmountable science and truths too difficult to grasp easily, all again created the all-too-common human truth, that we don't get along well enough with ourselves and each other, here on Earth, to act or choose in unison, on a global scale, even when facts and truthful analysis reveal the planet to be changing significantly. Not advanced enough, as a species, so that membership is denied in the community of planets and peoples who may or may not inhabit worlds like ours, even in our own Milky Way Galaxy.

Yet it needed to be tended, but we would not, or could not. And then we did not.

About this time, there would be a 'World Governments Conference on Planet Jupiter for Year 2418', held at EarthGlobe University in New Anchorage, Alaska. As such things go, it was a gathering of the best and brightest in the arena, those in-the-know about the Water Logic Wars at Jupiter. Jupiter Program management and survivors were up-front, the Space Authority and its committees, the Galactic Event Posture Task Force, the secretive powers also on-hand, and a wide swath of science, technology, astro-physics, space-travel, population, environmental, and survival experts, teachers, professors, workers and technicians, pilots, navigational, planetary researchers, alien-life specialists, many others. About 200 people were invited, men and women of all shapes and sizes.

Anchorage, Alaska, had recovered from natural disaster, by that era, including a complete rebuilding effort of a more continental urban city-scape, there in the cold, cold North, still a United States Regional Authority. The Anchorage area was laced with deep-core Earth-magma heat-transmitting energy rods, and heat-transfer stations, the world's first completely geothermal city. It made sense, as cold as it was.

Climate changes were not completely under control for the far North, creating a densely cold and snowy wonderland. By rebuilding over 100 years or more, the region was then an international marvel and resource for many projects and advanced facilities, energized by a system very similar to the one Dan Deveroux had found to be the 'cause-of-death', when the PPP needed him declared dead, in 2412.

The Earth's inner core lava-heat and magma, was penetrated by a series of numerous 20-mile deep copper-and-iron rods, each as big around as the tallest redwood trees back in California, to bring much-welcomed heat and warmth and energy to the surface. So the place was called New Anchorage.

The convention hall, known as the Earth Evolution Center For All People, was in the Southern corner of the city, nestled comfortably on a low-altitude, valley-type plateau, against higher mountains. It was a University campus environment, with course work, students, faculty, and many resources for the sort of work they did, and the academic-philosophical approach they favored.

Roadways and people-mover paths were supplemented with short-hop anti-gravity and levitation transports, like jets or helicopters with low-level anti-gravity assist and fast airspeeds along established routes. It was a busy place, but not particularly over-populated. They wanted isolation and safety, permanence, but not disconnection or lacking state-of-the-art for any and all improvements, depending on their goals and programs.

The trade-off was cold, and it was, and so were they all. Ice and snow were common.

The Evolution Center was designed like a large, upside-down fishbowl or serving dish, bright and full of colors and ringed around at the upper level by open-air view platforms and observation decks, also warm-heated gardens, and other services. It was a stadium, an indoor assembly hall, quite large enough for even 1,200 guests. Inwardly there were many seats, tables and screens and technological assists for the purposes of meetings, votes, speakers and public-address, tabulated opinions, security, and communications with the outside world. The current sessions on Planet Jupiter were broadcast to many other places in the world at that time. Deliberations were not available to the general public, and would not be available to the general public, even for many, many years. True enough, the leadership didn't want to generate any kind of a panic.

It was already the third day. It was the month of July. The convention would continue another two days. The speaker for the hour was Commander Martin Brandeis, large-scale deep-space transport pilot in charge of the 'Down'. Tanned, rested and ready, he looked older, and he was.

"Gentlemen and ladies of the global community leadership," Brandeis was speaking, then, at the 40-minute point in his presentation, scheduled much earlier. "Your various potentates and emissaries, those speaking other languages or of discontinuous origins, as a Jupiter Program pilot and commander for more than five years, assigned to the Planetary Transport Deep-Space Ship, the 'Down', my point of view concerning these events, bring to you but one central observation and also my plea to you: planet Jupiter and her moons, and all celestial objects in Earth's solar-system, or passing through it, to whatever extent the Sun has captured matter to within her gravity-well dimensions, are the property of, and under the complete authority, ownership and decision-making administration, of the people and powers and leadership of the humankind of planet Earth, in perpetuity, here, now, elsewhere and forever."

The audience applauded this statement with loud cheers and clapping for a moment.

Daniel Deveroux, and Angelo Mendoza were also in attendance. "He means we own the solar system," Daniel said.

"Good point, I guess. Who knew it was disputed?" Mendoza replied.

"This is all we've really been able to assess of things," Brandeis continued from the main stage podium. "Nothing else much can be recognized here, or among us, that would make much difference in our favor. What did the appearance of the Water-Logic Forms from within Jupiter's depths and high-pressure caverns of near black-hole density, what did they teach us? What did we learn? What is the lesson? What has really happened here? From all reports as you may know, this vast local planetary object and resource, Jupiter, is disputed for territorial and usage rights and control, by some other species of sentient awareness from another world somewhere in our own galaxy. And their claim goes 1,000's of years into the past. But our claim is even older. Mankind has ventured into space, into the galaxy, yes. But first here at home. If we don't own and control Jupiter and her moons and resources, then who does?"

He paused, and the groups of audience members and guests at the convention now relieved some of the boredom with another minutes' long cheer. Much hooting and hollering and noise overall, almost of a political-rally type, rather than the low-impact enthusiasm or emotional velocity of another cold scientific review.

Commander Brandeis was now very popular, and had many friends. *Wooooooooo! Right on! Yeee-aaaah! You got it! Tell the truth! Wooooooooo!*

Brandeis and others comprehended that the notion of material ownership rights, or the idea of possessing a thing, the ages-old story of human progress, could be reduced to a necessary theft, the dragon sleeping dull and stupid on his treasures. The very concept was probably rejected by most higher, more advanced consciousness, if any were found. So for the World Powers Assembly to declare these property rights, was similar to a declaration of war, a war between the men of Earth, and whoever might take our water-rights to planet Jupiter. A war that could last many centuries.

Brandeis and many here, were strongly recommending this stance on the matter at hand, as a Galactic Posture topic imperitive. Not that any off-world sources cared what we had chosen. Some had suggested he even had a future in politics.

Back within the ranks, at deck three, fourth row, seats 53 and 54, Daniel Deveroux and Angelo Mendoza, had taken the effort to attend the meeting, as interested parties. Home was home at last. The investigation was over, a journey from the very first assigned research, to that day, or more than four years, or 4.6 solar revolutions. Along with Deveroux and Mendoza, was a representative from Planetary Proficiency (not Vanessa Signo).

"That Brandeis, what an ego!" Deveroux was commenting, with only Mendoza paying any attention, and that in a sleepy haze of exhaustion. They were following the convention from session to session, the hearings and votes, like any of the guests.

"How many wars in the history warfare, have been started by ship captains??" Mendoza asked.

"A few, anyway," Deveroux answered him. "Maybe not as many as by public speech-makers."

"Here we have both in one man, then," Menodza observed callously.

They listened to more of what Martin Brandeis had to say. The conference had many speakers, but no one would be speaking for the PPP. They had no completed report to give. The truth, so valued and even worshiped, earned so often at such a high price, was but a footnote on this one. Off-world alien beings were now so common a concern, that the inter-being, or communications, were less of interest than whatever may truly constitute inter-stellar law, regarding star-system rights and planetary authority over inhabited star-systems.

The audience cheers for Commander Brandeis' speech were first more still, then to a quiet hush. The large audience-hall quieted down. He was making his concluding remarks.

Brandeis was finishing. "Thank you, I understand, thank you. Perhaps some of you have signed up for space-travel in recent years, or your parents, or your friends and family. Did you enjoy it? Was it something you felt to be a privilege, or a special time in your life? Did you find the experience enlightening? Did you ask yourself why? Yes, I'm a space-man. Yes, I know how my ship functions, I know how to get to Mars, I know how to reach the asteroid belt, how to navigate to Saturn, if needs-be. In my entire career, it never entered my thoughts that these local planets and moons, would be of significant value to others in the Milky Way Galaxy, besides the men of Earth. The vast star-to-star distances seemed much too far, for the resources to be realistically appreciated and applied, or even shared. It made no sense, so for me it didn't exist. Yes, I enjoy space-travel."

Deveroux squinted. "Huh," he mumbled. "He enjoys space travel. Huh."

"But I own Jupiter. I own the Earth. I own myself. And so do you, and so do we all. I run planet Io, and if anybody wants to know what's happening at Europa, they come to me. Gold mountains at Ganymede? Ask me, first, otherwise we call that stealing, if you like gold. Mysteries on Mars? Knock before entering, I run the place. I do, and so do you. So my plea again to the world assemblies: historic property rights and legal status of ownership and possessions as an Earth-born legal principle, must be extended to these local planets, as a communicated and asserted truth. To do otherwise, in the face of recent events at Jupiter, forfeits future eras to a venomous on-going dispute, by which our type of human animal will suffer loss and lack of critically needful resources. And that dispute, is become a war, and may even be considered ageless, without end, without beginning. Thank you, New Anchorage, thank you Evolution University, thank you world-assembly. Good night."

He left the speaker's desk podium, beneath a huge video-screen image of himself. The audience had dwindled for his speech to only two large groups of less than 50 people or so. His position was nothing new. More applause, cheer and shouts, greetings and laughter. *Wooo! Brandeis! Brandeis! Brandeis! Brandeis! Wooo-yeaaaaah! All right, Martin! Go for it!*

Brandeis was passionate, but had no real authority or decision-making power. He was well-informed, and deeply devoted to human rights as a galactic standard. The whistle-blower had come home, and he kept his whistle wet with wonderment. Our first steps as a child had never lost interest in the next and the next, and the next step to the future, the next step for Jupiter, the next step for Earth people, our land.

"Oh yeah, I'm a planetary man," Brandeis thought privately to himself. *"Is the Quantuum Mind paying attention here? I hope not."*

The he left the stage and podium as the EarthGlobe University conference management took a moment to prepare and introduce the next speaker.

Deveroux, and Mendoza, were still a bit dazed and confused by all they had been through. They finally arrived home, not aboard the 'Aerotica', but aboard the 'Down', after a ship-to-ship transfer, and then the voyage back to stationary orbit at Earth's moon. Either of them may have thought they had aged more years than actual real-time days and hours spent in space. It was only the stress, but the effect was telling on their appearance and demeanor. Tanned, rested and ready.

"So tell me, Mister Daniel Deveroux of Seattle, Washington. Did you enjoy space travel? Did you consider it to be a special time in your life?" Mendoza was joking with his partner, as they decided to change seats for the next round, and a break in-between.

"Yes, it was a special time in my life," Deveroux answered him blandly. "Very unique."

"Were you enlightened?"

"Uh, no. Sorry. I tried, I really did. Didn't happen for me. Sorry."

"Me neither," Angel-Face said blandly.

CHAPTER 56: Marta

Angelo Mendoza, Deveroux's more-or-sometimes-less devoted partner in their work for the PPP, associate researchers for as long as six or eight years, a team, and among the best, was sick, and tired, by the time he managed to at long last reach the sandy shores of his second wife Marta's comfortable home, in Belize, Central America. Their long chore of painful and not very successful team-work researching the Jupiter Program failures was over. His feet were planted as solidly as they would ever again be, back on Mother Earth.

He didn't always wish to know where his mind was, and what it was doing, following what would be his only lifetime deep-space journey. Something about working in deep-space did funny things to the hearts of even strong men, during that era. In his case, he was worn out, really exhausted. It was all too much, he had to admit to himself. The reasons they had failed were so much greater than themselves, as to be mind-boggling.

Not much ever seemed to be going on at his Belize home. He liked it that way, so did Marta and the kids. His peaceful little Mexican cabana on the coastal waters in the sun, he loved it there. It restored him, just what the doctor ordered.

The journey back from Jupiter was of course quite long, and then the required attendance at the convention in Alaska. He was still keen to follow the same affairs, many people were. Things settled out to a larger and easier to understand consensus: some strange machines or objects had popped out of the belly of planet Jupiter, caused a lot of trouble, and taken up long-term stationary orbits, with possible links to other-world Milky Way Galaxy alien 'neighbors'.

What anyone might be doing about it, over the long term, was also anyone's guess, and it was all guesswork, at Angelo's level. Which right about then was at ocean waves at sea-level. When they finally parted ways, after the New Anchorage Conference, Angelo seriously felt he would never work with

Deveroux again. They had been good friends, and successful researchers for a reasonable Earth agency that any serious person might certainly have considered to be helpful and good. But the voyage to Jupiter was too much. It was as if the entire effort was upside down: what had started for the PPP as an attempt to inform and identify various problems and failures, had expanded into a chance encounter with the galactic history of objects in space.

Additionally the stress and strain and deep-space environment had weakened them both, making them somewhat sickly by the time they were back on Earth. So it was a sadness and a sour good-bye for both of them, if only as friends.

"If you do any more work for the agency, I want you to find another partner, Angelo," Deveroux told him. "This is my last case for PPP. I'm through." Mendoza laughed, smiling. It wasn't hard to understand Daniel's choice.

"Well, it took a while, but at least you survived your own death," he said. "Maybe you'd best get back to Seattle and straighten that out. Maybe they'll bring you back to life, or get you proper papers, anyway."

"Rumors of my death are greatly exaggerated," Deveroux answered with a joke. So they parted on the best terms possible, and probably would see each other again, socially. But on a professional level, it was over. If PPP had to declare Deveroux officially dead to protect him from a revenge about the geo-thermal energy system corruption and conspiracy in the West five years ago, following that case, with the Jupiter scam, they may need to bury him even deeper, just to protect him.

And even his children and grand-children, if he ever had any. Maybe there was something to it, behind the scenes, that his buddy supposedly had ancestors who were from off-world. Menodza had come to feel over the years that Deveroux was a strongwilled and resourceful man, with a sharp and persistent mind. But he was also sometimes depressingly coy, evasive, emotionally unavailable.

"What happened? Angelo? What happened?" Marta really wanted to know, it had by then been three years apart for them. As one of two wives, they were often apart anyway. And she didn't understand his work, she was a much simpler person in life, mostly working the land and farm-work, and then with her friends and church and families. She also sometimes was a school-teacher. So later, after his return to their place in Belize and all the excitement about seeing him again, she asked, as if it was only a flat tire on his motorcycle, or broken taillight, instead of a planetary event. He enjoyed that about her, it forced him to simplify.

"What happened? How can you ask me that?" he said. They were resting together on one of the home's porch-patios, viewing the lands beyond. Angelo hardly recognized much of the place. The kids were grown now, and Marta had gained weight. "You know what happened. I went into the space-program on my work. And now I'm back. Do you have any more beers?"

"Yes, yes, plenty of beer," she said. "I'll get you one. Then tell me everything."

"I'm not sure that's possible," he said, then she went inside the home through a wooden door with screens for bugs. It was a hot, muggy day there in Belize, a favorite coastal hide-away for many who wanted to escape the urban world, like New York. Or other worlds. When Angelo was young, he studied at schools and colleges, looking at a military career, and also in technology. He had friends and lovers, a certain amount of money, he was healthy in general, and he managed to travel somewhat.

As a Hispanic, it wasn't unusual for him to encounter people from other cultures who felt he was 'a type of person'. Even by year 2,412, bloodline prejudice against Africans, Asian, Middle-Eastern, women and children, religious creeds, sexual preferences, obesity, and also any clones or genetically altered persons, or off-world peoples, was commonplace.

The Hispanic culture in South America had flourished over hundreds of years, by virtue of Earth-changes that left places such as the Amazon basin, to be among Earth's remaining havens of biodiversity and natural waters and forests. They had plenty to work with, in those regions, while other places, such as the Middle-East and North-Africa, had suffered tremendous dryness, hot and arrid, even un-livable. So as a Hispanic man, his work for the PPP agency was more discrete, less obvious, not intrusive or unexpected.

But in the Jupiter space-program, this had not been particularly true, except perhaps that he was well-liked and charming. So, for his own sake, and for Marta's sake, he spent the next few weeks there with her, and 'told her everything'. It was a great comfort. Marta was very intelligent and curious.

"Why did they kill them, then?" Marta asked casually, in her offhand way. In three years she had gained 20 pounds, but he always loved her body and found her appealing. She had dark skin and dark hair, she sometimes used henna on her hair, and her eyes always seemed to have tears behind them. They knew each other well, and the familiarity between them created a very easy way for them to enjoy co-habitation.

"To keep them quiet," Mendoza said. "To keep the secret."

"What was the secret?"

"They were scientists, they felt they could prove what was going to happen, big statues or rocks or giant machines deep inside planet Jupiter, that exploded to the outside, and went into orbit. They thought they were a million years old from other planets."

Marta seemed to lower her gaze. "You're crazy, Angelo, that's just stupid, to kill people because they think something is going to happen on a planet somewhere."

They were enjoying a modest lunch, with local fish, coffee, local bread, and rice with beans. "Then what is a good reason to kill people, Marta?" Angelo replied. "I didn't kill them. I don't like killing anyone if I can avoid it."

"Even if you could save the whole world? Like in the movies?"

"The world is not a movie, Marta! We all passed through a significant planetary event at Jupiter, and the leadership was both lying, and confused about what was really going on."

"What planetary event? What is that?"

Angelo was frustrated. He had gone over it so many times, with so many people, at so many levels of understanding or truthfulness. "Big giant rocks exploded from inside Jupiter!! I just told you!!"

Marta paused, drinking it all in, trying to imagine what he was telling her. "Oh," she said blandly. She was munching sensually on a bit of sweet-bread. "Did the rocks hit you in the head, my Angelo? You're crazy! That's a crazy story, I don't believe you!! They take you away and you go into space on some work and you come back crazy!! I think we should make that a law-suit and get an attorney for you."

He sipped at the hot coffee, a very dark brand from Columbia. It was amazing to him, that now back on Earth, the gravity and air and sunlight, the walking and talking and open-spaces, were somehow less successful at a sense of normalcy he craved, and had sorely missed on their journey to Jupiter. The coffee in Belize there with Marta was somehow 'different', than any food or drink 'up there'.

She was right. There was a certain sense of 'crazy', a sort of cognitive dissonance, but it tended to pass. Sad but true, Mendoza was not well, and he knew it.

"They had me put to sleep, too," he told her. "The head manager was angry and had us drugged so we would be unconscious while they shipped us back home like cargo. Like frozen animals. It was awful."

"You shouldn't have made him angry."

Angelo winced at the thought. Rage, or anger, was not something one might attribute to such as Menuda Wu, the Jupiter Program Space-Operations manager. He wasn't that type. Wu smoldered and schemed or plotted, detached and fully in control, coldhearted to a fault, not raging, not physical. Angelo's body kept the painful record of being drugged into suspended animation, against his will, with nothing he could do about it, and millions of miles from home. The memory was in his nerves and cellular-level limbic system, it still caused him another moment's worth of personal pain. That and other experiences.

"Well, he's dead, if that's any consolation," Angelo then told Marta, at that point. His coffee was now getting cold, but Marta had a hot carafe, and poured him a second cup, hot and steamy. Birds beyond the kitchen window chirped and played in the sunshine. The weather was a mix of clouds and cold, and warm sun.

"He's dead? The manager? Up in space?"

"Yes," Angel-Face answered, with no particular savor.

"What happened? You didn't kill him, Angel-Face, did you?"

"No, no, no, of course not. I think the judgment was that he died of a coronary heart-attack during one of his crazy sex parties."

Marta laughed. "What? Crazy sex parties!! I though these were astronauts and pilots and space-men!! Crazy sex parties!!"

Mendoza couldn't help laughing as well. There was an irony to it. Menuda Wu wasn't the enemy, and he wasn't behind all the trouble, or any conspiracy. But he was a very powerful person, as far as that program, and all the ships and bases and materials and wealth. And he had become personally corrupt. And now he was dead.

Rumors that the 'Ferrous-2' had carried a cargo item with altered DNA samples and methodology to begin human experiments on the astronauts, were also disturbing. So, good riddance Doctor Wu. He discussed this and much more with Marta, and had time with their children later.

One of their boys had left Belize during his absence, to fight in a secret war of some kind in China, and apparently was killed as long as a year before. Mendoza sensed himself fading into retirement. Why fight the powers, why bother the giants, why force the truth to carry her sacred burden for some new problem or difficulty, really, no matter how important it may seem? Truth is not forced.

The PPP was a good program, she only researched problems from a non-biased view, to help inform larger scale decision makers. But no one should ever have to be killed, or kill others, or be frozen like a drugged-up shipment of prime-beef on a 700-million mile deep-space run, or cloned, or handled so poorly, only for the sake of the truth. The truth of a matter is usually neutral, by itself.

The Mexican space-man stood on the shores of the Caribbean waters, later, there at Belize, listening to the sea-gulls and the waves, small buzzing sand-flies in his ears and his toes full of sand. He walked and picked up bits of wood, looking up at times to see the Sun again, and the blue sky and clouds. He dreamed of what he could recall of the sight of Jupiter, from the ship, then, and now so far away, so vast in his mind.

"Never swallow anything larger than your head, old man," he counseled himself, laughing. He threw the stick into the water, then walked on a bit. The seagulls laughed and danced in the wind over the waves.

CHAPTER 57: My Friend In Seattle

The Proficiency Agency had provided Daniel with another private jet, with the anti-gravity assist pods, to make his transition easier. The Pensacola was long gone, and the Seattle Regional Airspace Authority no longer had much interest in any violations. Deveroux was still tall and thin and somewhat gaunt-looking, in his black-and-tan business-attire (by this era a sort of pull-over ensemble for men, with pockets and a spandex-style, but loose and comfortable too). He was always athletic, but not husky or like a wrestler or line-backer. More like a runner, really. He had done some long-distance marathon runs, to be sure. His credentials as a pilot were limited, but the private jets were within his skill-set, and he loved to fly.

The airport at Ashland, Virginia, in the Washington, D.C. Authority, was big enough to handle all sorts of air traffic. The tarmac stretched East towards the Atlantic, and between were low mountains rimmed with pine-trees and granite stones the size of rail-cars, small lakes and creeks or rivers, roads and farms. It was all very green with plant life and every sort of Appalachian haunt, places he himself would never visit.

It was a modern facility, for a small airport: the antigravity aircraft were accommodated differently, much like helicopters, and most of the planes were hydrogen-jets, not prop-aircraft, long ago left behind as antiques. There were several long runways at various angles to each other. Deveroux's jet looked like a sleek, cleverly constructed steel-falcon or bird-of-prey, an eagle or hawk.

Of course it could not attain Earth-orbit at all, but was very fast for intercontinental Earth-globe travel. He could have reached London in only a few hours, but was headed to Washington-state. Back where he started. The jet was called a Presidio Formosa CrF4.

"Presidio F4, this is air-traffic control at the tower here, you can taxi to your runway for take-off, at the back of runway six,

please. Pilot, please taxi your aircraft and wait at runway six, and wait for clearance prior to exit. Thank you, pilot."

"Thank you air-traffic control," Deveroux said. "This is the Presidio Formosa F4 pilot, will comply."

"Thank you Presidio F4 pilot. Tower out."

The silver bird rolled easily towards the back-end or waiting area that approached runway six. Daniel was at the controls, and had no passengers. It was a very nice model, for a private personal aircraft, the PPP was good about things like that. He felt alive again, he loved to command an aircraft like this, it was a thrill.

The hydrogen-jets roared, the machine was tuned to warm-up, clean-burners. He checked and double checked the controls and measuring-levels, all seemed in order. He moved the jet into take-off position, then waited, thinking he might power down if the wait was very long. After a while the tower cleared his take-off flight path.

"Formosa CrF4, your flight plan is approved for take-off when ready. Have a good flight."

Deveroux was now alert, knowing what he needed to do. He had many flight hours, but it was always a thing to be very focused and aware. He bustled in the pilot's seat, getting comfortable. "Thank you Ashland. Starting my run. Best to you," he said into his head-set.

Then the Presidio's engines brightened and fired at his command, the ship pointed where she needed to go, gaining speed to more than 120 miles-per-hour in only a few moments, great gulps of air-mass forming faster and faster beneath the 60-foot wing span, flowing over the shape of the ship and curved wings as lower-pressure 'lift', until she was air-born above the Ashland Regional airport, and the lands below. Deveroux, now the pilot instead of the passenger, banked the ship north, climbing steadily to about 10,000-feet. After about 20-minutes he radioed back to the Ashland air-control to acknowledge his successful departure.

Not a space-ship. not the Abyss. Real breathable air outside. Clouds. The cool green hills on Earth beneath. Home.

Solitude suited Deveroux, but like most people, he didn't really like it. Somehow it seemed that over the past several years, there was very little that he really did like, or enjoy, and he knew as an experienced human being that his pleasures must balance with stress or discomfort, lest he become sad or morbid. And space-travel, or harsh adventures, certainly contributed to stress. He felt strong enough and competent. Angelo Mendoza seemed for some reason to have reacted worse to the journey than himself, physically, and when they parted ways, Deveroux felt his friend may need more doctoring.

Space sick, or home-sick?

By the time they had disembarked from the Earth-to-Moon orbital transport that connected to the 'Down', (parked in orbit around the moon), they were center-of-attention for a week with doctors and medical-advisors and de-briefing. This was standard, it was a lot like a required vacation, the facilities were nice enough, like a medical hotel or spa. A year or two in deep space could kill a guy, so they were extra-careful.

Seattle was ahead about three hours, from Virginia, in that type of air-craft. Deveroux settled in for the flight, adjusting the navigations and controls for automatic. Weather systems on that day, then late June of the calendar year, included storms over the Rocky Mountain regions, and dense cloud-banks from Tennessee to New Mexico. There was also a serious storm North West, mid-Oregon, and then more clear by Washington state. The Presidio Formosa CrF4 was a fine aircraft and could handle it easily with attention from the pilot and ground-assist.

Deveroux sighed, relieved, a long and heavy breath, somehow containing many months and many miles of personal trouble, with all that had happened. He was serious that the Jupiter scam research would be his last effort for the PPP. There were other jobs and forms of employment, such as teaching at the University in Seattle, and his small home on the tree-lined street there, and some of his friends he still felt he knew. He was a research-specialist, skilled at techniques for gathering information from a wide slate of public and non-public data-sources, and also skilled at attribution and confirmation. It was a certain area of education, so why not pass it on?

Daniel had started work for the Planetary Proficiency Program as a recruited entry-level data-manager. He fit their bill; middle-aged, healthy, single and able to travel, educated in the areas they needed, not compromised by commitments to military or political, and not deeply vested in corporate interests, or other agenda-driven groups. As a participating contributor to society and Earth-citizen, it was natural for him to sign up. The pay and perks were also good.

But he never really dreamed he would end up on such a duty as their recent try at 'fixing' the Jupiter Program, much less bouncing around the space-stations while a historic galactic event was unraveling. And in the end, that's all it really was: a big high-pressure explosion from within Jupiter, setting some oddball new forms into orbit around the gas giant. The facts were not clear as to whether the forms were truly of alien origins, or not, certainly an added bonus to the curiosity factor. But for himself, how could anyone have known? Who could have predicted or warned him?

The PPP only wanted to know why ships were crashing and the labor-unions were in rebellion. Out of his league? Over his head? It was an understatement before the idea even formed in his thoughts. So he could only let it go at last, without feeling he had failed himself, or failed at his best effort for others.

"Just that kind of guy, I guess," he thought to himself, with a humor. *"What a hero."* But to be sure, it was all too much, by then.

The so-called 'Jupiter plan' was the last straw for Daniel and the PPP, and he knew it deep down. Far too much, few other forms of employment would make those kinds of demands. A year or preparation and background, and all they could discern of any reliable sort was that some DNA-experiments were perhaps connected to Jupiter Program long-term goals and hotly disputed by the astronauts. No big deal, and eventually not even really very important.

"Mother Nature wins every battle," he reminded himself. *"Don't try this at home."*

Commander Brandeis really did most of their work for them, by the time the choice was made for he and 'Angel-Face' to be launched into space. Brandeis was a lot more of an insider, as far as Jupiter Program politics: the union disputes, the space-religion superstitions and traditions, the corrupt leadership and big-money deals from the top-down, the technological and science failures, an aging system of ships and docking platforms, and more.

But for Deveroux to make a journey into deep-space, it was a goose-chase from the start. And what they didn't know, finally came out, in spades, with the entire program rocked to its roots, many system failures and accidents, ruined equipment and gear, and deaths. And also the murders. Nothing Deveroux and Mendoza could have learned, had any possible chance of changing the course of events, it wasn't possible.

Like end-of-the-world predictions in the Bible. Even if you knew ahead of time, what real advantage was it, for the individual? Fear, and then some imaginary advantage, and then the madness.

The Presidio Formosa CrF4 danced a line among the clouds, a high-speed dream, a sleek silver idol of travel and speed. After a couple of hours, the Seattle Area Regional Air-Traffic Corridor came up on his navigations, from satellite. It was a sure shot to land the aircraft and begin to rebuild his life. Which was all he had in mind. It had all been too much. He was lucky to be alive, and he knew it, so he planned to take things slow and enjoy what remained. *"And many tales to tell besides,"* he thought.

Just about as he was beginning his landing-path and descent trajectory, computing the navigational from his current moment's position, the plane's radio-link alerted him to a call. *Beep-beep-beep!*

This continued a moment until he could answer. It was Vanessa Signo. "Daniel! There you are! Can you hear me? I'm in San Francisco! Are you there? Hello?"

"Well, well, hello Missy V, yes, I have your signal clear enough. I'm piloting a jet to Seattle, just about to land. How are you liking San Francisco? Missed you in New Anchorage at the Convention. What a bunch of egg-heads!"

Suddenly Deveroux could sense himself invigorated by talking with her. And both on the same planet! Vanessa was the only woman he had any intimacy with in several years, and that was unstable, since they were co-workers. Like two show-horses in a barn, they knew each other, and liked each other, and they knew they were 'supposed to' fall in love, like a planned thing, or 'just right'.

And it was maybe this aspect that deadened the thunder. Like two famous movie stars, everyone is certain they are in love, from the papers and the magazines. Yet they themselves are not so sure at all.

"Good news, Daniel," she said, on the radio-connection. "You're not dead any more. The agency has officially brought you back to life. When you reach Seattle, you'll have ID papers, residency, tax-status, passport and all the rest. You can pick up everything at the Immigration Office."

"Immigration? Like the US travel immigration office?"

"It was the best we could do on short notice."

He laughed. "From the land of the dead, back to the land of the living. I didn't know you needed papers for that. Well, that's great. I'll be at the University. What are you doing back in California?"

"Just work, for the agency, same thing, more cases, more researchers," she answered. "If I come to Seattle, maybe in a week or so, would you treat me to a hot-tub?"

"Sure thing, sure, sure," he said. The jet's flight-path was now moving into a pattern above the Seattle air-pad runways, some 20 miles out, so he had to position the plane to land. "I'm trying to land an airplane, Vanessa. I'll call you. Okay?"

The jet's engine's changed pitch as he adjusted the thrusters, and the aircraft dipped and began to bank. "Sounds good, Dev'," Vanessa said. "Just let me know. I want to see you. I'm sorry about everything, you know, this case at Jupiter was..I just..."

"We'll talk soon, dear," he said. "We'll take the hot-tubs in the snow. But not the Circulator this time. That's what killed Doctor Wu. Death by sex-toy. Not good. Releasing the call. Deveroux out."

The radio link went dead, a small green radio-activity light bouncing on his control panel was all that remained of her love, if indeed she loved him, and he thought she might.

And it was enough.

CHAPTER 58: Love Me, Love Me Not

Vanessa Signo was one hot babe, for year 2417-18 (Anno Domini, Common Era). By this day and date, she was approaching 50-years, and Daniel Deveroux was just a few years older. Her work was boring, but also placed her among the elite-class in Northern Hemisphere society: she had money, she had friends and family, she 'fit in' as far as her intentions and goals, both professional and personal. Signo was a positive contributor to the general public contract they all shared, co-existing in peace, work, play, and so on. By no means famous or a very important person or political-player, she mostly spent her years and hours as a case-manager for the Planetary Program Proficiency agency, assembling details on the assigned investigations (like researching a private-industry program to melt and then re-freeze huge tracts of frozen glacier ice in the Arctic Circle), and then presenting the investigation parameters in a sensible format that everyone could understand.

"When I get the blues or sad, I like music," she told her friends.

"You are music, Vanessa," the other woman replied.

She was also frequently involved in selection and final-determination on researchers to work on each specific case, (like Deveroux and Angelo Mendoza). It was a good life for 'one hot babe', but by this period in Western World civilization and history, there was a general slow-down in the popularity of romantic love, marriage and family, child-rearing and the so-called 'American Dream'.

Fewer adults trusted the whole idea of monogamous lifetime attachment, and fewer adults carried forward those long-term relationships, as traditionally understood. Only less than half of the Earth's population were even fertile, due to the 150-year battle to eradicate the AIDS virus. So by the same standards, or those of the so-called 'American Dream' (a house, a man and wife, jobs, work and wealth, children), perhaps only 20-percent of the general population walked through life in that way. Recreational sexuality was far more 'normal', recreational people, recreational relationships, recreational lives and lifetimes.

"Why does love grow cold, then?" she would ask, lifting a hot glass of brandy and coffee.

"More cowbell," her friend joked. "Get rhythm, when you got the blues."

Vanessa had a hard time expressing her feelings for Deveroux. He was a co-worker, very attractive to her in a male-female kind of way, they had spent time together in intimacy and 'romance', she enjoyed his company and he enjoyed her's.

And in a similar way, Deveroux was someone who 'fit in', he had money, social status, a place in society's sunshine zones, and the world by this period was far less suffering of total poverty, forms of enslavement, ruinous wars and criminal enterprise, devastating weaponry, despotic leadership or supreme-dictator mentality, and so on. The AIDS virus had been tamed, and other troubles as well.

"Love is a misunderstanding between two fools," Vanessa's girl-friend added philosophically.

"Good enough for me," Vanessa said. But she wasn't satisfied.

Tamed troubles like the Jupiter program and its failures. Which to be sure, had greatly troubled poor Dan Deveroux, almost two years on the case, and a year or longer in deep-space. *"Don't try this at home."*

In a very real way, the 'Jupiter Plan' was just that: much ado about nothing, a hot rock inside the gas-giant had exploded into orbit, with possible extra-terrestrial long-term involvement over water-rights, over several years. The PPP's work was mostly unrelated, perhaps by default or maybe ignorance of their goals, one investigation had led them into another, but an entirely different scenario.

Vanessa Signo had a hard time expressing her feelings about Deveroux. She wanted to, she yearned for that, she didn't know what it was, and most people she knew would call her foolish to 'fall in love'. It just wasn't in fashion, but there she was.

In San Francisco, at a popular club she enjoyed, Vanessa shared her thoughts with a friend, a husky, athletic woman, who worked in the under-sea housing and fish-food projects in the deep Pacific ocean, off Monterey. It was a chilly night, foggy as often the case for the City of Saint Francis, Vanessa had some chicken and spices, and a coffee-with-brandy. Her friend's name was Alisha, tall and bronzed, a woman of the waves and water, common to ships-at-sea, a beauty, but rough-hewn and sometimes spewing her words.

"He wants me to come up to Seattle, so I'll probably catch a flight by next Monday, so I can't do the bowling league awards ceremony or all that, probably for a couple of weeks," Vanessa was saying. The two women were part of a bowling-league when staying in San Francisco, it was a lot of fun. "So, sorry about that Alisha, just let the team know, it's no big deal."

"The awards ceremony?"

"Yeah, no big deal," Vanessa said. "Is it?"

"No, no big deal," Alisha said. She had a drink as well, the chicken dinner that evening was a fried sort, with spices and a thick batter-shell, very delicious. They munched away, white capped teeth ripping into cooked chicken flesh passionately, all smiles and eyes. "But he's a big deal, yeah? This guy in Seattle? Daniel."

Vanessa laughed, a bit strained. "He's a bigger deal than the bowling-league awards, for me, yeah," she said. "Other than that, what's the big deal about a big deal, if it's not a big deal? He just got back from a year at the Jupiter platforms, it was sort of rough on him. We've been close for a couple of years, so, sort of a reunion."

Alisha seemed impressed. "The Jupiter program stations? That's millions of miles away in deep space! He went up there, with your agency or whatever? What's his name?"

"Daniel Deveroux. He's a researcher. Really cute and sexy guy, too, I really like him, but..."

"Yeah, but-but-but, wait-a-minute, hold your horses Cupid, Romeo didn't come home for all that, but-but-but...you show all the signs of hooking up with this one for longer than just a week or so. True? False? It might affect your bowling score, that's all." Alisha was joking with her, woman-to-woman.

Vanessa was finishing up, the brandy-and-coffee were gone inside her animal self, with a deep-dark warmth and tonic feeling in her tummy. *Is that what love feels like? Could be, could be.*

"Maybe it's just me," Vanessa answered her. "But my instinct for hooking up longer than a week or so is...it's kind of important, higher in my thoughts, but also a failure to my heart, somehow, like a disappointment, but more important than other things that make me feel sad. An emptiness. A loneliness, but like it will never be satisfied, by him, or anyone, I guess. Like I somehow know love will fail me. Dan Deveroux or anyone else. And it makes me sad, so I want to find out, but I know somehow I never really will. Know what I mean?"

Alisha was also done with her food, they leaned back in their seats, the San Francisco city-scape was fuller with errors and mistakes than ever, plodding and dull, stupid and brilliant, gigantic, moving, cold, flushed with life, yet dead, machine-like.

"Of course I know exactly what you mean, and I'd be an idiot to pretend I could even begin to express it," Alisha said. "So why try? You want to fall in love, you stupid bitch, and you don't know how. Welcome to all that, I'll admit, it's a trashy business. Why even try? Go for it if the sex is good, I'd say, in any case. When did you ever understand that anyway?"

Then Alisha laughed, tipping back the dregs of a very cold beer she had ordered with her food, the second of two, a Japanese variety. Vanessa was laughing too. Alisha was known to her to often prefer the perversity of woman-to-woman sexuality, but they two never had. Vanessa liked men. Maybe it was genetic.

"Don't call me that."

"You want to fall in love, but you don't know how."

The San Francisco night embraced the morning, then a day, another day, another night, until Vanessa Signo later found herself in Seattle. Deveroux had settled in, now officially among the living, by writ-of-mandate from the secret methods of the PPP. No one even noticed, except those few friends he knew personally. The earlier faked death, the fake body and funeral-burial, the adjusted records and wipe-out of his life as a University professor (the house, vehicles, possessions and personal items), the gig teaching research methods to bored and uninterested students seeking careers in data-assembly and information-administration, all now a past dream.

Being that official records were not very much reviewed by other than tax-accountants and real-estate sales offices, the automobile accident he had supposedly suffered to the ending of his biological existence was a nothing, a mere cipher, a blip. But he still had some trouble with the idea of maintaining any actual relationships, it just didn't seem right. It was all computers.

So where have you been, Dan? Didn't you head to New York? About a year ago, they said you were, that is, I thought you were working for the space-program? Was that you? Am I wrong?

Glamorous, or the thrilling life of some sort of 'agent', or new-world cop, was not happening, it was much more of a social gaff, as if he had something to hide, and he did, but not for reasons anyone would find very appealing.

Yes, I was working for the space-program, sort of, in a second-level capacity on some research. Yes, I did some traveling, there was a mix-up, but it worked out, you know. I'm under restrictions about it all, you understand, thanks. How have you been?

Back with Vanessa in a pleasant Seattle-area bed-and-breakfast, it was a repeat of San Francisco, by then more than two years previously. Throughout the ordeal, Vanessa had been his link to reality, and home, moreso than Angelo Mendoza, or really anyone. At least the PPP cared about his well-being to the extent of keeping her in communications with what was happening with he and his pal, when they needed help, she was there, and she tended to come through, if possible.

He couldn't really see where she may have felt any similar affection or need, because on his part, he was millions of miles away, or in-transit, mostly complaining about conditions or the lack of cooperation from people like Menuda Wu, or various murders and crimes along the way, like being frozen to a deep-sleep state for the homeward-bound transit, prior to the revelation about the robotic-illusion alien-zodiac 'ships'. Against the rules? Illegal? Or just unpleasant?

The thing about it all was, she cared. She cared for him while he was going through all that. She cared. And he could feel that.

Was that why did Vanessa want to sleep in his bed and take his sexual pleasure as her own, again, after two or three years? Maybe she was dreaming about space-travel herself, or maybe she had grown fond of his lilting, stiff-necked, pseudo-sophisticated style, or maybe she lacked for men she trusted that way?

Just like her, in his maleness, the bottom-line inquiry he could not fathom in consideration of what people called 'love', was why? What did she get out of it? Why did she love him, or would, if she did, or, what was the value of a loving exchange, that would last or endure? From her side?

The question created an ego-circle, a snake with its tail in its own mouth. If she held the answer, and could tell him, it would only feed his pride, and not necessarily secure less rewarding forms of love, like if either of them became sick or ill, or grew old, or lost a leg, or had an accident causing an ugly facial deformity. Would you love me then? And if so, why?

Deveroux only eventually concluded that it was generally more pleasant overall to 'fall in love', than to be alone. They had a wonderful time together, again, at a tree-lined, rosewood and pastel-colored stone-masonry bed-and-breakfast hotel, away from the metropolis, very cozy and quiet.

No one called. No one contacted them by computer. No mysterious people dropped in. No anti-gravity helicopters overhead with line-of-sight spy-gear to cause fear. No huge global powers who wished him silenced. No urgent reports with priceless information that would change people's lives. No deft handling of events and people lost in confusion for some corporate cause or planetary loss.

"*Thank you, woman. That was great,*" *Deveroux told her softly.*

"*Mmmmmmm,*" *she answered, also softly.*

Deveroux was finally able to personally regard the Jupiter investigation as over, done, gone, in the past and not truly effecting him any longer, in the present. He'd come home, they'd done it, they'd made it, and the mess of it all was someone else's problem now, as he coolly stroked Vanessa's long, cream-and-gold hair, her sleepy head in the crook of his arm, naked against his chest in the queen-size feather-bed, her breath light, then heavier, her tender breasts of a movement, then a gentle cough, then she turned, sleepy from their loving, the daylight spent to afternoon and evening. Home. Peace, where love was even possible. Welcome home.

Vanessa blinked, and rested in his smile. "Hey there, good-looking," he said.

She just seemed to moan, again, sort of a purr. Ladyfingers searching for a cotton pullover.

He held her a bit, expanding the moment for them both. He considered offering some jokes about Doctor Menuda Wu's death, and the Group Sex Toy they called the Circulator. But somehow it just wasn't funny. That worthless asshole was dead, it just fell flat, his death was a justice, perhaps, but still a death, a brother's fall of some kind. He'd find other topics for them share.

Now that I'm alive again, he thought privately.

CHAPTER 59: The Proper View Is Lost

"Home is where you came from, and where you're going. But if you ever arrive anywhere else, and it's suitable or comfortable, a home is a home is a home. So welcome home," Daniel Deveroux, global programs efficiency researcher, in the Seattle-area Regional Authority urban area, 2,418AD/CE (to his friend, Seattle University Professor Richard D. Podliakov).

The proper view was lost. Even before they started. As a professional, and also a teacher of the same disciplines and practices, Deveroux faulted himself brutally for ignoring that much. Bias, pre-existing judgments, ignorance that skews the inquiry to the level of a fatal error in the helpful explanatory review (such as ignorance of the actual science, technology, mathematics and astro-biology involved), and qualities such as vain-glorious notions of heroics or adventure, had ruined his work on the Jupiter program for the agency.

The proper view was lost. And it had nearly cost him his life, and that of others, had his costly efforts for the Planetary Program Proficiency agency been taken more zealously, regarding Jupiter. The researcher's art, the skill of seeking and finding truth or non-truth, available or unavailable, reasonable or unreasonable, depended very specifically on 'the proper view', or 'the proper viewer'. He also taught this in his classes, when working as a University professor. When the proper view was lost, the research failed. And the truth would fail as well.

Afterwards, post-Vanessa, which he knew was important, for his overall well-being, (*'love is important for my health, thank you'*), Dan started to rebuild what he thought of as his life. Things about him, things more personal, things he had neglected or completely abandoned, or had to, even by necessity, like a wind-blown leaf somehow enlivened to search for the tree it fell from.

True enough, any other PPP research-investigator might have been assigned to Jupiter. Was Vanessa Signo really so cruel in her missing pieces, where what people called love had not found her, that she sent him into deep-space on a lark? Vengeance? Sort of a feminine beguilement, passing into a joke (to her, anyway)?

It riled him, there was no venting. So he put himself selfishly back into real life. There was no guarantee at all that work as a University staffer would be readily available. 'Information Gathering and Truth as a Research Data-Component' was a popular course, but the curriculum was very old, all the way back to Old School Communications and Journalism.

Routine, the sadhana of daily life, a boring blessing when things too difficult for the body, left pleasures and comfort depleted beyond exhaustion, was now a new friend. And he was thankful.

"A joke is a joke, Vanessa," he told her before they parted ways. "The agency needed me dead, for the good of the program, so the California Thermo-Electric Companies would drop the previous matter, that whole thing. There were a few they probably rubbed out, as the American Indians used to say. Rubbed out. But, a joke is a joke. And therefore, a levity, a non-sequiter, an apropos, a humorous aside that seems cruel, but isn't really true, and so is only..."

"Shut up, Daniel, hand me my legs," was her reply. "Gawd, man, you're just so fucking honest, it's like some kind of obsession with you."

Vanessa wore long, colored leggings, the same style as dancers would, or gymnasts, they draped her thighs and calves and dainty feet, not draped, but tight, hugging the flesh. And she was, and she was.

Why do you think my penis is so ugly? Sex-change surgeries were not for Deveroux, her attraction was an unlimited lust he never failed to enjoy, if he possibly could. And she knew it, of course, and he knew that she knew. Ugly. Penis. Beauty. Woman. Truth.

When Vanessa left Seattle, from the bed-and-breakfast they had together about a week, it was assured they would see each other again. And again. And they did so for the rest of their days. Eventually it dawned on them both with equal certainty: they were in love. It kept them both a long time, too.

Deveroux found his way again back to the company of Richard D. Podliakov, the mathematics professor and astro-biologist, a tenured Ph.D. at University in Seattle Regional. By this period, 400 years beyond the Bachelor of Arts and Master of Science degrees that once were the coded formality of illumination for the literate and well-read in the world, at the turn of the millennium, University functioned much more like elite camps, or long-term small societies, or like well-tended, peculiar monastics, wealthy villages devoted to learning, mostly so the participants could adequately pursue other things, including space-travel.

Gone were the days of frat-boy pranks and parties, and hot student-teacher affairs, or 'graduation ceremonies', rewarded by jobs. Podliakov, the younger, (Richard David) was the second son of the Elder Podliakov, one of the Old Men who set the tone for the early 'secret research' about planet Jupiter, that ended the lives of science and technology space workers George Philby and Montrose De Montrose, and maybe others. Murdered by such as Doctor Menuda Wu (deceased).

In terms of their social equation, the real reason Richard Podliakov enjoyed keeping company with Deveroux, was because of Daniel's ancestry as a descendant of Alex Deveroux, who was from another planet than Earth, a few mothers-and-fathers back. Alex had been brought to Earth from the Hydra star-system, as an observer, and also to assist with changing Earth history regarding terrorism, and warfare. But that was another story. And other loves, mothers and mothers ago.

A few Vanessa's back in the woodpile, Deveroux guessed, and not un-biased. *Are you unhappy and sad I'm having your baby? Is it ugly? Why? Don't you love me anymore?*

By then almost five full years beyond what Deveroux could recall of the man's shared wisdom about Jupiter and about the program collecting junk-stuff of the nearby Solar system planets, they scheduled a breakfast hour chat, again at the Yellow House Cafe, a favorite near the main campus.

PPP Jupiter Program Investigation Case Number 4436a (Verified), year 2412; a year of planet Earth background, two years (nearly) in space among the platforms and at the Alpha-base, a year's travel in-transit back from Jupiter to Earth, for both men, and by that time at least another solar orbit in recovery and de-briefing. Richard's previous warning and clue was simple: beware the secrets of the gas-giant. And he was right. And for the most part, Jupiter's secrets were still secret.

The breakfast was the same, two eggs, fried potatoes, with red hot-sauce, sourdough toast and butter (or margarine), and hot coffee. Deveroux added bacon, for that morning's meal, but Richard only wanted none at that time, no meat, a personal preference. Richard had also added his own breath-and-blood to year's passing, older, wiser, or not, yet to enjoy to cold North American air, each day he chose to live. A ruddy-looking white man, with wild, frilly and curly hair, grown long. He sometimes had a beard and sometimes he didn't. But his hair was turning white and gray. His head was larger than that of many men. He was Caucasian. An Earth man? Of course.

"Daniel, you freak of nature, an honest man, so sad, so sad," he said between bites of food. "Science adores you, natures abhors you. What does it mean? What does it mean?" Then he laughed.

"I am insulted," Deveroux said, sort of bland about it.

"Good, good, that's the best thing," said Richard D. Podliakov, Ph.D., his learned friend. "Back among the living, right? You made it! You've arrived! You've crossed over the river! And here you are! Insulted! Magnificently! Daniel Deveroux, the magnificently insulted space-traveler, my friend! Good, good to see you!"

Deveroux just ate his eggs, silently, from their talking. The Yellow House restaurant-cafe was busy, the usual student population and local teachers and University staff, the bus-boys and cooks, waiters and waitresses. He added some salt, and improved the red-hot sauce a bit, he felt. The toast was now soggy with other cooking fluids, watery egg leakage, oils and bits of unknown fried edible stuff, floating around the white-blondish plate in pools of other tasty edible salty liquids. *Stupid fucker is enjoying this,* he thought inwardly of himself, the Quantuum Mind experience still haunting at a far, far distance.

"Well, just wanted to see you again," Deveroux added. "For one thing, I spent time with your father, Eldon, he was at Ganymede for a while, and then at the main base. Before the rock broke out, I mean, before the thing happened, I spent some time with him. So I wanted to see you, with his greetings."

Both men were versed well enough in the global politics of the day, to have some conversational knowledge about the Big Deal at Jupiter that Year, or the year before. Distant gas giant planet, same as Earth in orbit around the Sun, a bit further out. Inside parts under great pressure explode into new orbits around Jupiter. Aliens from distant stars planned it all out, or knew about it before it happened anyway, they only really wanted water. Conflict with Earth powers, Earth-men, corrupt Earth leadership in near-planetary space. Many dead, much lost. Clones fight the battles, but there are no battles, the clones die, they know nothing. And it's all cool, it's all okay, it's normal.

Richard seemed displeased. "Well, how is he? My dad, I mean? At Ganymede base?"

"He's old, Richard, he's just old. Closing 90 years or more of age. He sends his greetings, of course. We mostly played guitars, or we learned to play guitars. One of the clones was the music teacher, I mean the clone's dad. The music-teacher and his clone. Amazing, in a way, they looked identical to each other, you couldn't tell one from the other. Your dad and the other science guides signed up, so we could discuss other things, the earlier research about planet Jupiter they were hiding. It was fucked up, we had just seen two men murdered on Alpha-base about then. By robots. But, at least we had guitar music classes."

"Robots don't kill people, people kill people," Richard answered.

"Guitars don't kill people, either, mostly," Deveroux quipped. "It was a way to conceal our research, the interviews."

Another moment. "Well, anyway, yeah, thanks," Richard said. Deveroux could observe that his friend was always like a child about his dad in that way, he could see the old man in the sky was dying soon. It was Eldon who showed him how to love education, the joy of learning. But the old man in the sky was dying. Again.

"You were right about Jupiter," Deveroux said. "But not so much about the stuff inside her. The forms. I think they were just rocks, or one big rock, not statues, not giant robots."

"Well, okay, maybe I was wrong," his friend answered. "Justice is served, I guess. Only the alien mind would seed a gas giant world's inner super-pressure dimension with little pieces of machinery they'd use 1,000 years later to get water. What if the machines broke up, or failed? I liked my dad's plan better."

"He had a plan?"

"Sure," Richard said. "Live free or die, sums it up. Great plan, too." Then he waved at one of the waitresses, more hot coffee flowed into his cup a few moments later. "Are you living free?"

"That's debatable. Also up-in-the-air if I'm still dead or not. I feel dead, dead inside. The journey was too much. I overestimated my physical ability, and my mind's stability. Space travel sucks, it really does, I don't care how fancy they dress it up. And it probably always will."

"But your own son, he's into space-work, yes?"

"My son? I don't have a son, Richard. I never had any kids."

They were finishing up their meal, a bit more coffee, they pushed back the plates on the oak-wood table. Richard was using the rest-room, and in the absent moment, Deveroux recalled what he meant. He did have a son. But he never knew him. There was a woman, years ago, they had relations sexually, and she was pregnant, but he wasn't informed and they were not for marrying each other. Daniel stood up by the table. It was good to see his friend. *A son? Into space-program stuff, technology? I never had any kids. Did I?*

As Richard's restroom run was over, he came through the doorway back into the main dining area. He handed Deveroux a paper card, with colored ink-and-pen artwork, it looked oriental, but the blank spaces had room for what he had written on it with a pen. He handed it to his friend.

"Here," he said firmly. "He's over across by the harbor towns, a place called Williamsburg. The guy's name is Arthur Morrey, I guess he never used Deveroux, must have been the mother. You can DNA test, but I checked. He's about age 35, by now, I think. He's studying off-world oceanic life-forms and sustainable distant planet water-worlds, something like that, all pretty egg-head. Your dick did it, and you're the dick that didn't even know it. Here's his info. Go find him. He's your son, after all. I ran into his records at the college while doing something else."

Deveroux took the pretty paper card, he looked at it casually, Richard had scribbled some information there, a name, a house number, a phone. Apparently how to locate his illegitimate son, from a 30 year's previous sexual moment, a person named Arthur. His son.

Daniel felt pretty good about it. *"I wonder if anyone ever told him about us? The family, grand-pa Alex, the Hydra legacy in the bloodline?"*

He saw Richard again later, they said their good-bye's for the breakfast that morning, then went their ways. Daniel had no idea his University mentor had planned to reveal anything about his past to him, that morning. But he was very glad that he did, in a way. Every man is glad to have a son, for some strange reason.

Maybe. Maybe not.

CHAPTER 60: Glad to Meet You, Son

Deveroux liked the idea of successful living. Still looking for Jesus? Look for human excellence, was his thought. He had never been dirt-poor, there were times in his life-passage when he handled things that way, traveling Europe as a student, for instance, or camping the wilderness on a survival budget. By this period, during his lifetime, much of global Earth poverty was 'relieved', by generous world powers, who eventually found it better to simply cave-in on notions of slavery and oppression as answers for labor-needs and law-enforcement.

Centuries passing, new and improved means to feed the poor, as planetary population was swelling to massive levels, some 17 billion mouths-to-feed, were best, even for the very wealthy. Weapons-free zones. Large-scale deep-ocean fish-farms occupied vast new industrial efforts, and provided monthly and yearly hauls of out-standing fish-protein for many.

PPP investigated those, too, problems with DNA rights and genetic ownership. For himself, the gift of a son, was something like a wealth, money he had never given, coins that were new, but unspent, not yet given, unknown but valued for worth and goodness. Grand-kids. Or just the enjoyment of knowing himself better through a son. So it was natural for him to wish this Arthur Morrey person well, and want to pass on some goodies, the booty of his days, some sense of being 'the provider'. It was also new to him.

"A son? Roger, I don't have any children," he remembered telling Roger.

The agedness of the cosmos approved. He wanted to give him something, as a father might, but he didn't really know what. Dan was able to spend more time checking out his treasures, left behind now far too long, in a storage, and also many items supplied by the Proficiency agency. The anti-gravity assist jet-aircraft among them, parked and serviced now again back at the same airport, the handling charges against him on the Pensacola model now long forgotten and dropped. Rotten thing, to die. Even worse, to die, and live on. Adding to the injury, to know one had died, and carry that knowledge along the way. Dead man walking, dead man talking, dead man flying an airplane.

'Don't try this at home', said the magnificently insulted space-traveler.

Deveroux strolled through the storage areas, a maze of large-style 'safes', or what banks once called 'lockers', strongroom vaults. These were big ones, walk-throughs, not the home-style aluminum sheds. They had excellent security, too. People using these were sometimes hiding things, it was true, but they served well for many items of a special nature like technological marvels, personal souvenirs, records, documents and books, weapons, yes, weapons, a few souvenir 'minooks' for fortune-telling favors, for some, not so much Daniel.

Deveroux kept no minook, he had come to see himself, or people in general, as a living minooks, as far as his spirituality. *Still looking for Jesus? Uh, no, no, sorry, I've already accepted, as a child. No, sorry. Seek human excellence instead.*

Look for the idolatry of gawdy riches, too, go ahead, don't be an ass. Anything, any thing at all, anything you want. Such his emotional thoughts, among his own treasures, there. He could enter by voice-command and voice-recognition, as well as finger-prints and a code. He had plenty of time, now. It eventually dawned on him that what he was seeking was a gift for his son.

But now, after the voyage, he was returning to life, instead of letting go of life. Death and rebirth. Let the bad ideas and bad memories die, let the sub-conscious renew the man, let the aged cosmos chuckle a bit. Romance, the whole idea of loving Vanessa. Too old to bear a child? Maybe not, maybe not. The things he owned, his suits and clothing, his athletic gear, a set of 'epee'

French swords, foils, and sword-fighting gear, the masks and vests, gloves, (without gloves you could do nothing in the sport, true, true. The gamers would whack the hands by accident very painfully), all his collections of high-tech gizmos, some specialty items, some inventions, some PPP creations used in his work for them, and a few official immigration files related to the Earth-passage of Alex Deveroux (deceased), the agreements with the civilizations at the Hydra star-system, and more, all these things were his.

Proof of his off-world lineage was of course quite sensitive, but it was all legal and for the good of everyone. Hydra star-world men and women were identical to humans of Earth, no one really knew why, or cared much. They were much more advanced, however, and had mastered a technique of inter-stellar arrivals and departures. They also had no violence or war among them, and at some point chose to approach Earth, to both learn about evil, and also assist and help with terror-wars going on at the time Alex Deveroux had immigrated to Earth. Alex loved a woman, during his day, an athletic dancer named Misha, of dark-skin and certified Earth-born lineage. At the time he worked for governments to remedy terrorism. Alex Deveroux eventually was drawn into a terrifying trap, and died there. But that was another story.

He could hear foot-steps, moving toward the locker-vault, from outside, it was a long metallic hallway. He knew it was the man (no boy, or child), Arthur Morrey. Previously, the info on Podliakov's card was applied almost effortlessly to leave a voice message:

"Message for Mr. Arthur Morrey, of Williamsburg, at this number, I believe. Please confirm to meet me briefly if possible, this Thursday, at the Bay Stone Underground Vault, Unit 203, just South of Port Gerald's Point, at noon or about mid-day. I will be there, my name is Daniel Deveroux, with University and a government program. I believe we are related by DNA bloodline,

to about 30 years ago, previously unknown to me. That is, with all due respect, I'm your father, we've never met, unless it's untrue or a mistake. I have some information and family details to discuss. Confirm if possible to this number. Thank you."

Deveroux's hand-phone soon after received the confirmation he wanted, also a voice-recording: *"Yes, thank you, got your message, Mr. Deveroux, confirming your schedule at Bay Stone Underground Vault at noon Thursday. I've also been told about you and my mother, from other sources, from my birth-year in 2,381. I was also contacted by your associate, Dr. Podliakov. A pleasure to meet you sir, if it's true, a rare thing. Thank you, see you Thursday. Arthur Morrey, Williamsburg General Public Communications, 20917b. Cid Bixi Mimim."*

A voice-recording can sometimes do justice for a truthful reproduction of the human voice. Deveroux could hear Arthur's voice, as he got the message. The voice was male, sort of 'sandy-sounding', not wheezy, but breathy, somewhat girlish or young-sounding, but firm and clear, strong. He more or less understood the process.

The air first pushes to Arthur's unit from his mouth, to an inner a piezo-element, vibrating against an electronic magnet, translating to a quantifiable digital or math equatable signal, mere electrons. The same signal is stored temporarily on a semi-permanent iron-oxide based computer chip, sustaining the sequence as tiny atomic alignments, at the molecular level, formed by precision magnetics. Iron-oxide is particularly good at this. The signal sequence is perfect, and can be repeated without error as a radio wave, frequency-modulated through the air-space, to satellite or a ground-based tower-repeater, then arriving by code-identity to Deveroux's hand-phone unit. His hand-phone then also stores the identical signal sequence of tiny atomic alignments, and so on. He may touch a button, his device recreates Arthur's voice, new puffs of air rise to his ear external to the device.

He had never heard Arthur's voice, of course. Yet, knowing that men and women have always recognized one another by voice-patterns, also a legal form of identification, also a social form, like birds or even barking dogs, or lovers, the human voice and speaking sounds arising from electronics this way were often a disappointment. Not the same, not the same at all. For just an instant he tried to recall the sound of Arthur's mother's voice, from more than 30-years past, and he knew he could not, and never would. A woman, a moan and a sigh, a lie on the lips of a lost lover's hot passion, gone, gone, only this man to show, a new life. *Cid Bixi Mimim.*

Now the foot-steps of a grown man, in the steel-cave hallway just beyond unit 203 at the Bay Stone Underground Vault system. Arthur or whoever it was seemed to be wearing heavy boots, snow-boots or hiking boots. *Maybe it was raining again outside?*

The units had a hard-door entryway, Deveroux left his open while inside viewing his treasures. A dragon counts his money the same way. Then Arthur was there, with a boyish grin and smile. He was a big guy, tall, husky, more pinkish or Caucasian looking than Deveroux, his dad, who looked more Eastern, darker-of-the-flesh. He did have snow-boots on.

"Is it raining?" Deveroux said. The boots had bits of mud and water still on the toes and soles.

"Yeah, a bit heavy weather, some snow too. I'm Arthur Morrey. Mister Deveroux? I was sent a message..."

He paused. Daniel could only grin back at him. "Yeah, Morrey, yeah. I'm your fucking dad. How you doing?" Then he laughed, and Arthur did as well.

"Fucking good, fucking okay, nice to meet you," Morrey said. "Dad." They stood there, then hugged like men, strong, shoulders blocked. It was a moment, often repeated, reunited with a lost relative, father, mother, sister, brother, even over many years. Now Deveroux could see him more clearly. He was collegiate, not a geek or another egg-head, he wore athletic gear, and a funky hat. His hair was brown-reddish, and he had a chin-hair beard, with a wide forehead and deep-looking eyes, a somewhat weak-chin. Not much at all like Doctor Menuda Wu, of the Jupiter Program leadership in deep-space. Long-term space-travel will do that to a man, I guess, Deveroux was thinkng. But not this one, and I hope perhaps never much space-travel, for one I care about anyway.

They exchanged formalities, small-talk, this and that. Deveroux was showing him some of the items he had in the storage, including Alex Deveroux immigration papers and records, but those went on and on. By now some other clients who used the underground vaults were moving around beyond where his unit was, next to another, and another, and another, all so similar, identical, but not the same. No two things are identical.

"So, my mentor, Podliakov, from University, mentioned that you work on off-world deep-telescope analysis of oceanic planets, like distant star-systems with oceans and possible life? Is that true?" Deveroux asked him.

"Yes, that's right, it's a special area," Morrey answered. We call it Porphorps, or Porphopolis Worlds. It's fascinating, but the distant-star system detection from Earth-telescope and wave-analysis is exceptionally detailed and challenging. It takes years, we scan vast regions, then analyze for light-wave variables and other factors. But they exist, we're fairly sure. It means a lot, in a way, but it's rather dull. I like it, though, I mean, I like the work."

Deveroux paused, thinking what he meant. Off-world deep-space planets with oceanic life forms. Got it. That's easy to understand, isn't it?

"Home is where you're going, and also it's where you come from, son," he said. "We need to talk about some family matters. I have some inheritance stuff, too. And then just maybe, you know, get to know each other. I was very shocked to hear about, you know, having a kid, I guess you'd say. Not that you're a child, I mean, but, you know..."

"Sure, sure, Daniel, no worry. Dad. I'm all grown up. Sure."

They hugged again, shoulder-to-shoulder, more like brothers than father and son. It was not different, and it was not the same.

Cid Bixi Mimim.

THE END

Cover art "Starship" by David Schleinkofer | artist20@verizon.net

ABOUT THE AUTHORS

Tom Luong | Film Producer | Director
tvluong1@hotmail.com | 951-660-6010

Tom Luong is a Vietnamese American born in his native country just after the Vietnam War ended and raised mainly in California. Tom was born in November 11, 1976. His parents were refugees of the war and fled Communist Vietnam in the late 70's and was awarded sponsorship with a relative to live in Orange County in Southern California in late 1981. Like many Vietnamese that fled the Communism, his Dad (Mike Manh Van Luong born in 1949; Mom, Nancy Ngat Thi Le born 1954) was a POW during the war and this affected Tom in many ways. To understand about how his Dad felt during the war, Tom joined the US Army at one point and underwent basic training and was later deployed to South Korea. Tom later went to film school to hopefully make films someday about the war. After working with Julian Phillips on the first movie script, he expanded to writing books. Tom went to many colleges to gain a thorough understanding of the physical world and has a BS degree in Aerospace Engineering from Cal Poly Pomona in 2001. He completed film studies in 2008 and directed his first feature film "The Grounded" in 2011. Tom is a futurist and likes to make movies about the future of Human existence.

Michael Julian Phillips | Author | Film Screenwriter | Producer
pog777@inbox.com | www.kumaskitchenfreelancemedia.webs.com

Michael Julian Phillips is a 54-year old retired journalist-artist, a Fifth-generation Californian whose career has included many projects for film-video, stage, numerous children's books, newspapers and magazines, and music. Julian was born in 1957, at Saint Mary's Hospital in Long Beach, California, the son of John William Phillips, an Army Radio Man who served in World War 2 and the Korean Conflict, and his young German born bride, Brigitte, an orphan girl whose childhood included witness of Hitler's murderous rule, and Russian military incursion into East Germany. Julian's grandfather, Edward Julian Phillips, known as 'Jules', was a Salinas area farmer and crop scientist, who was schoolyard pals with famous California writer John Steinbeck. Julian has worked in freelance and small newspapers since about 1973, and was editor of a countywide weekly tabloid newspaper, year 2000 to 2003, in San Luis Obispo County, California, where he spent most of his life, and grew up on ranches and farms. The writer earned a BA degree in Journalism/Communication from San Jose State University in 1981. Julian has been married for 25 years to his wife, Carol Lynn, and his son Preston Laverne Phillips is now an art student in San Francisco.